THE KISH CYPHER

The Story of KLJN for Unconditional Security

THE KISH CYPHER

The Story of KLJN for Unconditional Security

Laszlo B. Kish

Texas A&M University, USA

World Scientific

NEW JERSEY • LONDON • SINGAPORE • BEIJING • SHANGHAI • HONG KONG • TAIPEI • CHENNAI • TOKYO

Published by

World Scientific Publishing Co. Pte. Ltd.
5 Toh Tuck Link, Singapore 596224
USA office: 27 Warren Street, Suite 401-402, Hackensack, NJ 07601
UK office: 57 Shelton Street, Covent Garden, London WC2H 9HE

Library of Congress Cataloging-in-Publication Data
Names: Kish, Laszlo B., author.
Title: The Kish cypher : the story of KLJN for unconditional security /
Laszlo B. Kish (Texas A&M University, USA).
Description: [Hackensack] New Jersey : World Scientific, [2016] |
Includes bibliographical references and index.
Identifiers: LCCN 2016019952 (print) | ISBN 9789814449458 (hardback : alk. paper)
Subjects: LCSH: Computer security--Technological innovations. | Data encryption
(Computer science) | Public key cryptography. | Thermodynamic cycles. |
Electronic noise. | Kish, Laszlo B.
Classification: LCC QA76.9.A25 .K546 2016 (print) | DDC 005.8--dc23
LC record available at https://lccn.loc.gov/2016019952

British Library Cataloguing-in-Publication Data
A catalogue record for this book is available from the British Library.

Desk Editor: Amanda Yun

Typeset by Stallion Press
Email: enquiries@stallionpress.com

Printed in Singapore

This book is dedication to the unbiased seekers of true knowledge;
and my family, teachers and colleagues.

Foreword
by Peter V. E. McClintock

Laszlo Kish's book relates to one of the most important and pressing problems of our age — security of information.

Since time immemorial humans have needed to communicate with each other in confidence, passing on or receiving secrets without them getting picked up by the wrong person. In an earlier era the two communicators, Bob and Alice, just had to ensure that they were together alone, out of earshot of the potential eavesdropper, Eve, in order to keep their conversation to themselves. These days, however, Bob and Alice are probably not in physical proximity and may be communicating via e.g. a telephonic, radio, or optical link, but they still want to preserve the confidentiality of their exchanges. Eve is nowhere to be seen but, to be on the safe side, it must be assumed that she may somehow be able to intercept the communication, whence the need for it to be encrypted.

Of course, Bob and Alice need not be real people at all. In a closely similar way, communications related to log-on credentials, credit card usage, bank transfers, and so on, all need to be encrypted to prevent confidential data being intercepted and used, for example, in frauds or for other criminal purposes. It turns out that the encryption and decryption processes themselves need not present too much difficulty, and there are some standard methods that provide very high security. They typically rely on a private key known only to Bob and Alice that is used both for encrypting and decrypting the message. Using a 256-bit key, for example, a message encrypted by the advanced encryption standard (AES) scheme would take billions of years to decode by brute force using state-of-the-art computing resources. The really difficult problem is how Bob can convey the private key to Alice (or vice versa) without it being intercepted.

Apart from Bob writing the key on a piece of paper and personally handing it to Alice in a sealed envelope — which is seldom practicable – unconditionally safe key-exchange protocols, or ideas for such protocols, are few and far between. The most fashionable idea at the moment is based on quantum information processing. Although at least one leading expert has challenged its security claims, the general view seems to be that it is inherently safe, albeit difficult and expensive to implement.

Several years ago, however, Laszlo Kish proposed a completely different kind of key-exchange scheme based on classical statistical physics. It too should be inherently safe, as far as anyone can ascertain, because it is based on the fundamental properties of electrical noise, a subject on which Kish is a world expert. The presence of an active eavesdropper on the line can be detected unequivocally and it seems, at first sight, that the scheme could be implemented and developed more easily and at lower cost than its quantum rival. This protocol, known by a variety of names such as "Kish cypher" or "Kish key distribution" scheme, forms the central topic of the book.

But the book is much more than a dry scientific apologia for a key-distribution protocol that seems to have been inexplicably undervalued up to now. It has a strong autobiographical strand and is addressed, in part, to young scientists so that they will be able to appreciate what real science is really like, including the ideas, comradeship and (by implication) politics and petty jealousies that are part and parcel of all human enterprise. The current political backdrop, with a US Senate bill under consideration that would effectively prohibit unbreakable encryption, adds a topical twist to what is already a very interesting book. I commend it warmly to all its readers.

Peter V E McClintock
BSc, DPhil, DSc, CPhys, FInstP
Emeritus Professor (Physics)
Lancaster University, UK
Lancaster, May 2016

Preface

With all those cyber attacks and compromised personal and corporate data around us, communication and computer security has been becoming an issue of importance. Perhaps, the scientifically most interesting and challenging issue of communication and data security is the distribution of the secure encryption key among the communicating parties. Until 2005, only quantum physics had offered unconditional security for the key distribution and commercial quantum communicator pairs utilizing this technique had very high price (around $100k). In 2005, a new invention[1] of mine introduced an inexpensive competitor utilizing classical statistical physics. This scheme is the subject of this book.

Kish cypher was the Wikipedia-name of the unconditionally secure Kirchhoff-law-Johnson-noise (KLJN — my preferred version) secure key exchanger, which has been running with other names, in chronological order: Kish cipher; classical key exchange system based on amplified Johnson noise; Kish key distribution (KKD), etc. The intention to write this book about KLJN was originally triggered by the fights of Supporters and Deniers (Quantum-supporters) on the 'Kish Cypher' Wikipedia page. Almost everything was incorrect, including the word "cypher" in the name. Most importantly, the impact of the good-willed Supporters was not better

[1]Adrian Cho, "Simple noise may stymie spies without quantum weirdness", *Science* **309** (2005, September 30), 2148.

than that of the enthusiastic Deniers. Three books with "Kish cypher" title and several other books with the same chapter title were published by simply printing out the Wikipedia text of the day and, of course, their text and quality varied according to the fights on Wikipedia. After unsuccessful personal efforts in 2012 to get the Kish cypher wiki page deleted, I decided to write down and publish the truth about the KLJN scheme and signed a contract with World Scientific.

Right then, an exciting and intensive period of deeper understanding and fresh results began, which is not over yet. This is the reason why the book has been so much delayed. Even though the Wiki page has been deleted with my strong request/support in 2013, the mission to provide a more accurate account of the KLJN system persists even more because, year after year, fundamental mistakes have been published by professionals even in high-profile journals. The related debates and discussions are actually good because they indicate a lively field of research with many open questions. Hopefully, due to this book, more scientists and engineers will join these efforts.

After several trials, I decided that it is the best to follow a chronological order based on published papers with added notes, corrections and history. Chapters 2–5 roughly go with such a scheme. The prehistory begins a few years prior to KLJN with showing the thermodynamical limits of Moore's law. After the Moore's law myth was nullified, as a natural continuation, the next step was proving that quantum computing is fundamentally less fashionable for general-purpose computing than classical binary computers. This was successfully done in 2004 and the dogma that quantum computers are superior to classical computers for general-purpose applications was eliminated.

But there was yet another myth to address: the claim that only quantum encryption can be unconditionally secure. Various attempts in that direction finally resulted in the classical physical KLJN scheme, which killed that myth. Eventually, it has turned out that the KLJN system itself is potentially more important than the dogma it annihilated. This book also briefly shows another "fallout" of the KLJN concept — the new computing scheme, *noise-based logic*, that aims at dreams that quantum computers may never realize.

By 2015, KLJN has the most robust and general proof of unconditional security (Section 5.1) based on the *continuity of functions* in classical physical linear and stable non-linear systems. It should be noted that this foundation is more robust than the *no-cloning theorem* (which is the foundation of quantum security) because quantum measurement theory is a dynamically changing field and many surprises may come. For example, now we have weak quantum measurements and their yet-to-be-explored implications, too. On the contrary, the continuity of functions mentioned above has been with us since Newton; robust and unchanged.

The annotations in this book have an important role in enriching the reader's experience. They can carry information taken from popular science, hard science, history, my personal anecdotes, or other miscellaneous information sources. These are distinguished from the main body text by the frames used to identify them, and are embedded throughout the work. The scientific notes aim to help scientists, engineers and students to follow the published papers. Deep understanding of the papers has not been easy due to the strongly multidisciplinary nature of KLJN research, which includes *statistical physics*, *circuit noise theory*, *stochastic signals*, *computer science*, *information theoretic security*, etc. As a result, even professionals with narrow background or superficial understanding of some of these subjects can produce fatal mistakes.

The notes of historical and/or personal nature help readers who are interested in knowing how ideas evolve to have a more complete picture of the short history of KLJN.

We avoid page numbers for locating references within the book. Figures and tables within papers reproduced in this book are numbered within their section. Each paper also has its own References section. With the exception of Figs./Eqs./Notes/etc. within papers, the remaining in this book are numbered with their section number added at the front. For example, Eq. (1.1.2), Note 1.1.3 or Fig. 1.1.1 are found in Section 1.1. Whenever such citation is misunderstandable, we give more specific data. The running titles of sections help to quickly locate these sources without the need of using page numbers.

Readers will also encounter references to 'Alice', 'Bob' and 'Eve' throughout this book. These are the commonly accepted forms of representing person/computer/system 'A' and 'B' (the communicating parties at the ends of a communication channel) and the Eavesdropper.

There are various ways to read this book. A few possibilities:

- First of all, I strongly recommend for *all readers* to read Chapter 1, which is a brief introduction to noise and its beauty, and to the fundamentals of security.
- The readers most deeply interested in this topic, with plenty of time, can read the whole book as it is.
- Those short of time can start with the Table of Contents and the Subject Index and cherry pick from them.
- For the few who are already familiar with the KLJN scheme and want to learn more about its history and additional aspects, I recommend to go through the notes embedded in the text first.
- For those who not only know the KLJN principles but also want to build unconditionally secure key exchangers, the final chapters can be useful.

Finally, an important disclaimer: The described experiences with foundations and organizations (NSF, Wikipedia, arXiv, etc.) are *my personal experiences and their interpretations are my personal opinion, which do not represent the official opinion of Texas A&M University*. My general observation and personal opinion are that foundations and organizations have people with generally good intentions but the rules and structure may produce situations and pressure where the correctness of the resulting actions are questionable.

I wish you an enjoyable reading. Comments and constructive criticism are very welcome.

Laszlo Kish
KLJNbook@gmail.com

About the Author

Laszlo B Kish is a physicist and professor of Electrical and Computer Engineering at Texas A&M University. His activities include a wide range of issues surrounding the physics and technical applications of stochastic fluctuations (noises) in physical, biological and technological systems, including nanotechnology. He is particularly interested in creating and exploring original "high-risk" ideas and to challenge celebrated scientific dogmas — pursuits that have triggered international discussions at various occasions, since 1986. His earlier long-term positions include the Department of Experimental Physics, University of Szeged, Hungary (JATE, 1982–1997), and Angstrom Laboratory, Uppsala University, Sweden (1997–2001). During the same periods he had also conducted scientific research in short-term positions, such as at the Eindhoven University of Technology (Netherlands, 1986, 1997), University of Cologne (Germany, 1989, 1990), National Research Laboratory of Metrology (Japan, 1991), University of Birmingham (United Kingdom, 1993), and Uppsala University (1989, 1990, 1992–1995).

Professor Kish received his MS in Physics from Attila József University (JATE), Hungary, 1980; and Doctoral degree in Solid State Physics, at JATE in 1984, while establishing noise research there. His mentors then, were Miklos Torok, Laszlo Vize and Imre Hevesi. He received a Docent degree (habilitation) in Solid State Physics from Uppsala University, Sweden in 1994, and a Doctor of Science (Physics) title, from the Hungarian Academy of Sciences in 2001.

He received the Honorary Doctor (Honoris Causa) titles from Uppsala University (2011) and the University of Szeged (2012). He was the recipient of the Benzelius prize of the Royal Society of Science of Sweden (2001).

He founded the journal Fluctuation and Noise Letters, and several conferences with a flavor to challenge established beliefs: Unsolved Problems of Noise (UPoN, 1996–); the Fluctuation and Noise Symposia (SPIE, 2003–2007) and Hot Topics in Physical Informatics (HoTPI, 2013).

Contents

Dedication v
Foreword by Peter V. E. McClintock vii
Preface xi
About the Author xv

1. Preparations 1
1.1 Noise in Physics: The Voice of Infinity 6
1.1.1 Additive, stationary noise 7
1.1.2 Finite/infinite measurement information 7
1.1.2.1 Transients 8
1.1.2.2 Periodic signal 9
1.1.2.3 Arbitrary deterministic signal 9
1.1.2.4 Pseudo-random number generation . . . 10
1.1.2.5 Noise 11
1.2 Noise Essentials 12
1.3 Gaussian Noise 19
1.4 Thermal Noise: The Universe is Talking about Itself 24
1.5 Noise Measurement at Its Best 28
1.6 On Secure Communications and Key Exchange 33

2. The Dawn (2002–2004) 49

2.1 The Thermal Noise Death of Moore's Law of Computer-chip Miniaturization 52
2.2 Who Wins the General-purpose Computer Race, Classical or Quantum? 68
2.3 The Kish–Sethuraman Cipher 74

3. The Birth of KLJN (2005–2008) 81

3.1 Birth Pains: Noise-free, Random Kirchhoff-loop (RKL) Cypher 82
3.2 Pre-KLJN: Zero-energy (Stealth) Communication . . . 85
3.3 Henry Taylor's Notes about the Stealth Communicator 96
3.4 Enter the Dragon. The First KLJN Paper 99
3.5 Terminating the Man-in-the-middle and Other Hackers 115
3.6 The (Bergou–) Scheuer–Yariv Attack and Its Correct Derivation by Kish–Scheuer 127
3.7 The Hao Attack and the Response to It 138
3.8 Telecloning and the First KLJN Network Application 147
3.9 The Szegedin Whisper Project: Experimental Verification of KLJN 165
3.10 Securing Computers, Instruments, Games, etc. 184

4. Developments and Fall-Out of KLJN (2009–2014) 191

4.1 To Beat the Quantum Computer Dreams. Noise based Logic and Computing 195
4.2 Liu-attacks and Responses. Circulators and a Deeper Analysis of KLJN 206
4.3 The Outstanding Fidelity of Alice and Bob and How to Make It Better 233
4.4 In the Ring: Response to Charles Bennett and Jeff Riedel 252
4.5 In the Ring: Non-validity of the Landauer Principle, and Leo Szilard's Mistake 299

4.6 The Analyses Triggered by the Gunn–Allison–Abbott Attack 307
4.6.1 Waves or only handwaving? What does physics say? 309
4.6.2 The correct analysis and the failure of the Gunn–Allison–Abbott attack 327
4.6.3 The experimental attenuator artifact in the Gunn–Allison–Abbott attack 342

5. Cream on the Cake 353

5.1 General Security Proof of the KLJN Key Exchange . . 354
5.2 Nullifying the (Bergou–) Scheuer–Yariv Attack and the Second-Law Attack 365
5.3 Enhanced KLJN Schemes 375
5.4 Physical Unclonable Function (PUF) Hardware Keys 395
5.5 Unconditionally Secure Smart Grids 409
5.6 Unconditionally Secure Vehicle Communication Networks 411
5.7 Design Rules-of-thumb for Security 413
5.7.1 General directives 414
5.7.2 Random numbers and noises 416
5.7.3 Privacy amplification 417
5.7.4 Enhanced usage of the shared key (new method) 417
5.8 The Most Recent Developments 418
5.8.1 The cable capacitance attack 418
5.8.2 The current injection attack 418
5.8.3 Vadai–Mingesz–Gingl: perfect security at non-zero power flow 419
5.8.4 The Random-Resistor–Random-Temperature KLJN scheme 420
5.8.5 More light on the flaws in the Gunn–Allison–Abbott (GAA) “directional coupler” attack 420
5.8.6 A new, successful transient attack (Gunn–Allison–Abbott) 423

5.8.7 Deficiencies in the Gunn–Allison–Abbott transient attack paper 425
5.8.8 The ultimate cause of all my debates about the physics of noise and information 429

6. Epilogue 433

Subject Index 435

Chapter 1
Preparations

The Kish Cypher — as the Kirchhoff-law-Johnson-noise (KLJN) secure key exchanger has been (erroneously) nicknamed — is the focus topic of this book. The KLJN is a classical statistical physical system (of 4 resistors and related measurement and communication systems), which is able to generate and share an arbitrarily long and unconditionally secure binary key between two communicating parties so that any information leak towards an eavesdropper is negligible. The KLJN system is described and analyzed in this book at various depths at different places; Section 1.6 provides a condensed summary of various aspects of secure communications and key exchange, and also outlines the KLJN system.

One of the most important components of KLJN is the *thermal noise* of the resistors (i.e. *Johnson noise* — or its artificial, amplified version in more advanced systems). It is thus essential to have a brief introduction to *noise* in a book dealing with the Kirchhoff-law-Johnson-noise (KLJN) key exchange. And, of course, the basic notions of secure communications and key exchange are also needed for readers who are not familiar with them. Chapter 1 is a simple introduction to these issues.

1.1. A characteristic feature of the history of noise research is the numerous occurrences of erroneous interpretations and models, starting from the beginning, such as Nyquist's erroneous way of obtaining the Johnson noise formula (to explain thermal noise at low frequencies) in his famous 1928 *Physical*

(*Continued*)

(*Continued*)

Review paper. The errors usually originate from ignoring or misunderstanding some of rules of the physics of mathematics of the noise, while surfacing at other times simply from the mistreatment of the physics. The history of the KLJN scheme is no different; see later in this book, where various kinds of mistakes by different scientists are rectified. Some of these errors are fundamental; for example, quantum informatics founder Charles Bennett's assumption about the eavesdropper's unlimited information (see Section 4.4) or the recent Gunn–Allison–Abbott attack papers (see Sections 4.5 and 5.8.7). Section 5.8.8 explains the generic source of these types of errors. Some other works are only technically wrong, such as Scheuer–Yariv's wire resistance attack (see Section 3.6 for the correct derivation).

Everybody "knows" Kirchhoff's loop-law (for 'the sum of voltage in a closed circuit loop') and everybody has "heard" about electrical noise (at least, as *statics* from a radio). Thus, when the KLJN scheme was picked up by the news in 2005,[‡] many people felt self-confident about this seemingly simple system; and, because they thought they understood it, they did not believe that it could be secure at all.

Security is intricately puzzling, thus it needs something as mystic as "quantum weirdness", doesn't it? Can a bare wire that is freely exposed to the public, with measurable voltage and current, offer security at all? The answer is *yes* — and, to understand *why*, the reader needs to be aware of a few basic properties of *thermal noise* (i.e. *Johnson noise*) and *noise measurements*.

1.2. The thermal noise of a resistor is its Johnson noise: stochastic voltage fluctuations on the open-ended resistor and stochastic current fluctuations in the resistor with shorted ends (see Section 1.4). The randomness of thermal

(*Continued*)

[‡]A. Cho, "Simple noise may stymie spies without quantum weirdness", Science 309 (2005) 2148; http://www.ece.tamu.edu/~noise/news_files/science_secure.pdf; see also:

D.J. Palmer (2007) Noise encryption keeps spooks out of the loop. New Scientist, issue 2605 (2007) p.32, http://www.newscientist.com/article/mg19426055.300-noise-keeps-spooks-out-of-the-loop.html

"Quantum Cryptography Outperformed By Classical Technique", MIT Technology Review, 2012, June 14, http://www.technologyreview.com/view/428202/quantum-cryptography-outperformed-by-classical-technique/

(*Continued*)

noise is based on the great number of degrees of freedom and the randomness of scattering of the electrons in the resistor. The randomness of thermal noise can be deduced back to an axiom of thermodynamics: if this noise was deterministic then the Second Law of Thermodynamics could be violated. Though the Second Law of Thermodynamics is an axiom, it has been vindicated in experiments and observations at a very large scale, similarly to the Energy Conservation Law.

Noise is the least known key feature of the security of KLJN and — don't forget — noise (*quantum noise*, temporal or spatial, with its rules of conditional-probability) is also behind the "quantum weirdness", which is foundation of quantum-physics based security.

1.3. Quantum noises, with the exception of shot noise, are irrelevant for this book. Shot noise is a thermal noise triggered quantized fluctuation, which can be used in the noise generators in the KLJN system. Quantum noises can originate from the uncertainty principle (which is due to the statistical meaning of the intensity of wavefunction and the non-commuting nature of conjugated physical quantities such as position-momentum, frequency/energy-time, etc.) and/or the quantized values of the physical quantities.

1.4. It is important to note that, *contrary to some quantum-promotion jargon talking about provable randomness of quantum number generators*, their randomness cannot be proven (except that of shot noise, which is thermally activated, thus its randomness is due to classical physics) because that is itself an axiom: the Born-interpretation axiom meaning that the squared absolute value of the wavefunction is the probability density of finding the system in that particular state at quantum measurements. However, experiments with limited statistical sizes do not contradict this assumption of quantum randomness while classical randomness has also been present at any of these experiments. Up to the time of writing this book, there has been no better mechanism to generate truly random numbers than to use thermal noise, which is a classical physical noise. Although at the microscopic level both quantum (electron-phonon scattering) and classical randomness (with astronomically large degrees of freedom) play a role, together they synergistically enhance stochasticity.

Johnson noise (thermal noise; see Section 1.4) which is utilized for KLJN belongs to classical physics: *thermodynamics* and *statistical physics*. One of the key foundations of the unconditional security of KLJN is the *Second Law of Thermodynamics*, which claims the impossibility of building a perpetual motion machine (of the second kind). It is also essential that this noise is a *Gaussian* process (see Section 1.3), and that measuring the thermal noise in a system can communicate essential information about the system without introducing any perturbation into the system (see Section 1.5). This property is only characteristic of classical physics.

Figure 1.1. Claes-Göran Granqvist holding a book from his valuable collection of old books (in 2011). He is full professor (former division head) at Uppsala University and one of the international key players with wide expertise about material science, material optics and nanotechnology and their applications for energy savings, etc. He is a longtime colleague and friend. Our first collaborative paper addressed a noise-phenomenon: to explain the lognormal distribution of nanoparticle sizes at condensational growth where the famous old models failed. We used the noise (Brownian motion) in the drifting of particles to introduce a new mechanism[§] for lognormally distributed particle sizes, and this model has also found applications in different nano systems, such as bacterium detection and neuron-firing statistics.[**] Since then, our collaboration has become much wider and, among others, included fluctuation-enhanced sensing, noise measurements, the Maxwell demon, and KLJN with the related debates (see later in this book), too. The most advanced security proof can be seen in Section 5.1.

[§]J. Söderlund, L.B. Kiss, G. A. Niklasson, C. G. Granqvist, "Lognormal size distributions in particle growth processes without coagulation", *Phys. Rev. Lett.* **80** (1998) 2386.
[**]L.L. Kish, J. Kameoka, C.G. Granqvist, L.B. Kish, "Log-Normal Distribution of Single Molecule Fluorescence Bursts in Micro/Nano-Fluidic Channels", *Appl. Phys. Lett.* **99** (2011) 143121; E.A. Kish, C.G. Granqvist, A. Der, L.B. Kish, "Lognormal distribution of firing time and rate from a single neuron?", *Cognitive Neurodynamics* **9** (2015) 459–462.

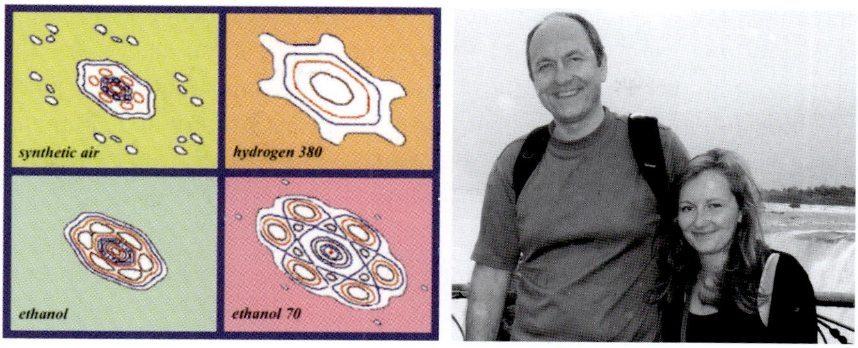

Figure 1.2. Left: Chemical fingerprints, information extracted from the non-Gaussian components of conductivity noise measured in a single resistive gas sensor.[††] It is the bispectrum of the noise of the sensor in different ambient atmospheres. Numbers indicate parts-per-million concentration. Right: Janusz Smulko (with his daughter Michalina at the Niagara Falls in 2011) whose research led to the figures on the left. Janusz is the lead researcher on higher order statistics in Fluctuation-Enhanced Sensing. He is a longtime colleague and friend. He is vice rector for research and head of department at the Technical University of Gdansk (Poland), and is also a leading expert of advanced noise-based statistical methods in corrosion detection and sensing. He recently entered to the KLJN field by applying advanced statistical analysis to study Eve's measurement information.

Noise is often viewed by the mainstream of engineers as a trash-to-be-removed. There is gold (i.e. useful information) in that trash; therefore it is advisable to dig into that "trash". See Figure 1.2 about *Fluctuation-Enhanced Sensing*.

1.5. I started research on Fluctuation-Enhanced Sensing back in Sweden in 1997 and received the Benzelius Prize for this activity in 2001 but no research funding. Interestingly, this research topic was the only one for which I received Governmental funding in the USA. It was from the Army Research Office (via the small business, Signal Processing Inc., headed by Chiman Kwan). Grant applications to the defense sector have the advantage that the researcher must convince only a single person, the Program Manager, instead of a committee, like at NSF. The more revolutionary the idea the higher the chance that at least one committee member will not support it.

[††]From: J.M. Smulko, L.B. Kish, "Higher-order statistics for fluctuation-enhanced gas-sensing", *Sensors and Materials* **16** (2004) 291–299.

In the next few sections, we present a very brief introduction about those aspects of the noise in the KLJN system, which are essential for the understanding of its basic security aspects. Most of this material is taken from the author's graduate courses for engineering, physics and computer science students: "Fluctuations and Noise in Electronics and Physics" (ECEN 771); and "Low-Noise Electronic Design" (ECEN 658), at Texas A&M University, USA.

Finally, in Section 1.6, we briefly address some of the basic concepts of secure communications, too.

1.1. Noise in Physics: The Voice of Infinity

The measured value of a physical quantity has an additive component that is randomly fluctuating in time. This is called (additive) *noise*, which is the generic nickname for *stochastic* processes in physics, electronics and information technology. *Stochastic* means a random time-dependent probability distribution or pattern (of future amplitude values) that, with the exception of some statistical parameters or expected values, may not be predicted precisely.

> **1.1.1.** Discussion on the origin of the noise is out of the scope of this book. Here we only mention a few causes, such as random interactions (collisions, etc.) of charge carriers, fluctuations of system parameters, and measurement effects including the probabilistic effects in quantum and classical physics. In everyday life, the "*static*" sound and image of old off-station televisions are good examples of this noise. It feels like the system is whispering or showing something "mysterious" that is continuously changing, virtually without repetitions, *infinitely* — a phenomenon that has inspired various art pieces, including music, pictures, and even horror movies like *White Noise*.

In general, only a *statistical* (i.e. *probabilistic*) description can be given for a noise. During a measurement process, the noise and the measured values of physical quantities emerge as signals, thus one can use the word (stochastic) "signal" even for the noise-components. Here, we deal with classical physical systems and neglect quantum effects unless it is stated otherwise.

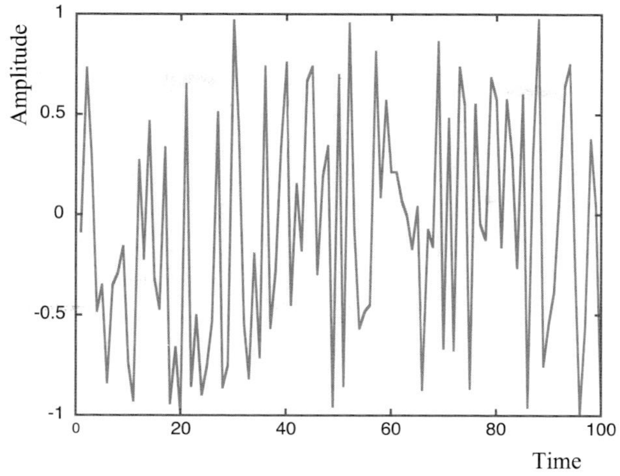

Figure 1.1.1. Computer-generated white noise (band-limited).

1.1.1. *Additive, stationary noise*

Mathematically, the value $U(t)$ of a physical quantity can be given as:

$$U(t) = \langle U(t) \rangle + U_{\mathrm{d}}(t) + U_{\mathrm{n}}(t), \tag{1.1.1}$$

where $\langle U(t) \rangle$ is the time average (mean value), $U_{\mathrm{d}}(t)$ is a deterministic component, and $U_{\mathrm{n}}(t)$ is the (additive) noise of the $U(t)$ quantity, a stochastic fluctuation, where, in this book we automatically assume that the noise has zero average value unless we indicate it otherwise. *Deterministic* (the opposite of stochastic) means that its amplitude can theoretically be predicted from past measurements and a mathematical model, such as the 50/60 Hz powerline interference, a deterministic evolution (e.g. the time-dependence of the temperature of a cooling warm body).

1.1.2. *Finite/infinite measurement information*

Suppose a *practical* (physical) measurement system, which means finite measurement speed (finite sampling frequency) and finite accuracy (finite bit resolution).

Figure 1.1.2. Julio Gea-Banacloche. (Courtesy of J. Gea Banacloche). He is a theoretical physicist with quantum optics focus, and full professor and head of the Department of Physics at the University of Arkansas, USA. He has been assistant editor for quantum informatics for *Physical Review A*. We first met in 2003 at the first Fluctuation and Noise Symposium in Santa Fe, where we fought on the same side to challenge unrealistic dreams about quantum computers. In 2004, in a joint project, we compared the minimum energy dissipation of quantum and classical logic gates versus their error probability. This led to the conclusion† that an ideal general-purpose quantum computer would dissipate more heat by several orders of magnitudes than an ideal classical computer and killed the old dreams of desktop quantum computers replacing our PCs (see the paper in Section 2.2). It was a great pleasure to collaborate with Julio, who not only deeply understands the physical phenomena but who is also able to work and sharply analyze creative pictures/models outside of his own field.

1.1.2.1. *Transients*

Mathematically speaking, the time integral of the absolute value of a *transient signal* $U_t(t)$ (see Fig. 1.1.3) is finite, $\int_0^\infty |U_t(t)|dt < \infty$. A hypothetic, *infinitely long* measurement sequence with *finite* amplitude resolution results in only a *finite* amount of non-zero binary data; i.e. during the infinitely long time scale, there is only a finite duration when the sampled signal is distinguishable from zero. The physical measurement information about a transient signal can be compressed into a *finite* number of bits even when the observation time is approaching infinity. The precise scaling is a

†J. Gea-Banacloche, L.B. Kish, "Comparison of Energy requirements for Classical and Quantum Information Processing", *Fluctuation and Noise Letters* **3** (2003) C3–C6.

J. Gea-Banacloche and L.B. Kish, "Future directions in electronic computing and information processing", Invited Paper, *Proceedings of the IEEE* **93** (2005) 1858–1863.

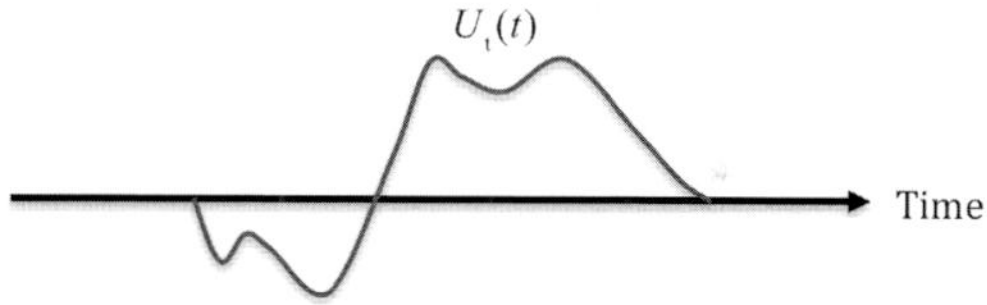

Figure 1.1.3. A transient signal $U_t(t)$ can be deterministic or stochastic. It can be distinguished from zero only in a finite time period. Therefore, its practical information content (relevant for finite accuracy and speed of recording) is finite.

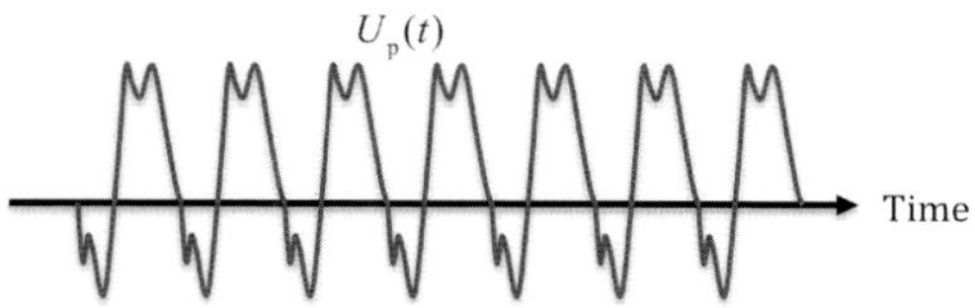

Figure 1.1.4. Periodic signals are deterministic because their future can be predicted. The practical measurement information is finite even for durations approaching infinity (see the notes about logarithmic divergence).

logarithmic divergence due to the need to also store the numbers describing the boundaries of intervals with zero values; however, that is practically considered finite for very large time intervals.

1.1.2.2. *Periodic signal*

The same is true for a *periodic signal* $U_p(t)$ (see Fig. 1.1.4), which is always deterministic, because that is equivalent to the infinite number of replays of a transient signal equivalent to its time function during a single period. Thus the practical measurement information of a periodic signal is also *finite* even over measurement durations approaching infinity. The situation is again a logarithmic divergence, similarly to the transient signal. In this case, the bits describing a single period, and a binary number describing the number of periods, must be stored. The latter one results in a logarithmic divergence with increasing duration.

1.1.2.3. *Arbitrary deterministic signal*

This observation is true even for an *arbitrary deterministic signal.* Why? The future amplitude of a deterministic signal can be given by a finite

Figure 1.1.5. Sunil P. Khatri. (Courtesy of S.P. Khatri). He is a computer-engineering expert and a full professor at the Department of Electrical and Computer Engineering at Texas A&M University. Our collaboration started with exploring the logic space of *Noise-Based Logic* (NBL), which has a similar logic depth as that of quantum information systems (see Section 4.2.1). Then a number of other joint explorations in NBL followed (see e.g. Section 4.2.3). Later he implemented correlator-based NBL for the Boolean Satisfiability (SAT) Problem; a very original scheme. He has also been an ally in debunking Landauer's principle.

amount of measurement information from the past and a mathematical model. The information describing the mathematical model is finite and that together with the measurement information from the past form only a finite amount of information. Thus the practical measurement information of any deterministic signal is also *finite* even over infinitely long measurement duration.

1.1.2.4. *Pseudo-random number generation*

A particular computational (*non-physical*) scheme is a computer's random number generator. The information content of the random number sequence is not more than the number of bits needed to describe its generating algorithm. A historical example is the infinite, never repeating sequence of random digits of the number π.

$$3.14159265358979323846264338327950288419716939937 5\ldots$$

Even though Lambert's theorem proves that this infinite sequence cannot come to an end or repeat, the related information is *not more* than the number of *bits needed to describe its generating algorithm.* Thus the information in the infinite number of random digits of an irrational number is also *finite.*

1.1.2.5. *Noise*

In the physical measurement system, a noise will produce an infinite amount of measurement information over infinite time because its future amplitudes cannot be predicted and it virtually never repeats itself.

> **1.1.2.** Even if a noise time function may seem to repeated in a finite interval t_0 at a given measurement accuracy, one can increase t_0 up to the point where this experienced repetition breaks.
>
> **1.1.3.** It is easy to see that, *philosophically speaking, everything is noise.* Deterministic behavior exists only for a limited time period in the future. For longer times, measurement and model inaccuracies, instabilities, etc. prohibit accurate prediction, and only statistics remains. An everyday example is the weather forecast, which models the weather as a deterministic process but fails with this goal; the longer the term the more random the deviations are. Science has been made possible by the fact that colored noises like deterministic physical processes have a "long-time" memory (correlation time) through which they "forget" their former amplitudes. Fortunately, for a short-enough duration, the abstraction of deterministic science exists. Of course this "short-enough period" can mean a very long time interval, especially when operating on scales such as geological or astronomical time scales, depending on the system.

Here, *Physics* with its infinite number of degrees of freedom is the "random number generator". No algorithm exists to generate the same random sequence because this is *not a pseudo* generation of random numbers, *but a real* generation of *truly* random numbers. Thus, though the practical measurement information of a physical noise is finite during a finite-time measurement, this information converges to *infinity* over an infinitely long measurement duration.

1.2. Noise Essentials

In this section, we present some simple but important definitions, rules and facts that are essential to understand the KLJN scheme and its security.

Physical independence. Independent but identical physical noise sources at identical conditions will produce statistically independent random signals. Mathematicians call these signals "independent realizations of the same stochastic process".

> **1.2.1.** Independent noise sources mean that, regarding the fluctuating physical quantities (*fluctuations*), the physical interaction between these entities is negligible. For example, two unconnected resistors at a distance in separate, magnetically screened faraday cages can be considered independent Johnson noise sources (see Section 1.3) because the interaction of their fluctuating electrons is negligible. However, removing the faraday cages will allow interactions via capacitive, inductive or radiative coupling, and the noises may become significantly correlated depending on the separation, geometry, resistivity and frequency band.

Below are a few important facts about the statistics of noises.

Stationary noises. Their statistical properties do not depend on time. All the noises in the KLJN system are stationary.

> **1.2.2.** Strictly speaking the stationarity claim is true only for the *correct* realization and analysis of KLJN. In some of the incorrect attack schemes (e.g. see Section 4.5.2) time derivative and related integral operations can potentially result in nonstationary $1/f^2$ noise components due to integrating parasitic noise and that can contribute to experimental artifacts.

Mean (average). The simplest statistical description is the *time average* (mean value) $\langle U_n(t)\rangle$, which is meaningless for our noises, as defined above, because their mean value (per-definition) is zero.

$$\langle U_n(t) \rangle = 0 \tag{1.2.1}$$

Note: another type of average is *ensemble average*, $\langle U_n(t_0) \rangle_N$ where the average is taken at a fixed instant of time t_0 over a large ensemble of N identical systems where the noises are *independent realizations* of the same stochastic process. Ensemble averages have not yet been utilized in KLJN; however, they may be so in some multi-line arrangements.

Ergodicity. A noise process is ergodic if time-based statistics and corresponding ensemble-based statistics result in identical values. (All statistical measures use time or ensemble average to get their final result; see their definitions above). If a noise process is ergodic then it must be stationary too; while the opposite statement is not necessarily valid.

> **1.2.3.** A closed thermodynamical system is always ergodic and stationary provided a sufficiently long observation time is available to do the correct time average.

Mean-square. The next simplest quantity is the *mean-square value*, which is the time (or ensemble) average of the square:

$$\left\langle U_n^2(t) \right\rangle = U_{n,rms}^2 = U_{n,eff}^2 \tag{1.2.2}$$

where the *root-mean-square* (RMS) *value* $U_{n,rms}$ and the *effective value* $U_{n,eff}$ of the noise have the same well-known meaning as for AC signals.

> **1.2.4.** If $U_n(t)$ is voltage or current, its heating power is the same as that of a DC quantity U_{DC} of the same amplitude value: $U_{n,eff} = U_{n,rms} = U_{DC}$. This is why the jargon often incorrectly uses the term "noise power" for the mean-square value even though it does not have a power unit, and its nominal value is only equal to that of the power with a 1 Ohm resistor load.

Amplitude density function. The (first-order-) probability density function $g(U)$ of the amplitude provides the finding probability $P(U_0, dU)$ of the noise amplitude around an arbitrary U_0 amplitude level within an infinitesimally small interval dU is:

$$P(U_0, dU) = g(U_0)dU, \quad \text{where } dU \to 0. \tag{1.2.3}$$

1.2.5. Amplitude density is a one-dimensional density function, just like the one-dimensional mass density (with mass/length unit) of a wire. Imagine a wire with varying diameters to picture its dependence on the location. For laymen readers the only important feature to keep in mind is that it is proportional to the probability of finding the noise amplitude around U_0. After introducing the Gaussian amplitude density and declaring that the noises in the KLJN are such, this function will become practically unimportant except at the calculations of the fidelity of the key exchange, see Section 4.3.

Power density spectrum (noise spectrum). The amplitude density function is totally "dead" for stationary processes because it does not have a time variable; thus there is no information in it about the dynamics of the noise process. One of the simplest and most useful statistical measures that provides information about the dynamics of the noise is the power density spectrum (noise spectrum) $S_u(f)$, which is the density of the mean-square amplitude versus frequency.

1.2.6. Example. Suppose we record a noise. If we play the noise back 100 times faster, the *amplitude density function will not change*. However the spectrum will go through a transformation of $S_u(f) \Rightarrow S_u(f)(100f)$. For example, if the maximum of the spectrum $S_u(f)$ was at the frequency of 1 Hz, this maximum will now be at 100 Hz. The noise spectrum gives information about the "dynamics" of the noise.

In Figure 1.2.1, a simple measurement scheme illustrates the meaning of the noise spectrum.

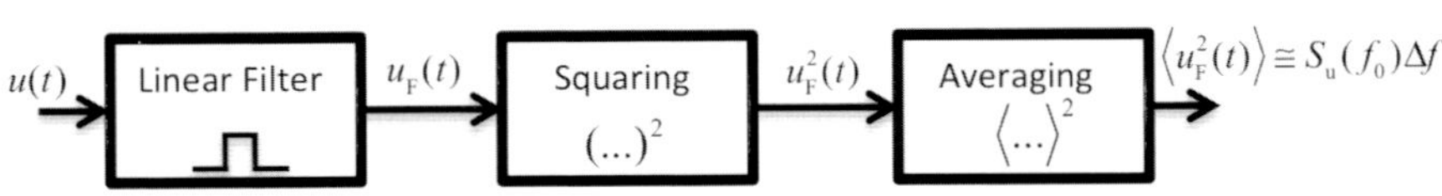

Figure 1.2.1. Measurement scheme illustrating the meaning of the noise spectrum.

The $u(t)$ noise goes through a linear filter with a square shape transfer function, center frequency f_0 and the *infinitesimally narrow* bandwidth Δf. In the filtered noise $u_F(t)$ only frequency components in the Δf band around f_0 remain. The measurement output provides the mean-square of the noise within the Δf band around f_0. If $\Delta f \to 0$, we obtain the density of the mean-square versus frequency

$$S_u(f) \equiv \lim_{\Delta f \to 0} \frac{\langle u_F^2(t) \rangle}{\Delta f}. \tag{1.2.4}$$

As a consequence, the power-density spectrum of a sinusoidal oscillation is infinite at the oscillation frequency and zero elsewhere (a Dirac pulse), and the mean-square amplitude of a narrow-band noise (sufficiently narrow band within which the variations of the spectrum are negligible) is:

$$\langle u_{f,\Delta f}^2(t) \rangle \simeq S_u(f)\Delta f. \tag{1.2.5}$$

The mean-square of the total noise and that of the noise band-limited at frequencies x (lower limit) and y (higher limit) respectively are:

$$\langle u_{tot}^2(t) \rangle = \int_0^{\infty} S_u(f)df \quad \text{and} \quad \langle u_{x,y}^2(t) \rangle = \int_x^y S_u(f)df. \tag{1.2.6}$$

A further essential feature of the power density spectrum is that it does not have any information about the phase of the noise. For example, the noise spectrum does not change when the noise passes through an amplifier that has a frequency-dependent phase shift, which completely changes the time function, provided the absolute value of the amplification does no depend on the frequency. As a consequence of zero phase information, not only do the infinitely many different realizations of the same noise process lead to the same noise spectrum, but infinitely many different noise processes (even with different amplitude density functions) can also have identical noise spectra.

1.2.7. The noise spectrum is different from Fourier spectra in the sense that it cannot be inverted to regenerate the original time function of the noise. Fourier spectra always form a pair (such as amplitude and phase spectra, etc.), while the noise spectrum is single and contains no phase information.

(Continued)

(*Continued*)

A similar statement can be given about the amplitude density function: Not only do the infinitely many different realizations of the same noise process have the same amplitude density, but infinitely many different noise processes (even with different noise spectra) can also have identical amplitude density function.

Spectra after linear operations on the time function of the noise. As a consequence, a linear, frequency-dependent complex amplification $\mathrm{A}(f)$ of the noise amplitude results in the multiplication of its spectrum by $|\mathrm{A}(f)|^2$ as:

$$S_\mathrm{u}^*(f) = |\mathrm{A}(f)|^2 S_\mathrm{u}(f), \tag{1.2.7}$$

where $S_\mathrm{u}^*(f)$ is the resulting spectrum (see Fig. 1.2.2).

1.2.8. As a consequence, time derivative of the noise has a modified spectrum that is multiplied by $4\pi^2 f^2$ and the time integration yields a spectrum divided by the same factor.

Correlation functions. These are only important for certain argumentations at a few places in this book. The *cross-correlation function* $R_{1,2}(\tau)$ of two stationary noises $u_1(t)$ and $u_2(t)$ is the average of their time-shifted product:

$$R_{1,2}(\tau) = \langle u_1(t)u_2(t \pm \tau)\rangle = \langle u_1(t \pm \tau)u_2(t)\rangle, \tag{1.2.8}$$

where τ is the time shift. For independent noises $R_{1,2}(\tau) = 0$ (remember: noises are defined to have a mean of zero in this book unless specified differently).

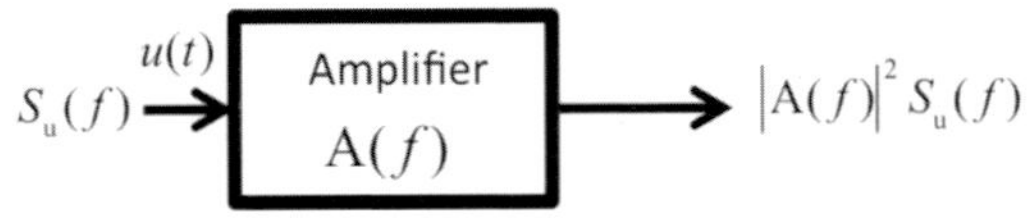

Figure 1.2.2. Impact of linear amplification on the noise spectrum. Only the absolute value of the complex amplification matters; its phase shift is unimportant.

Figure 1.2.3. Peter Heszler (2005). He was a photonics scientist, and a close friend. Peter died in 2009 at the age of 50. We were young scientists in 1986 when we had the first international showdown. It was against a celebrated theory of the time: the *quantum 1/f noise model.*† We were supported by a senior quantum-electrodynamics-guru, Peter Hrasko from Budapest, who taught us the related tools, thus our denial became much deeper than condensed matter theory arguments — our starting points. The noise community supported our view and I became an invited speaker for a public debate at the International Conference on Noise and Fluctuations in 1987, in Montreal, Canada. It is funny how the debates in my life started with a quantum noise myth; the present existence of the KLJN system is itself a denial of a quantum (informatics) myth; i.e. *that only quantum informatics can achieve unconditional security.*

The *autocorrelation function* is the cross-correlation function of the noise with itself:

$$R_{1,1}(\tau) = \langle u_1(t)u_1(t \pm \tau)\rangle. \tag{1.2.9}$$

At zero time shift, the autocorrelation function yields the mean-square amplitude. The autocorrelation of stationary noises converges to zero for increasing time shift:

$$\lim_{|\tau|\to\infty} R_{1,1}(\tau) = 0 \tag{1.2.10}$$

that is, the noise becomes independent of itself for sufficiently long times. The jargon uses the notion of "*correlation time*" within which the noise is still correlated to itself (above a certain level such as the fraction 0.5

†L.B. Kiss, P. Heszler and P. Hraskó, "Non-validity of the quantum 1/f noise model", *Proc. 9th Internat. Conf. on Noise in Physical Systems*, ed. C.M. Van Vliet, World Scientific, Singapore (1987) 373; also, a much shorter version in: *Journal of Physics C* **19** (1986) L631.

Figure 1.2.4. Peter Hrasko is a theoretical physicist, an outstanding teacher, the former director of a theoretical group of the Central Physics Research Institute of Hungary, and emeritus professor of the University of Pécs, Hungary. His current activities include scientific skepticism by debunking false scientific claims. He recently published several books including one on statistical physics, and another one on relativity. In 1985, when I told him about my condensed matter physics based doubts regarding the quantum 1/f noise models, he immediately said that the mistake is much deeper; according to quantum-electrodynamics, the fundamental effect is non-existent.

or $1/e$ of its mean-square). The correlation time τ_{cor} of band-limited white noise — the essential feature of the KLJN scheme — is proportional to the reciprocal of the bandwidth B of the noise and is in the order of that:

$$\tau_{\mathrm{cor}} \sim 1/B. \tag{1.2.11}$$

Sampling theorem. Nyquist was the earliest to deduce the sampling theorem‡ and he did that without emphasizing the aspect of reconstructing the signal after sampling. This form is essential for the security analysis of KLJN key exchange:

> **In a communication channel with B frequency bandwidth, *at most* $2Bt_{\mathrm{w}}$ independent samples can be sent during a time window of t_{w} duration.**

‡H. Nyquist, "Certain topics in telegraph transmission theory", *Trans. AIEE* **47** (1928) 617–644.

In another words, a band-limited white noise (such as in the KLJN scheme) has at most $2Bt_{\mathrm{w}}$ statistically independent samples. This law is the uncertainty relation between observation time and measurement accuracy which progressively limits eavesdroppers' information at non-ideal situations of the KLJN system; where, if Eve (our hypothetical eavesdropper) could have infinite time, she could extract the key bit. However she cannot achieve that because both the effective noise bandwidth B and the active time window t_w of the key exchange are controlled by Alice and Bob (the communicating parties).

1.2.9. This is a point of frequent misunderstanding even by computer scientists and quantum physicists. Several experts thought (see, for example, Section 4.4, the debate with Charles Bennett) that we pose a conditional technical limit on Eve's abilities — that is that KLJN offers only conditional security. The real situation is far from that; Eve's instrumental cababilities are unlimited while her information is still converging to zero. Eve's measurement system is assumed to be unlimited; she can theoretically have infinite *measurement-instrument-bandwidth* and an infinitely long observation time; however, her *actual signal* is band-limited by B and time-limited by t_{w}.

Measurement of a band- and time- limited signal by an instrument of infinite bandwidth for an infinitely long time does not offer more information than executing the measurement with an instrument that has the same band/time limits as the signal. In conclusion, when she goes beyond B with her measurement bandwidth and goes beyond t_{w} with her measurement time, she gains no extra information: *she expands her measurements only with empty frequency and time domains.*

1.3. Gaussian Noise

The most widely occurring and the most studied stochastic processes are the Gaussian ones. Their amplitude distributions are Gaussian (normal):

$$g_{\mathrm{G}}(U) = \frac{1}{U_{\mathrm{rms}}\sqrt{2\pi}} \exp\left[-\frac{U^2}{2U_{\mathrm{rms}}^2}\right], \qquad (1.3.1)$$

where U_{rms} is the rms (effective) value. Figure 1.3.1 shows the "bell-shaped" Gaussian curves with various rms values and zero mean.

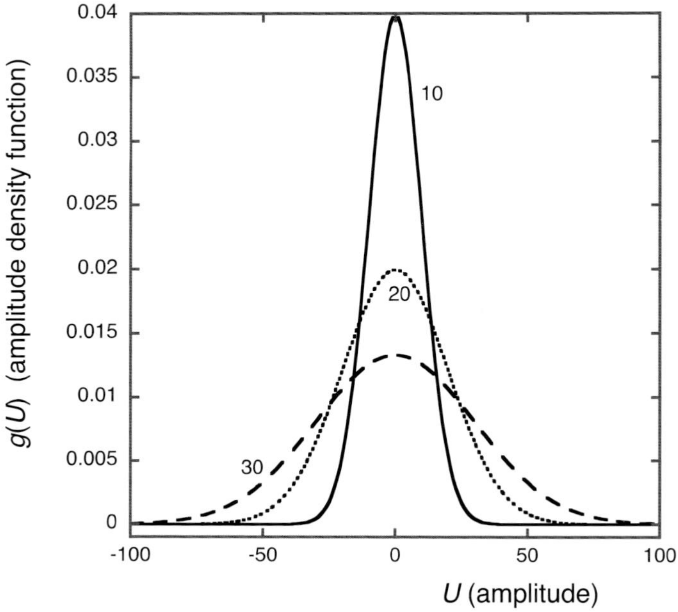

Figure 1.3.1. Gaussian amplitude density functions. The numbers indicate the rms (effective) amplitudes.

The frequent occurrence of Gaussian (-like) stochastic processes in nature are explained by the Central Limit Theorem of probability theory: it states that the distribution of the sum of a large number of independent random variables with finite variance and identical statistical properties converges to normal distribution. Macroscopically measurable noises (such as thermal noise) are often the sum of a great number of independent elementary fluctuations; thus Gaussian noises usually dominate in linear statistical physical systems with macroscopic size.

Linear operations on Gaussian noises result in Gaussian noises. As a consequence, *linear combinations*, *time derivatives* and *time integrals* of these types of noises are also Gaussian; and when a Gaussian noise passes through a linear amplifier (see Figure 1.2.2) the output noise will also be Gaussian, even if the amplification and its phase shift are frequency-dependent. But the sum of a Gaussian noise and a non-Gaussian signal will be non-Gaussian.

1.3.1. This is a very important aspect of the security of KLJN and can lead to experimental artifacts and related false claims about cracking the security when significant spurious non-Gaussian voltage/current components exist in the channel. See the analysis of an unsuccessful attack in Section 4.5.2. Non-linearities also cause information leaks. All these facts indicate that practical KLJN systems are no different from quantum key distribution in the sense that the practical realization of a system with near-to-perfect security is a serious design challenge requiring a wide range of experiences and a deep understanding of all the fine details of the system.

Independence versus cross-correlation. Another essential fact is that zero cross-correlation of two Gaussian noises (of zero mean) guarantees that these noises are statistically independent.

1.3.2. In the ideal KLJN system, the current and voltage have zero cross-correlation — a requirement of the Second Law of thermodynamics. These noises are Gaussian and thus statistically independent even though the current is a result of the voltages generated there. Without this feature, a serious information leak would exist.

Another consequence of this feature is that the time derivative or time integral of a Gaussian noise is statistically independent from the original noise. For proof, see the random phase oscillator formalism at the end of this section.

This feature is also essential in the analysis of the unsuccessful attack attempt in Section 4.5.2.

The tails of the Gaussian distribution and its spectrum determine the threshold crossing statistics of high thresholds, including the error probability and fidelity of the KLJN key exchange.

1.3.3. The combination of current and voltage statistics yields astronomically low error probabilities of KLJN due to their independence (see Note 1.3.2 above and Section 4.3).

No hidden features (higher-order statistics) in Gaussian noises. It is also an essential feature of the perfect security of the ideal KLJN scheme that any Gaussian process is fully characterized by its power density spectrum. Thus even the amplitude density function is not needed (out of knowing

Figure 1.3.2. Zoltan Gingl (2012) is head of the Department of Technical Informatics at University of Szeged, Hungary. Zoltan was originally my PhD student at the Department of Experimental Physics and, in the 1990s, he took over directing the Noise Lab, which I founded in the 1980s. He went into the direction of digital informatics, which is a powerful tool for the KLJN, too. Zoltan has a wide range of expertise including noise theories and experiments, digital and medical informatics, noise generation and analysis, and more. His research interest endured even when he became an Associate Dean, and he (with Robert Mingesz, his former PhD student), made the first mathematical proof[†] of the uniqueness of Gaussian noise for the security of KLJN.

that it has a Gaussian shape) for the full statistical characterization of these noises because the rms value directly follows from the noise spectrum, see Equation 1.2.6.

1.3.4. One can say that Gaussian noises are the most "honest" noise process: nothing remains hidden after measuring their noise spectrum. For example, the existence of the bispectra in Figure 1.2 is due to the non-Gaussian components of the noise carrying extra information that cannot be seen in the power density spectra.

Another interesting feature of Gaussian noises is that their time dependence can be described by the random-phase-oscillator formalism[‡] (of S.O. Rice). One can envision a division of the frequency scale into the infinitesimally

[†] Z. Gingl, R. Mingesz, "Noise Properties in the Ideal Kirchhoff-Law-Johnson-Noise Secure Communication System", *PLoS ONE* **9** (2014) e96109.

[‡] G.A. Korn, T.M. Korn, *Mathematical Handbook for Scientists and Engineers*, McGrawHill, New York, (1968); Dover, New York (2000).

narrow frequency bands of Δ and at the center of each band an oscillator with random initial phase to represent the noise. The peak amplitude $u_i(t)$ of the oscillator in the i-th band provides the power density spectrum around the frequency $k\Delta$:

$$S(i\Delta) = \frac{u_i^2}{2\Delta} \quad \text{where } u_i = \sqrt{\Delta 2S(i\Delta)} \tag{1.3.2}$$

because $f = i\Delta$, we get:

$$u_f = \sqrt{\Delta 2S(f)} \tag{1.3.3}$$

Thus for $\Delta \to 0$, the time function $U(t)$ of the total noise is:

$$U(t) = \sum_{i=0}^{\infty} u_i \cos(2\pi i\Delta t + \varphi_i) = \sum_{i=0}^{\infty} \sqrt{\Delta 2S(f)} \cos(2\pi ft + \varphi_i), \tag{1.3.4}$$

where $f = i\Delta$ and φ_i is the random initial phase $0 \leq \varphi_i \leq 2\pi$ of the i-th oscillator. A different realization of the noise (such a thermal noise with the same spectrum coming from another resistor) has an independent infinite random set $\{\varphi_i\}$ of initial phases.

Thus the infinity included by a Gaussian noise has two aspects: the infinite number of oscillators with different frequencies and the infinite set of random numbers in $\{\varphi_i\}$. The random initial phase of the oscillators and the Central Limit Theorem guarantee, that $U(t)$ will indeed be Gaussian noise.

The time-derivative of $U(t)$ is given as:

$$\frac{dU(t)}{dt} = -\sum_{i=0}^{\infty} \sqrt{\Delta 2S(f)} 2\pi f \sin(2\pi ft) \tag{1.3.5}$$

thus the cross-correlation of the original Gaussian and its derivative is zero:

$$\left\langle U(t)\frac{dU(t)}{dt} \right\rangle = -\left\langle \sum_{i=0}^{\infty} \Delta 2S(f) \cos(2\pi ft) 2\pi f \sin(2\pi ft) \right\rangle = 0, \tag{1.3.6}$$

which, due to the "*independence at zero cross-correlation*" theorem for Gaussian processes (see above) implies that the Gaussian noise amplitude and its velocity are statistically independent Gaussian fluctuations.

1.4. Thermal Noise: The Universe is Talking about Itself

Thermal noise is the random fluctuations of the physical quantity around its mean value due to the existence of thermal motion in the system. Boltzmann's *Energy Equipartition Theorem* says that, in a system of thermal equilibrium, at the classical physical limit, each (thermodynamic) degree of freedom carries $kT/2$ mean thermal energy, where k is the Boltzmann constant (1.38×10^{-23} Joule/Kelvin), and T is the absolute temperature.

1.4.1. The Energy Equipartition Theorem is fundamentally interrelated with the Second Law of Thermodynamics. The number of thermodynamic degrees of freedom (TDF) is equal to the number of generalized coordinates on the second power in the equation describing the total energy of the system in consideration. Examples:

(i) When restricted to a one-dimensional motion, and rotation is excluded, a ball hanging on a spring has TDF $= 2$ because the total energy that can vary with the motion of the mass is $E = 0.5Mv^2 + 0.5Dh^2 + Mgh$, where M is the mass, D is the spring constant, h is the deviation from the vertical equilibrium position and g is the gravitation constant. The third term in the energy does not contribute to the TDF because it is linear in a coordinate.
(ii) A capacitor has TDF $= 1$ because $E = 0.5CU^2$, where C is the capacitance and U is voltage on it.
(iii) An inductor has TDF $= 1$ because $E = 0.5LI^2$, where L is the inductance and I is current passing through it.
(iv) A parallel LC resonator is naturally TDF $= 2$.
(v) An electromagnetic wave with a single frequency and single polarization has TDF $= 2$, and with N different frequency modes it has TDF $= 2N$ and its total thermal energy is:

$$E = 2NkT$$

This fundamental result proves that there are no electromagnetic waves in a short cable at low frequencies because there is not enough thermal energy in the cable to satisfy the energy equipartition law (see the *no-wave* proof about an unsuccessful attacks in Section 4.5.1.

(*Continued*)

(*Continued*)

To put these systems into thermal equilibrium, they must be coupled to a heat reservoir in equilibrium. For example, the systems in (ii)–(iv) can be connected to the two terminals of a resistor, which is kept at temperature T.

Thus any physical quantity that is coupled to thermal motion is subject to thermal noise. Various types of thermal noises are also responsible for exchanging thermal energy with the environment. Because virtually everything is at least at or beyond the $\approx$2.7 Kelvin background radiation temperature within the Universe, thermal noise is everywhere and it is manifested in many different ways. The Universe is "talking"† about itself everywhere, and has been doing so since its origin.

1.4.2. The discovery of the 2.7 K cosmic background radiation — or cosmic microwave background, is a thermal noise; specifically the thermal radiation left over after the Big Bang — earned the Nobel Prize in Physics for Arno Penzias and Robert Wilson in 1978 for discovering it, and for George Smoot and John Mather in 2006 for analyzing its structure. On the other hand, Johan Johnson and Harry Nyquist, two Swedish-born scientists, have never received a Nobel Prize for their discovery and explanation of Johnson noise, despite its much higher impact on science and technology.

The characteristic feature of thermal energy and thermal motion is randomness. The light coming from the sun and the stars is dominantly a thermal radiation (i.e. thermal noise) of the electromagnetic field. The Brownian motion of a particle in a fluid is the thermal noise in the velocity of that particle. The thermal noise in the velocity of atoms or molecules in the solid body is their thermal motion, which is manifested via random oscillations (phonon waves) around their average position. The higher the temperature the stronger are these fluctuations and the higher are their rms velocities. A hot plate burns our finger because these vibrations propagate through our skin and destroy the weak organic structures.

The thermal noise voltage and current (Johnson noise voltage and current) of a resistor is due to the (random) thermal motion of charge carriers. The

†Here Stanislaw Lem's excellent science fiction novel *His Master's Voice* comes to mind where the noise detected by the radio-astronomers contained high-impact information after successfully decoding it.

Figure 1.4.1. The two Bell Lab scientists of Swedish origin who first identified and described the thermal noise of resistors in 1927–28, two years after quantum physics was founded in 1926. Left: John Bertrand Johnson (1887–1970); Right: Harry Nyquist (1889–1976). It is an interesting situation that the secure KLJN scheme, which is classical-physics based, was also discovered later than quantum key distribution (QKD). Note: QKD could have already been discovered in 1926 and KLJN in 1928.

Johnson noise voltage is measured on an open-ended resistor, while the current in a shorted resistor. The power density spectra $S_u(f)$ and $S_i(f)$ of the voltage and current, respectively, are given as:

$$S_u(f) = 4kTR \quad \text{and} \quad S_i(f) = \frac{4kT}{R}, \tag{1.4.1}$$

where R is the resistance.

> **1.4.3.** It is obvious from seeing these equations that we are talking about a deeply fundamental physical phenomenon, not only because Boltzmann's constant k and the absolute temperature T are there, but also because, besides the resistance, there is nothing else in it about the resistor. Its material, its way of conduction (conductor, semiconductor, electrolyte, etc.) its shape, volume, structure, etc. are all irrelevant. The necessary condition is to follow Ohm's law with its resistance R (or the real part of its impedance) and must be in thermal equilibrium at temperature T; nothing else matters. Johnson's pioneering discovery [*Nature* **119** (1927) 50] was about this very fact. Nyquist was the first one who gave a thermodynamic derivation of Johnson's empirical formula [*Physical Review* **32** (1928) 110]. Note, this celebrated derivation is not without problems, even though the correct result is obtained. Some of the key

(*Continued*)

(*Continued*)

steps, such as an implicit assumption that the noise is white in the low-frequency limit, cannot be vindicated without extra proofs that are missing there.

A personal memory: during the fall of 1973, in a college at Kazincbarcika (Hungary), I got into an intensive debate about this issue because an engineering student dorm-roommate claimed that, with advanced material technology, the Johnson noise of resistors could be reduced. He was obviously wrong; however, at that time I did not have enough knowledge of physics to back up the correct view, thus I claimed only that the Johnson formula looks like a fundamental lower limit thus material technology cannot be of any help.

If the resistance is independent from the frequency then the noise spectra in Equation 1.4.1 are also frequency-independent, which means we have *white noises*. As a consequence, in a frequency bandwidth B, the mean-square noise voltage and current are given as:

$$\langle u^2(t)\rangle = 4kTRB \quad \langle i^2(t)\rangle = \frac{4kTB}{B} \tag{1.4.2}$$

1.4.4. White noise with infinite bandwidth is unphysical for various reasons including infinite speed of processes and infinite mean-square. Thus there must be a cut-off at high-enough frequencies — however this is well beyond our range of interest. Practical cut-off frequencies are due to parasitic shunt capacitances and serial inductances; frequency-dependent resistance due to the mean free time of charge carriers; or other inertia in charge transport properties of the material, etc. A fundamental cut-off is due to the "freezing out" of the thermodynamic degrees of freedoms at frequencies $\sim kT/h$ and beyond ($h = 6.6\times 10^{-34}$ Js is Planck's constant), where quantum phenomena take over.

Note, the author and his coworkers currently debate[‡] the popular results about the behavior of the spectrum in the quantum domain but this question is irrelevant for KLJN due its low frequency band.

[‡]L.B. Kish, G.A. Niklasson, C.G. Granqvist, "Zero-point term and quantum effects in the Johnson noise of resistors: A critical appraisal", *J. Stat. Mech.* (2016) 054006. http://arxiv.org/abs/1504.08229

The functional forms of Equations 1.4.1 and 1.4.2 are required by the Second Law of Thermodynamics. As a consequence of these equations, there is a power flow P_{12} between two parallel-connected resistors with resistances R_1, R_2 and temperatures T_1, T_2 given as:

$$P_{12} = 4k(T_1 - T_2)\frac{R_1 R_2}{(R_1 + R_2)^2}B. \tag{1.4.3}$$

In accordance with the Second Law and Equation 1.4.3, the heat always flows from the warmer resistor to the cooler one and there is no net heat flow when the two temperatures are equal, that is, when the resistors are in thermal equilibrium with each other.

> **1.4.5.** These facts are important to guarantee the perfect security of the ideal KLJN system with uniform temperature and have an important role in totally eliminating the famous cable resistance-based attack of (Bergou–) Scheuer–Yariv (see Section 5.2).

1.5. Noise Measurement at Its Best

Physical measurements can be *active* or *passive*.

Active physical measurement analyses the response due to an excitation (see Figure 1.5.1). The perturbation of the system is significant and can unfavorably interfere with the information obtained about the system.

Passive physical measurements, see Figure 1.5.2, record the values of the physical quantity of interest while avoiding perturbation as much as possible. In a system of thermal equilibrium, they observe the spontaneous fluctuations around the mean value.

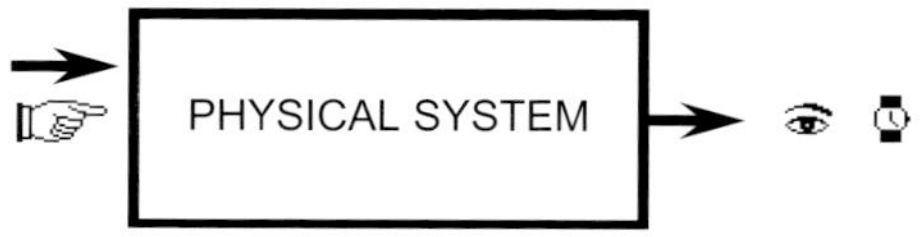

Figure 1.5.1. Active physical measurement. The system is perturbed and the perturbation is measured, recorded and analyzed.

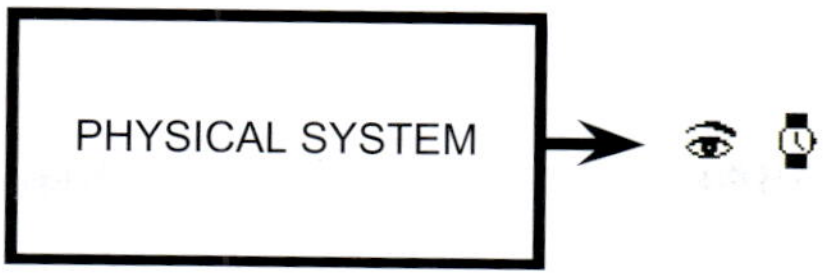

Figure 1.5.2. Passive physical measurement with minimal perturbation of a system: analyzing its spontaneous fluctuations.

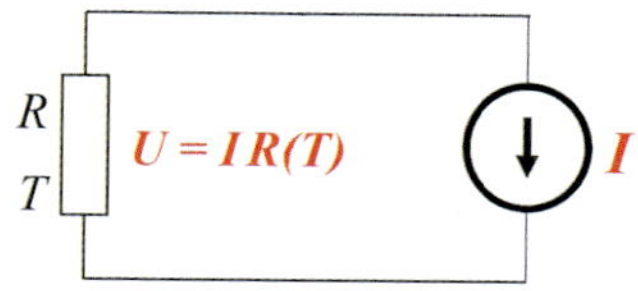

Figure 1.5.3. The usual scheme of resistive thermometry. The measurement current is heating the thermometer thus causing the worst type of perturbation.

Naturally, noise measurements can be done in either active or passive ways; however, no doubt, the latter ones represent "noise measurements at their best". Two examples are shown below.

Figure 1.5.3 shows the usual way of *resistance-based thermometry*. The resistance of a temperature-dependent resistor is calibrated versus temperature, and a lookup table is generated. During temperature measurement, the temperature corresponding to the actual resistance is found out from the lookup table. However, typical resistance measurements are taken using a calibrated current drive and Ohm's law to determine the resistance (see Figure 1.5.3). That results in $P = I^2R(t)$ heating power generated at the most sensitive point of the measurement system — within the thermometer — leading to errors.

The most advanced way of resistance-based thermometry is shown in Figure 1.5.4. Both the thermal noise voltage and current are measured, and by solving the system of equations given by the Johnson formulas, both the temperature and the actual resistance can be determined without heating the thermometer. Note: In practice they often use only the voltage noise and a temperature-independent resistance; however, the scheme in Figure 1.5.4 is superior to that, because it is a first-principle-measurement with no assumption about sensor details.

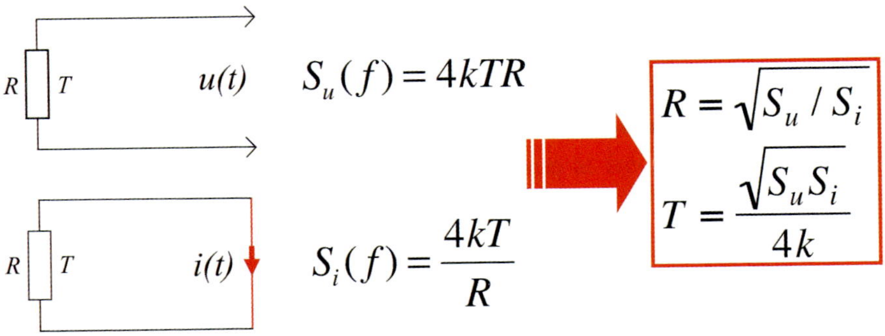

Figure 1.5.4. Resistor-based noise thermometry at its best. Without injecting energy into the system, both the resistance and temperature are determined by passive measurements.

> **1.5.1.** It is proper to mention the KLJN operation in this aspect, too. The scheme requires the determination of the total serial resistance in the communication loop in such a way that does not give out other information to the eavesdropper (Eve). This process utilizes the thermal noise of the resistances in the loop (it can be enhanced by noise generators to emulate a much higher temperature). Simple passive noise measurements inform Alice, Bob and Eve not only about the total loop resistances but also about the values of the used resistance (without specifying their locations, whenever they are different). Any active type of measurement, which injects energy into the loop, would give out information to Eve. Thus the fully secure system must have additional protection against active attacks, see, for example, Sections 3.4, 3.5 and 5.8.2.

The second example is about a famous fundamental application of noise in the field of spin glasses, see Figure 1.5.5, which are disordered magnetic materials with frustrated interactions between the atomic magnetic moments, e.g. created by a distribution of ferro- and anti-ferromagnetic interactions. Thermal noise measurements above the "glass melting" temperature show that glass melting is a dynamical phenomenon where the "glass lifetime" is strongly temperature dependent. This effect could be studied by thermal noise measurement of the magnetization, where the white noise to 1/f noise crossover frequency showed the reciprocal time constant of the glass state. The cut-off of the distribution of time constants, also referred to as the relaxation time, increases as the length scale of spin correlations increases, approaching the glass transition from above and should diverge at the glass transition.

1.5.2. The famous 1/f noise (flicker noise, pink noise, excess noise, etc.) has spectrum that is roughly proportional to the reciprocal of the frequency. It is peculiar in the sense that, in a linear system it is impossible to generate it without the incorporation of a large system of properly distributed time constants that "graphically" synthesizes the spectral shape from bandlimited white noise. Due to this reason it has not been easy to generate correct models to explain various 1/f noise phenomena, and the history of 1/f noise is characterized by many papers with mathematical and/or physical mistakes and the related debates. Even single-fluctuator-based linear models (seemingly without the distributed time constant) have those in a hidden way. For example, the "Transient-diffusion 1/f noise model" [L.B. Kish, J. Hajdu, *Solid State Communications* **72** (1989) 799] that generates 1/f noise in the surface current via diffusion of defects away from the surface (and seemingly avoids this constraint), inherently contains an infinite distribution of time constants due to the increasing diffusion times for greater distances away from the surface.

At a specific temperature above the glass transition temperature and for time scales larger than the relaxation time, spin correlations at length scales larger than the correlation length "melt"; the spin correlations are shaken away by thermal agitation. Magnetic noise measurements are performed without external magnetic field excitations and are thus preferred over active magnetic relaxation measurements because even the smallest excitation will prevent the relaxation time/correlation length to diverge at the glass transition. The noise spectra in Figure 1.5.5 are not smooth lines because of the finite measurement time. The statistical errors (i.e. roughness, or "hairs") seen on the noise spectra are typical at noise measurement and, to have absolutely smooth curves without statistical error, an infinite measurement time would be required. This phenomenon is due to the *uncertainty relation between the measurement accuracy of noises and the length of the time window*. The same mechanism progressively reduces the eavesdropping information in non-ideal KLJN situations where there is non-zero information leak, and preserves the unconditional security even then (see Sections 4.4 and 5.1 for examples).

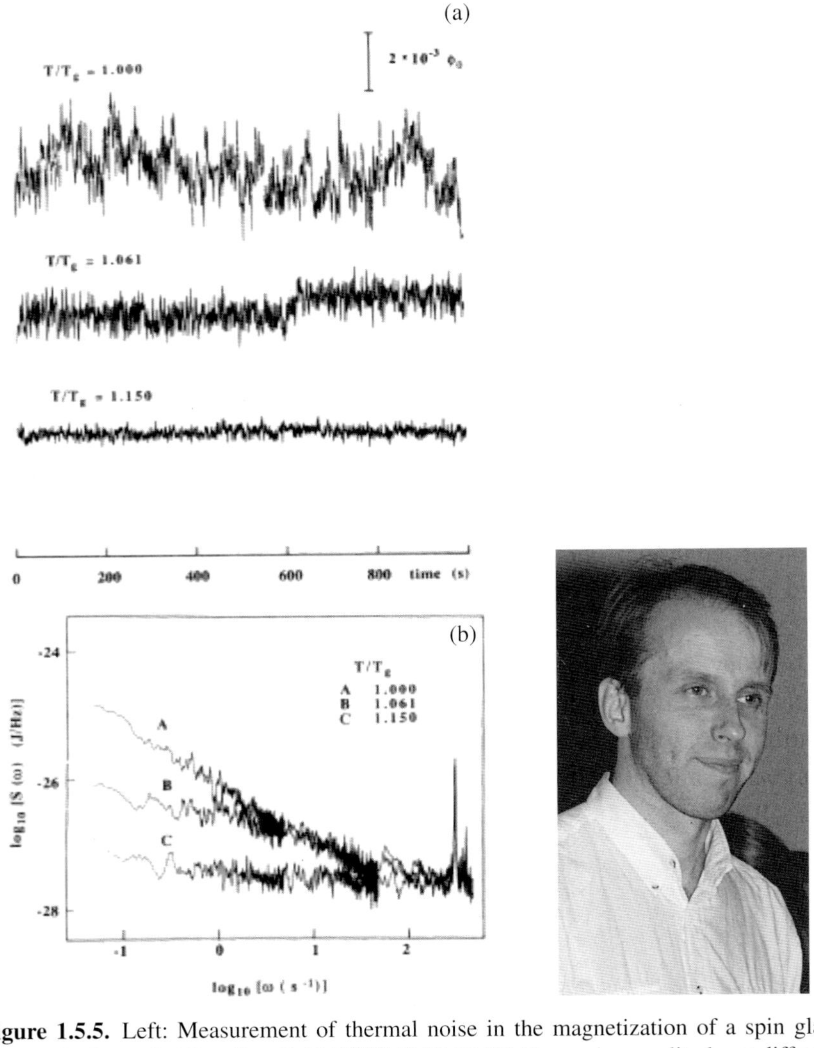

Figure 1.5.5. Left: Measurement of thermal noise in the magnetization of a spin glass [P. Svedlindh, *et al.*, *Phys. Rev. B* **40** (1989) 7162–7166]. Top: noise amplitudes at different temperatures relative to the glass melting temperature. Bottom: the corresponding power density spectra. Right: the leading author, Peter Svedlindh from Uppsala University, Sweden (September 19, 1989). Peter Svedlindh is full professor at Uppsala University and a leading magnetism and low-temperature physics expert (also a longtime friend and colleague). We got in touch due to this paper of his, and we had conducted joint research on the noise in high-temperature superconductors. We experimentally and theoretically proved the existence of percolation by scaling analysis of the measured noise intensity; an idea that was resisted by the mainstream of that time.

1.6. On Secure Communications and Key Exchange

Secure communications are probably as old as humanity. There has always been confidential information exchange between two persons that a third one is not supposed to understand. In this section, we discuss a few basic aspects of communication and security.

1.6.1. Related intriguing questions:

- Do animals of the same species purposefully use secure communications?
- Does biology use secure data exchange?
- How about the brain?

My daughter, Eszter, is a neural scientist; and I recently asked her if it could be said that there was secure data transfer in the brain. The conclusion of the discussion was that perhaps, the inhibitory signals and related coding provide such a function. Secure communications may be part of the scheme of how the conscious mind stays unaware of the information flow within the subconscious mind.

Non-secure data exchange. Figure 1.6.1 shows a non-secure communication scheme where Eve, our eavesdropper, has access to the information channel in which the data exchange is not encrypted. Therefore, Eve has full access to the exchanged information.

What does security mean? It can mean various situations, such as:

(i) Security by making Eve's access impossible. See Figure 1.6.2. The channel may be secured by force (security personnel with sensors and alarm; etc.) or by physical means, such as robust encaging. The latter case belongs to the problem of *tamper-resistance*. This class of security is not the subject of the book.

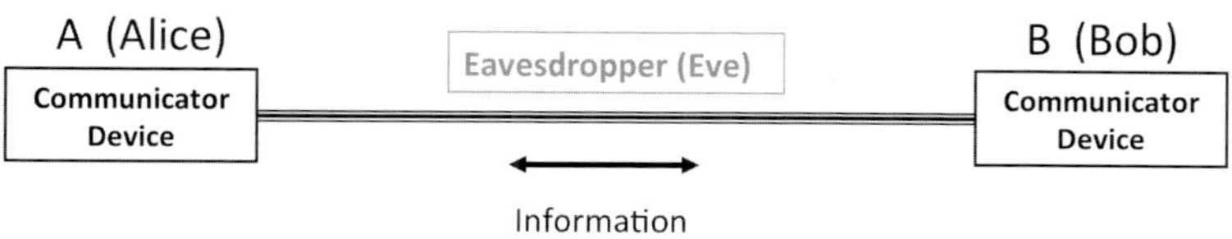

Figure 1.6.1. Non-secure communication scheme. No data encryption and Eve has access to the information channel.

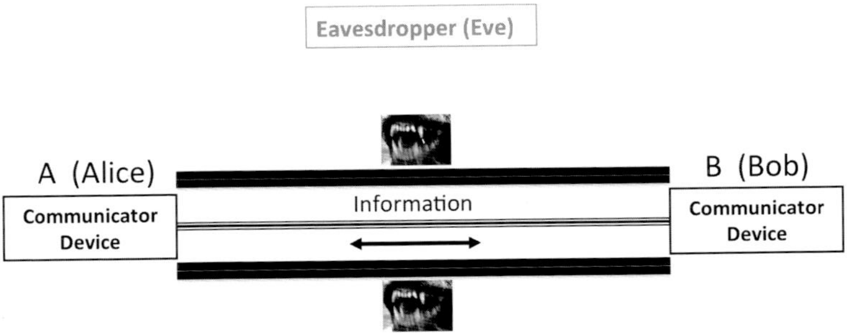

Figure 1.6.2. Eve is prohibited from accessing the information channel. This type of security belongs to the larger class of tamper resistance.

(ii) Security by measurement theory and/or information theory. The eavesdropper has access and can do measurements on the channel but the laws of physics do not allow extracting the communicated information from the measurement data. The KLJN key exchange is an example for this type of security. The related basic definition of (perfect) secrecy of communication channels is*:

> *Eve is limited only by the laws of physics and her knowledge about the communicated data does not improve by her monitoring of the communications.*

(iii) Security by eavesdropping detection. The eavesdropper can extract the communicated information — however, when that happens, it disturbs the channel so that the sender and receiver discover the eavesdropping activity. Quantum key distribution (QKD) and the security of KLJN against invasive (active) attacks are relevant examples.

(iv) Secure key and encryption. Alice and Bob have a secure key, which is unknown to Eve, and use one-time-pad cipher to encrypt and exchange information, see Figure 1.6.3. An essential question is: how do Alice and Bob share a key securely in order to be able to use this scheme? (see "Secure key exchange" below).

---

*C.E. Shannon, "Communication theory of secrecy systems", *Bell Systems Tech. J.* **28** (1949) 656; G. Gilbert, M. Hamrick, "Secrecy, Computational Loads and Rates in Practical Quantum Cryptography", *Algorithmica* **34** (2002) 314–339.

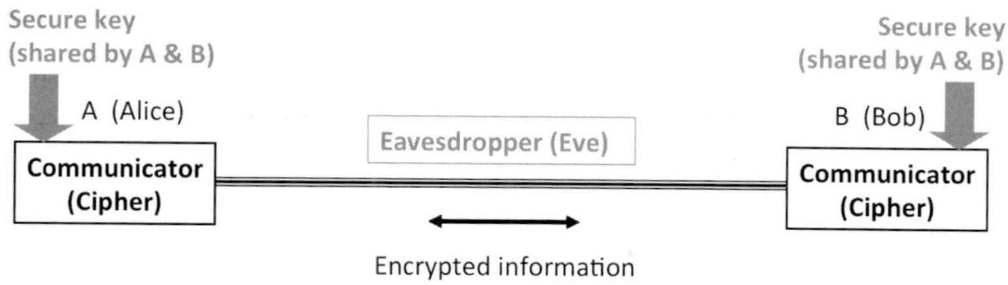

Figure 1.6.3. Example of encryption-based security: symmetric key encryption scheme. The "key" question: how do Alice and Bob share the key securely before this scheme can be applied? This is the focus topic of the book.

In the rest of this section, the reader can find some basic definitions about security, which are useful to get the KLJN concept.

Kerckhoffs's principle. It says that a perfectly secure communication system should stay perfectly secure even if all the details about the system become public information — *except the key*. Shannon had differently formulated the same statement later on as "the enemy knows [all details of] the system" except the secure key (Shannon's maxim), and security should endure even in this situation. When we are aiming information-theoretic security (security derived from information theory), we must suppose that Eve knows everything about the system, except the momentary key. She knows the blueprint of the system, the parameters of the building elements, the protocol, the cypher, etc, but the key.

Secure key exchange (or secure key distribution). This is a secure communication protocol to generate and share a secure key between Alice and Bob. Everyday examples are today's secure computer connections, which are software based and *basically not secure* (conditional security). Only a *physical secure key exchange* can offer unconditional/information-theoretic security and they belong to points (ii) and/or (iii) above. Our answer to this question is the KLJN key exchange, which offers unconditional (information-theoretic) security (see below).

Conditional security in key exchange. In today's software-based secure communications (tools we use when connecting to our bank via the internet), before the secure data exchange can start, Alice and Bob must generate and share a joint secret key through the communication channel while the eavesdropper (Eve) is supposedly monitoring and recording the related data. Everyday examples are all software-based key exchanges (RSA, Diffie-Hellman, etc). Alice and Bob typically generate, manipulate and share random numbers and their combinations. Finally, this process results in identical key bits "popping up" with Alice and Bob. (Note: there are also asymmetric key systems, but the basic features [including conditional security] are the same). Eve can record the communicated data and has all the information to crack the key, therefore, *from an information-theoretical point of view, these schemes offer absolutely zero security*. Then security exists only when her computational abilities are limited but it is only a question of the time in which she will know the key. Because new algorithms and computing solutions are continuously researched, today's software-based secure communications are potential time bombs and offer *non-future-proof security* (see below).

Information-theoretic security for the key exchange means that information theory (and measurement theory, if it is physical system) guarantees the security of the key exchange even if Eve's measurements extract the

Figure 1.6.4. Eleni Diamanti is an experimental quantum information scientist. She is a tenured researcher at CNRS, Paris. Her most relevant research for KLJN is the effort to integrate a quantum key exchange system onto a chip (a system-on-chip). So far, only KLJN can claim this advanced feature. While a quantum chip will always be about 1000 times larger due to the greater photonic wavelengths, a successful integration of even a single system on a chip would be a major breakthrough, because today's quantum units are bulky; and it would make major applications (hardware keys, securing chips, digital money, etc.) on a large scale possible.

Figure 1.6.5. Renato Renner is a leading quantum key distribution (QKD) researcher. He is full professor in theoretical physics, at ETH, Zurich. He is involved with other issues too; such as thermal demons and quantum random generators. Renato, Horace Yuen and Osamu Hirota have been in a high-profile debate† about the security of QKD, and this is exciting because science progresses most efficiently via debates. (Renato and I have different opinions‡ about some aspects of demons/Landauer's theorem, and about the question on whether classical or quantum random generators are more random, and whether quantum randomness is theoretically provable or only an axiom in quantum physics). In the picture, Renato is preparing to act in a performance at the wedding of his brother-in-law in Korea. (Courtesy Renato Renner).

maximum amount of information allowed by the laws of physics. Thus, for example, nearly infinite computational power or measurement sampling rate do not offer any way to crack the key. Neither future quantum computers nor noise-based logic processors can crack such a key. With simple wording: for Eve, no accessible information exists about the key in the extracted signal/information.

Future-proof security of the key exchange. Unconditional/information-theoretic security is future-proof because even the advancement of technology cannot crack the key by further analyzing the signal recorded in

†R. Renner, "Reply to recent scepticism about the foundations of quantum cryptography", http://arxiv.org/abs/1209.2423 [11 September 2012] and H.P. Yuen, "On the Foundations of Quantum Key Distribution — Reply to Renner and Beyond", http://arxiv.org/abs/1210.2804 [22 October 2013].

‡L.B. Kish, C.G. Granqvist, T. Horvath, A. Klappenecker, H. Wen, S.M. Bezrukov, "Bird's-eye view on noise-based logic", *International Journal of Modern Physics: Conference Series* **33** (2014) 1460363. http://www.worldscientific.com/doi/pdfplus/10.1142/S2010194514603639

Figure 1.6.6. Vadim Makarov, a leading quantum hacker, in Korea. Formerly, he led the quantum hacking lab in Norway where they hacked and completely cracked all commercial quantum communicators, and provided patches to fix the vulnerabilities. Currently, he is a research professor at the University of Waterloo, Canada, and Head of the Quantum Hacking Lab there. He is one of the most knowledgeable experts in the hardware (photonics, electro-optics, etc.) aspects of quantum security and their weaknesses. (Courtesy of V. Makarov). See more at Figure 5.4.1.

the past. For example, conditional security is not future-proof if it is based on Eve's limitation of computational power. Eve can record the whole key exchange and eventually crack the key when her slow computers are able to do the job or when proper quantum (or noise-based logic) computers will be available for her.

On quantum key distribution (QKD) schemes. There are many clever QKD-related schemes (starting with Stephen Wiesner's [1970s]; Charles H. Bennett and Gilles Brassard's [1984]; and Artur Ekert's [1990]). They are claimed to offer unconditional (information-theoretic) security. Even though this claim is currently debated as a practically reachable goal, I am convinced that, under proper conditions (such as slow speed and proper anti-hacking expansions) they will be able to offer unconditional security, probably even at the strict requirements given by Horace Yuen.

The information bits are carried by single photons. Here the *no-cloning-theorem* of quantum physics, which is one of the quantum uncertainty principles, is the theoretical foundation of the security. It means that a single photon state cannot be cloned without error (noise). If Eve captures

Figure 1.6.7. Horace P. Yuen pointing out why quantum key distribution has not yet been proven secure.[§] Picture taken at the first international conference of Hot Topics of Physical Informatics (HotPI), November 11, 2013, Changsha, China. He is a full professor at the Northwestern University, Chicago, and is a recipient of the prestigious Quantum Communication Award of Tanajawa University, Japan. His activities involve a broad range of topics in quantum optics, such as quantum communication theory, quantum cryptography, the foundations of quantum physics and new quantum devices. He is one of the most knowledgeable experts in the theories of quantum security and security in general. He is the inventor of the KCQ (keyed communication in quantum noise approach) approach. (Courtesy of He Wen).

and measures the photon, it gets destroyed and she must regenerate and reinject it into the channel, otherwise this bit will be considered invalid by Alice and Bob. However, due to the no-cloning rule, while Eve is doing that, she introduces noise into the channel — thus the error rate in the channel will become greater than without eavesdropping. Therefore, by evaluating the error statistics, after analyzing a number of transmitted bits and their errors, Alice and Bob will discover the eavesdropping. Without getting into further detail: in the *ideal case*; single photons with proper randomness in their polarization (or other properties), a correspondingly set up detector system, and an authenticated classical communication channel, are used to generate/exchange the secure key. Multiple photons,

---

[§]H.P. Yuen, "Essential lack of security proof in quantum key distribution", http://arxiv.org/abs/1310.0842 [22 Oct 2013].

channel noise, non-ideal building components, etc., compromise security. Moreover, quantum hackers, such as Vadim Makarov, have so far used only the simplest vulnerabilities for attack. However, according to Horace Yuen and Osamu Hirota, even the idealistic scheme has unproven aspects of security (for more information, see the paragraphs on "Statistical distance" and "Known plain text attack" later in this section).

The Kirchhoff-law-Johnson-noise (KLJN) key exchange scheme. It is the focus topic of this book. The KLJN scheme offers unconditional (information-theoretic) security (see Section 5.1 for the most general security proof). In a KLJN system, our communicators, Alice and Bob, are connected by a pair of wires (one of the wires can be the ground). If they shortcut the wire pair then a wire loop is formed. However, instead of shortcutting, they terminate the wire pairs by a randomly selected resistor with temperature T and with resistance unknown by anybody else; say R_A at Alice's end, and R_B at Bob's end. Thus the total loop resistance is the sum of the two secret resistance values:

$$R_{loop} = R_A + R_B \tag{1.6.1}$$

The system is fully defended against external voltage or current excitations (see, e.g., Sections 3.5 and 5.8.2); however there are spontaneous current and voltage fluctuations in the loop due to the Johnson noise of the resistors. Everybody, including Eve, is allowed to execute passive current and voltage measurements in/on the loop. Then from the measured Johnson noise current and voltage, the total loop resistance R_{loop} can be determined. Then Alice determines Bob's resistance by the $R_B = R_{loop} - R_A$ operation while Bob determines Alice's resistance with $R_A = R_{loop} - R_B$.

Eve can determine even the values of the connected resistances but, when they are different, not their actual location. The only directional quantity is the current-voltage cross-correlation, which is zero due to the Second Law of Thermodynamics prohibiting power flow in thermal equilibrium.

1.6.2. Note, KLJN can also offer ***secure analog data transfer*** if resistors with continuum (and arbitrary) values are used because Alice and Bob can determine the resistance at the other end in a secure way whenever they are not equal, see, e.g. Section 5.8.4.

(*Continued*)

(Continued)

1.6.3. KLJN was named the "*Kish Cipher*" by Bruce Schneier on his blog site around 2006; and the "*Kish Cypher*" by Derek Abbott around 2008 when he created a Wikipedia page with that name; the Wiki page was deleted in 2013 (at this author's request) during intensive fights between *supporters* and *objectors* (most of whom were *quantum-supporters*) where even the supporters were often wrong (a situation that triggered the contract to write this book). However, a key exchanger is very different from a cipher/cypher; thus the name is incorrect and misleading. Bruce was not completely wrong because, in the original paper, KLJN was proposed together with a simple one-time-pad cipher (see Section 3.4): the randomly generated key bits were numbered and Alice sent the index numbers to Bob to form a secret message. However, KLJN very quickly took the shape of a secure key exchanger, and by 2007 it had taken on its current form — thus the name of the Wiki page was completely improper. If we were to use the Author's name in the naming of KLJN, the correct one could have been the Kish Key Distribution (KKD), which Derek Abbott has been using recently. However, analogously to QKD, an alias without family name is better, especially since there are now 11 different systems (see Sections 3.4, 5.3, 5.8.3 and 5.8.4). Ten different versions are by Kish (iKLJN, MKLJN, KKLJN, iMKLJN, iKKLJN, KMKLJN, iKMKLJN, RRKLJN, RRRTKLJN, and the original KLJN), and one by Vadai–Mingesz–Gingl (VMG–KLJN). In conclusion, the usage of KKD is inaccurate and misleading.

It is important to note that, similarly to QKD, KLJN is perfectly secure only when the idealistic mathematical model holds. In the non-ideal (practical) case, there is information leak (see Chapters 4, and 5) while the unconditional security can still be maintained by various techniques, including privacy amplification (see Sections 5.1–5.3 and 5.8). Against invasive attacks, further extensions are needed (see Sections 3.5 and 5.5).

1.6.4. During the recent years, cyber attacks, hackers, whistleblowers, etc., brought secure communication needs into focus. While governments have announced heighten information security initiatives in response, in fact, not much happened. Neither QKD nor KLJN has a market yet. The headlines and much hype we have seen about QKD in the media are either promotions (like using it for communicating the Swiss voting results, or securing a key exchange between two banks), or due to the high-profile cracking of its commercial

(Continued)

Figure 1.6.8. Robert Mingesz in the Noise Lab at Szeged, Hungary, during the "*Szegedin–Whisper*" project** (see Section 3.9) making the first experimental demonstration of the KLJN key exchange. Behind him, Zoltan Gingl, the Lab Director. Robert is the leading author of experimental works on KLJN. He is currently Assistant Professor at the University of Szeged, Hungary. Be careful if you recommend him as a referee: he is just too smart and will not miss any of the weak points in your paper!

(*Continued*)

versions. KLJN is even more behind, partially because its life begun 31 years later — not even a commercial KLJN-based product exists yet, just research papers and some debates about its security (often due to reasons of limited expertise, ignorance or bias). If personal data security would be as important as to develop a telephone network in Alexander Graham Bell's time, we would have already seen small businesses popping up to create chip realizations of KLJN and cheaper versions of the currently bulky and expensive versions QKD: e.g. non-clonable (PUF-based) credit cards utilizing KLJN; ultra-secure computers; etc. However, while Big Brother is getting bigger everywhere he may not be too interested in promoting unconditional security on a large scale because information-theoretic/unconditional security cannot be cracked — not even by the National Security Agency (NSA), USA.

**R. Mingesz, Z. Gingl, L.B. Kish, "Johnson(-like)-Noise-Kirchhoff-Loop Based Secure Classical Communicator Characteristics, for Ranges of Two to Two Thousand Kilometers, via Model-Line", *Physics Letters A*, 372 (2008) pp. 978–984.

Figure 1.6.9. Ferdinand Peper (2013, Courtesy of F. Peper). He is a senior researcher (with tenure) at NICT, Osaka, Japan. His main field is unconventional computing schemes — however his range of interests includes unconventional informatics and the brain. He attended the INCF noise conference in Tokyo (2007) where we first met. Later, our collaboration was centered on noise-based logic, where Ferdinand solved the difficult problem of making a non-squeezed AND noise-based logic gate.[††] Also, we co-published the first paper about utilizing power lines for KLJN.[‡‡]

Eve's successful guessing probability *p* of a key bit. This quantity is relevant for a physical secure key exchange. In the case of perfect security, $p = 0.5$, equivalent to a perfectly random coin; which means Eve has zero knowledge about the key. Perfect keys need also to have IID (Independent and Identically Distributed) statistics. If the key is not IID then p is not enough to fully characterize the security. Then measures based on *statistical distance* between the real and perfect keys are used (typically in quantum communicators — see segment on **Keys with independent and identically distributed (IID) statistics** for more information).

Key with independent and identically distributed (IID) statistics means that the probability distributions of the key's bit values represent independent, identically distributed random variables. This property is easily destroyed by error correction and high-efficiency privacy amplification protocols, which are essential for quantum encryption (see

[††] F. Peper, L.B. Kish, "Instantaneous, non-squeezed, noise-based logic", *Fluctuation and Noise Letters*, **10** (2011) 231–237.

[‡‡] L.B. Kish, F. Peper, "Information Networks Secured by the Laws of Physics", Invited paper, *IEICE Transactions on the Fundamentals of Communications, Electronics, Information & Systems*, **E95-B** (2012) 1501–1507.

Figure 1.6.10. Pao-Lo Liu at the La Rambla Fountain in Barcelona (2010, Courtesy of P.L. Liu). He is a very original thinker and a full professor at The State University of New York at Buffalo, USA, with a background in photonics. He proposed the idea that the resistors in the KLJN system can also be emulated by circulators (see Section 4.1), which are active elements. He was the first one who tried to (unsuccessfully) crack the KLJN scheme by utilizing the delay effect in the steady state. His most interesting contribution is his own key exchanger, which is basically a computer model of the KLJN scheme realized by linear systems theory in the wave-regime; thus the signals propagating in the two directions are separated. Pao-Lo now says his system is crackable in the transient limit but it is secure in the steady-state even though the signal components propagating in the two opposite directions can be extracted and separated (on the contrary of the situation in a KJLN system). Thus, directional couplers (which actually do not exist for KLJN) would not be able to assist Eve (see also Sections 4.6.1–4.6.3 about an unsuccessful attack). With Zoltan Gingl, we have not been able to crack Pao-Lo's key exchanger in the steady state mode though we tried hard. Neither has Tamas Horvath succeeded with that goal. More on this in the next two figures.

below). The simplest XOR-pairs based privacy amplification, which is sufficient for KLJN (see below) keeps the IID property, thus KLJN keys are typically IID (especially because no error correction is needed).

Brute force attack. If Eve knows the cipher and the protocol, she can test all the possible bit variations in the N bit long key to crack an encrypted message. Because each bit has two possible bit values, there are 2^N different key possibilities.

Uniformly distributed key (perfect key). When, during the brute force attack, each key possibility has the same probability, the key is uniformly distributed and perfect. This probability is 2^{-N}.

Figure 1.6.11. Zoltan Gingl and a bottle of Paulo red wine, which we bought to acknowledge the Pao-Lo Liu key exchanger after our intensive but unsuccessful effort to crack it (taken in the author's house in College Station, 2010).

Statistical distance measures between the actual and the ideal keys. Note: this is not needed for KLJN; however, it is useful for showing KLJN's superior security compared to the security of existing quantum keys. For IID keys (see above), the p value provides a full characterisation of the security of the key. However, in quantum communicators, the IID property is usually destroyed by aggressive error correction and privacy amplification. Thus they evaluate the security by a statistical distance measure called the *trace distance*. To compare the security of KLJN and quantum keys, it is desirable to calculate the equivalent statistical distance measure for KLJN, namely, the *total variation distance*[§§] between the probability distribution characterizing Eve's extracted key and that of a perfectly secure key of the same length. In order to assess the security of the shared key, one must compare the probability distribution of successfully guessing each possible key sequence of the N-bit-long key extracted by Eve, encompassing 2^N different sequences, with that of the perfectly secure key having uniform distribution. The total variation distance Δ of the probability distribution characterizing Eve's extracted key and that of a perfectly secure key of the same length is

$$\Delta(E, I) = \max_{j=1,\ldots,2^N} [P(E_j) - P(I_j)], \tag{1.6.2}$$

[§§]J.S. Rosenthal, "Convergence rates of Markov chains". *SIAM Rev.* **37** (1995) 387–405.

where E and I indicate Eve's extracted key and a perfectly secure key, respectively, and $P(E_j)$ and $P(I_j)$ are the probabilities for correctly guessing the jth version of Eve's key and of the perfectly secure key, respectively.

1.6.5. The trace distance, the statistical distance measure used in QKD, is a topic of heated debates between leaders in the field. Though the measure is valid, its interpretation differs. Horace P. Yuen[***] and Osamu Hirota[¥] point out that the interpretation of the situation by Renato Renner and his coworkers is incorrect. Without entering into details, we mention that practical KLJN can easily offer less than 10^{-500}, which is close to perfect security for the given key length (see the Chapter 4, Section 4.4).

Known plain text attack. Eve may launch a key-inverting attack when, e.g. due to Kerckhoff's principle, she knows part of the text that is encrypted. This attack is particularly dangerous when there are correlations within the key (non-IID key), such as in quantum keys.[***]

Man-in-the-middle attack. This is an invasive (active) attack. Eve breaks the line and installs two communicators to conduct two separate communications with Alice and Bob. She shows up to Alice as Bob and to Bob as Alice. (Note, the expanded KLJN system is naturally protected against this type of attack — see Section 3.5).

Privacy amplification of the key. Privacy amplification is a software tool (based on hashing) that is basically an error enhancer. Suppose, on the original key, Eve has a successful bit guessing probability of $p_1 > 0.5$ (error probability $(1 - p_1)$). Alice and Bob runs the key through a privacy amplification protocol (Eve must do the same and she can do that due to Kerckhoffs's principle). The new key will be shorter and Eve's related error probability $(1 - p_2)$ will become greater because $0.5 < p_2 < p_1$ due to the

[¥] O. Hirota, O.: *Incompleteness and Limit of Quantum Key Distribution Theory*. http://arxiv.org/abs/1208.2106, [26 Aug 2012].

[***] H.P. Yuen, "Simple explanation why QKD keys have not been proved secure", http://arxiv.org/abs/1408.4780v1, [20 August 2014].

Figure 1.6.12. Tamas Horvath (2010) is a tenured senior computer scientist at the Fraunhofer Institute for Computer Science in the Bonn area in Germany. He is an old friend from Szeged. He analyzed the issues of the XOR-based privacy amplification for KLJN and has had several important contributions for noise-based logic, which is the "fall-out" of KLJN. He brought abstract mathematics and rigor to the projects. He also failed to crack Pao-Lo's KLJN-mimicking scheme (see the caption of Figure 1.6.10).

privacy amplification. There are two problems with privacy amplification:

(a) Because it is an error enhancer, it will increase the error probability of Alice and Bob (if they had a non-zero error probability). This poses serious implications for QKD because of the huge error probabilities of Alice and Bob there. The error correction protocol that must be used before privacy amplification in QKD gives out information to Eve.
(b) Unless the inefficient but simple and "clean" way (XOR-ing bit pairs) of privacy amplification[†††] is used, it can introduce correlations between key bits. This is a new source of information for Eve. These serious problems with QKD have low visibility in the literature and only security experts, like Horace Yuen and Osamu Hirota, are discussing them.

Hacking attacks against physical key exchangers. Because, in accordance with Kerckhoffs's principle, Eve knows the systems except the keys, she can utilize the non-ideal features of the building elements to extract information. Security proofs are always based on mathematical

[†††]T. Horvath, L.B. Kish, J. Scheuer, "Effective Privacy Amplification for Secure Classical Communications", *EPL (formerly Europhysics Letters)* **94** (2011) 28002-p1–28002-p6

assumptions about the behavior of building elements and any such element will deviate from these assumptions in some way at proper conditions. Such non-idealities lead Vadim Makarov to totally crack all the commercially available quantum communicators in many different ways by simple means, for example by blinding the detectors and then measure into the systems. And what has been seen up to now is only the tip of the iceberg because those systems are very complex with many non-ideal components. This is the situation that offers a treasure for hackers to exploit. On the contrary, KLJN is relatively simple; still, hacking attacks must be considered very seriously.

Chapter 2
The Dawn (2002–2004)

Some important personal history follows here. In 2001, I left my non-permanent position at Claes Granqvist's Solid State Physics Division at Uppsala University Sweden for a tenured (permanent) faculty position at Texas A&M University, College Station, USA.

> **2.1.** I had made it clear already during my hiring interviews that I will leave the well-paying topic of nanotechnology because I want to return to noise research which I felt was more fundamental and interesting. Concerning research money, fluctuation-enhanced sensing (FES) for chemical agents was the envisioned source (we had two Swedish patents for it). I was originally promised a multi-year collaboration project by a sensor company in Bangor; however, the director for electronics who was my actual collaborator left before the project was realized. In the new situation, I naively submitted a research plan to said company at their request without a signed contract/agreement to protect my rights. Then, after I helped them to hire an inexperienced postdoc with a noise-based PhD, they turned down further collaboration with me — probably in the hope that they can realize the projects without me, which (I suppose) seemed the more economical way. The project failed, and I "washed" my (still empty) hands.

Having a permanent position and no grants and their respective research plans to satisfy, I felt energized to take up new topics.

2.2. Here, I am sharing my *personal opinion* about research funding but much of these hold for confidential paper peer reviewing systems, too. The pressure to get research grants in order to be able to function as a scientist and to get a tenured faculty position is killing science itself. When greed takes over, personal connections, salesmanship, politics and political correctness enter; causing damage. In such a system, it pays better to be a part of the mainstream and follow its mediocre goals and do its incremental research. The typical National Science Foundation (NSF) grants (*although there are always some good exceptions*) are for projects at the top of the mainstream. In the best cases, they are technically of the highest possible quality, but their potential to transform science is doubtful. There is no funding for new Einsteins: those ideas are regarded as too high-risk. The sizable committee of experts, even with good intentions, automatically provides the unavoidable filter of mediocrity: there will be always somebody there who does not like the breakthrough ideas of today's Einsteins. Thus objection-free, "*safe*", and technically high-level proposals have higher chances of getting funding instead. But let us assume the improbable; that today's Einstein gets funded. I know excellent scientists and their students who have been broken/destroyed by large research funding; I have seen them overwhelmed by the pressure to get more money, by the difficulties of spending it, and the need to manage a large group — all tasks of a small-business CEO, not of a *breakthrough scientist who, most of all, would need a pressure-free life and quality time for deep thinking*. In such cases, the students are supervised by postdocs or senior students instead of the supervisor who has the knowledge, skills and experience to share. And the research output (typically in leading journals and magazines) is not surprising: the mass-production of shining papers with sophisticated technical stuff and 3D color figures, which are highly cited at the moment (citing a high-profile journal looks good in the reference-list of a paper), but with no true impact on the course of future science, and which were likely to be forgotten in a decade or two.

In Figure 2.1, Derek Abbott (in his house) is enthusiastically showing me a paper (December 14, 2001). At that time, he was excited about Moore's law of miniaturization, quantum computing, and quantum communications. It was then that I got "infected", and the result was the first paper of this chapter, challenging the fundamental limits of Moore's law.

The second paper, with quantum physicist Julio Gea-Banacloche, proved that quantum computers could never win over classical computers for general-purpose applications; thus what remains are special-purpose computers built for specific purposes — just like analog computers.

Figure 2.1. Derek Abbott in 2001. Derek is a full professor of electrical engineering at the University of Adelaide, Australia, and a longtime friend. He has a strongly multidisciplinary interest spanning across many engineering and science fields. We ran some conferences and symposia together, often for the purpose of challenging the status quo and popular dogmas. He is always enthusiastic about some idea, and has on several occasions triggered my own scientific interest — including those challenging the quantum claims and developing the KLJN. However, we are at a standoff regarding several aspects of KLJN, see Chapter 4, and Sections 5.8.5, 5.8.7, and (most importantly) 5.8.8.

And even then the unanswered question remains: who is the winner for special-purpose computing: quantum or (non-Turing-type) classical computers? My opinion is that classical computers utilizing noise-based logic will be the winner, but only the future will tell.

This period was the dawn of the KLJN. The Moore's law project was about challenging Moore's law with energy dissipation arguments. The next step was to do the same with quantum computers. All these can be summarized as *exploration of the error-speed-heat dissipation triangle*.

2.3. Continuing on sharing my *personal opinion* on research funding, the grant proposal requirements of the NSF and similar agencies are simply ridiculous: *The applicant must present a plan about what he/she will "discover" during each year!* Those who founded these grant formats and expect the predictions to be fulfilled on schedule have certainly no understanding of how scientific discovery happens! First rule: you cannot plan it. You must seek the truth and, from 100 ideas, you will find that 99 are bad or trivial. If you are lucky,

(*Continued*)

(*Continued*)

the remaining 1 idea in the hundred will change the course of understanding science today by, for example, killing a popular dogma or presenting a new effect. An anecdote reportedly said that *Albert Einstein* was also annoyed by this situation and had responded with the following: "***If we knew what it was we were doing, it would not be called research, would it?***"

2.4. The creation of the NSF Foundation was proposed by *Leo Szilard*, who said that the agency should pay *summer-salaries* for interested university science faculty who would then use this time to pursue ***unidentified ideas*** to revolutionize science. Things today work in the opposite fashion: the NSF committees typically turn down unidentified ideas as high-risk and limit summer salary to a fraction of the full summer. (In the US system, universities usually pay faculty for 9 months/year; the summers are unpaid. In Canada and Europe, faculty members receive a 12-month salary and about 30–40 workdays' worth of paid vacation).

After these, the next logical step was to challenge the uniqueness of quantum physics for secure communications. The third paper of this chapter, introducing the Kish–Sethuraman cipher, was the first attempt.

2.5. While the Kish-Sethuraman cipher probably cannot be built by classical physical means, it was an intensive effort at the time and the first exposure to security.

2.1. The Thermal Noise Death of Moore's Law of Computer-chip Miniaturization

Brief summary:

2.1.1. Contrary to the claims of the industry and the media, Moore's law was broken a long time ago. The most important reason is as follows:

The maximum clock frequency has not been utilized since 2003 where it was 3 GHz. Today (2015), we are around the same clock frequency instead of the

(*Continued*)

(*Continued*)

12–15 GHz that would be easily achievable if heat were no issue. It is like a Hong Kong pocket radio in the 1970s marketed as an "18 transistors" radio where 12 transistors were only used as diodes.

2.1.2. In this section, for historical reasons, we present our analysis of the fundamental physical limits of processor miniaturization when the speed of the transistors is fully utilized. This analysis directly led to challenging general-purpose quantum computing on a similar basis, from where the next natural step was secure communications and KLJN.

2.1.3. Note: Moore's law has not yet been broken due to these physical limits; we are far from that. There are non-ideal features in the chip, including noise from cross-talk and limited trans-impedance of MOS transistors, which result in similar effects as that shown in the error-speed-heat triangle described below.

The issue here is the error-speed-heat triangle, which supported the questioners of some former claims of quantum informatics about heat-dissipation-free computing. The paper below (invited by Jamal Deen), was the second one in line on this topic. Though it was the first paper[1] that became famous, the present paper contains an expanded message because of its considerations about quantum gates; why they cannot be used in a small-error operational mode. While the deduced fundamental limits are hard, the predictions turned out too optimistic because Moore's law broke next year when Intel chips were either calculating correctly or being overheated. The main conclusion of the first paper was that Moore's law (in a single core) couldn't be followed beyond the limit unless the clock frequency was decreased, thus Moore's law would have lost one of its main motivations: faster transistors. At that time the clock frequency was 3 GHz and processors were single-cored. After that the clock frequency was first decreased, later slightly increased and now it is still at or below 3 GHz. To avoid errors, the power supply voltage was not reduced as much, which goes against energy efficiency. Today processors are multi-cored — through they cannot be fully utilized due to problems with parallelism — and the

[1] L.B. Kish, "End of Moore's Law; Thermal (Noise) Death of Integration in Micro and Nano Electronics", *Phys. Lett. A* **305** (2002) 144–149.

Moore's law issue is no longer a question of science or practicality: it has become strongly political in a hope to convince funders to further invest into miniaturization. Miniaturization to these extremes is good for profit, especially because chip lifetimes are getting progressively reduced due to the high electrical fields. Microprocessors (the logic board) are becoming like light bulbs; in need of replacing every few years if you want to keep your computer. This is one example of the many in the history of technology where greed results in decreasing the useful life span of a product instead of improving it. Much later, in 2009, these thoughts expanded by facts about the brain led me to introduce *Noise-based logic* (see Chapter 4.2) which has the ambition to reach the same goals that quantum-computers aim to reach — but much earlier — and at a small fraction of the cost via classical physics, utilizing random noise. We have yet to see if this is possible; however, the success of KLJN with beating the quantum communication dogma is very encouraging and maybe it is the sign that classical stochastics can do as much or more than quantum informatics.

2.1.4. A Note: The references in the following paper will be found in this form: e.g. [1], [2], [3], etc., with full reference information at the end of Section 2.1

Paper: Moore's Law and the Energy Requirement of Computing versus Performance[2]

Abstract. It has recently been recognized that speed, noise and energy dissipation are strongly interrelated entities. Following Moore's law of miniaturization, at sizes below 40 nanometers, physics will impose fundamental and practical limits of performance by shrinking the noise margin, increasing and quickening noise, and increasing power dissipation. It is important to locate the fundamental aspects of the problem, explore relevant practical problems and possible solutions, and investigate this situation not only in microelectronics (CMOS, etc.) but also in single-electron-transistor-based nanoelectronics, and quantum informatics applications. The energy requirement of running classical and quantum logic gates is compared.

[2]L.B. Kish, "Moore's Law and the Energy Requirement of Computing versus Performance", *IEE Proc. — Circ. Dev. Syst.* **151** (2004) 190–194.

1. Thermal noise and the Rice equation

Thermal noise is an omnipresent small voltage fluctuation on resistors. It has been thought that thermal noise will never be an issue in digital electronics. This view has been re-evaluated and changed recently [1,2]. On a parallel resistor-capacitor (RC) unit, the effective thermal noise voltage U_n is given as

$$U_n = \sqrt{kT/C}, \tag{1}$$

and the bandwidth f_c of this noise is

$$f_c = \frac{1}{2\pi RC}. \tag{2}$$

Gaussian noise processes, such as thermal noise, can cross large amplitude levels provided sufficiently long time is available. In a logic circuitry, noise amplitudes reaching beyond the noise margin U_{th} cause false bit flips, which can result in bit errors. For a single, band-limited noise process, the mean frequency $\nu(U_{th})$ of bit errors can be obtained from the Rice formula [1,2]:

$$\nu(U_{th}) = \frac{2}{\sqrt{3}} \exp\left(\frac{-U_{th}^2}{2U_n^2}\right) f_c, \tag{3}$$

where U_n is the effective noise voltage and f_c is the bandwidth.

So, what happens during miniaturization? In CMOS technology, the resistors are two-dimensional conductors, so the resistance stays (roughly) constant. Therefore, the supply voltage has to be decreased to keep the electrical field and dissipation at an acceptable level. The capacitances are decreasing. Together, these effects yield the following trends:

(i) A shrinking noise margin because U_{th} is only a fraction of the supply voltage.

(ii) A growing noise because of Eq. (1).

(iii) A growing bandwidth (see Eq. (2); quickening of the noise).

All these phenomena, **i**, **ii** and **iii**, act toward radically increasing the frequency of bit errors via Eq. (3). To have a feeling for the nature of this problem, on Fig. 1, the bit error frequency versus the noise margin normalized to the noise voltage is shown for different bandwidths (clock frequency) and differing numbers of transistors. The practical limit of

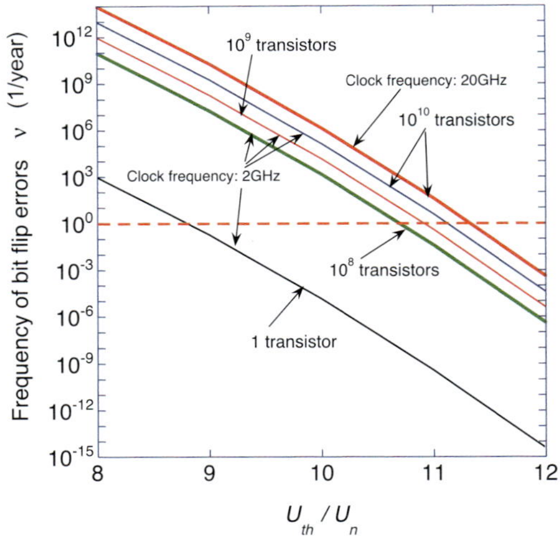

Figure 1. Bit errors versus the ratio of noise margin and effective noise voltage [1,2].

usability is a certain U_{th}/U_n ratio when the bit error rate is around 1 error/year. Even a 10% decrease of the U_{th}/U_n ratio compared to its critical value yields an error rate increase to 10^5 when we consider all transistors in a modern PC ($3 * 10^9 - 10^{10}$ transistors).

Using these results and a prediction of the evolution of capacitance and noise margin using present trends, recently, a prediction of the end of Moore's law was published [1,2], see Fig. 2.

The shrinking noise margin (logic threshold voltage) and the increasing noise margin required by the increasing and quickening noise pose conflicting requirements which would stop miniaturization in 6–8 years if the trends of last years continue.

In the rest of this article, we outline the objectives that need further analysis and give some initiatives for extended research in the future.

2. Refinements of the predictions for CMOS technology

It has been pointed out in [1,2] that the prediction was based on strong approximations in order to keep generality and because of lack of information about certain device parameters. Since [1] was published, new information from microprocessor makers about some previously unknown parameters have emerged, and some new efforts have been made to reduce

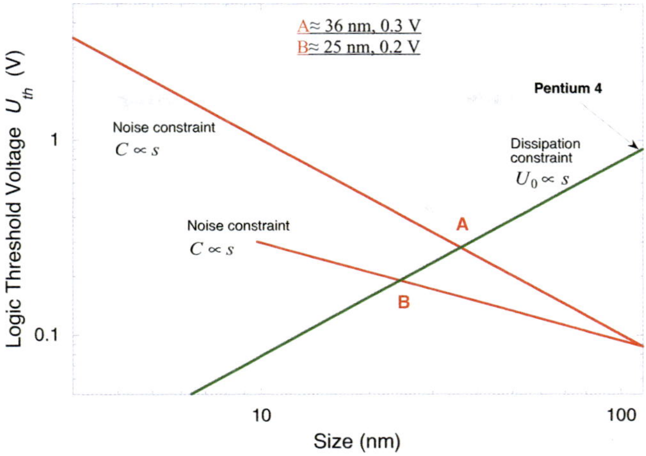

Figure 2. Prediction of the end of Moore's law [1,2]. The technology faces a difficult problem when the upper limit of the noise margin set by the dissipation/field constraint and the lower limit of the noise margin required by the noise/error constraint cross each other (between points A or B, depending on the evolution of gate oxide thickness).

the supply voltage with a rate *less* than previously supposed. These efforts are however *controversial* because many large-scale users, for example in aviation electronics [3], would like to *increase* the rate of supply voltage reduction in order to improve the device failure rate which has been steadily growing due to high electric fields in chips.

Therefore theoretical efforts have been made on refining the prediction for the bit error problem due to thermal noise. The refined model takes into the account the following aspects:

- Further noise margin decrease due to the opening threshold voltages of the P and N type MOSFET transistors.
- Somewhat reduced noise due to parallel gate circuits.
- The various supply voltage reduction strategies (predictions are controversial).
- The fact that the internal supply voltage and noise margin of microprocessors in 2002 was already less than supposed in [1].

Interestingly, the preliminary investigations indicate that the size range, where the problems begin, remains the same because the different corrections act in different directions and the effects compensate each other. Thus, the conclusions obtained in context of Figure 2 remain the same.

3. Single-electron transistor based microprocessors

The next question is the bit error situation of microprocessors based on single-electron transistors (SETs) [4]. SETs utilize tunneling, which is not dissipative, but the processes coupled to it are dissipative. The electrons at the tunnel junction of a closed SET can be excited by thermal energy fluctuations to the energy level where tunneling can occur and a single electron can cause a single bit error in a SET. Similarly, an open SET can be temporarily shut down by thermal fluctuations. So, the tentative expectation is that SETs will have similar nature of bit error characteristics as CMOS. However, the picture is more complex [4]:

- Instead of a single capacitance, three different capacitances influence the bit errors: the two tunnel junction capacitances and the quantum dot capacitance (gate capacitance).
- The noise margin cannot be increased arbitrarily by increasing supply voltage. It has a practical maximum which is equal to the voltage difference between a totally closed and totally open transistor. Higher voltages cause multiple electron operation mode and large noise and dissipation.
- There are two different working ranges versus the quantum dot size. A larger sized, Coulomb blockade controls the current transport, but with smaller sizes (<10nm), quantum confinement effects dominate.

The preliminary studies [4] show that the requirement for small quantum dot size becomes much more difficult to satisfy when not only the DC characteristics but also bit errors matter in a microprocessor with 10^8 or more SETs. To have a SET based microprocessor, the characteristic quantum dot size has to be around 1 nm.

4. Ultimate limits of energy dissipation versus performance in classical and quantum computing

The ultimate and most fundamental questions are related to the power requirement of classical and quantum information processing. If CMOS and SET fail, it is a natural question if quantum computing and quantum information can help and how. Because, so far, the existing quantum computing architectures are not practical, the only question we may be able to answer is the ultimate limits of performance. Performance includes error rate (accuracy), speed (bandwidth) and power dissipation. It is very important to take temperature into account, and to compare the ultimate performance limits of classical and quantum computers at the same temperature [5].

Recent studies [5, 6], see in the Appendix, show that quantum computers have problems at high accuracy due to Heisenberg's uncertainty principle. Classical computers perform much better as far as accuracy is concerned; however, quantum computers can balance this deficiency by a greater speed [5], see in the Appendix. For a comparison of the energy requirements of a classical and a quantum gate when they run at the same clock frequency, see Figures 3 and 4 [2, 5].

Recent studies [7] comparing general-purpose quantum and classical computers indicate that, even by using error correction, quantum computers will dissipate at least 100 times more energy at the same performance.

However, it is important to note that the ultimate focus of this study, in the future, will not be the accuracy (error rate) of classical and quantum computers but the energy requirements of processing Shannon-information

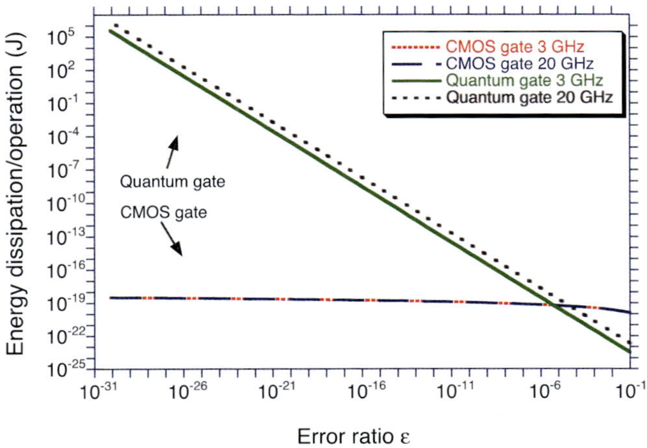

Figure 3. Minimal energy dissipation of a single logic gate [2, 5, 6], classical (CMOS) and quantum, versus the error ratio of the gate. The quantum gate result is for zero temperature. See the derivation in the Appendix.

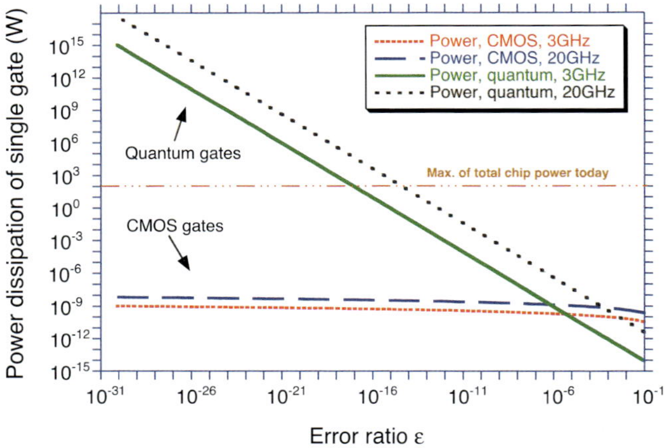

Figure 4. Minimal power dissipation of a single logic gate [2, 5, 6], classical (CMOS) and quantum, versus the error ratio of the gate. The quantum gate result is for zero temperature. See the derivation in the Appendix.

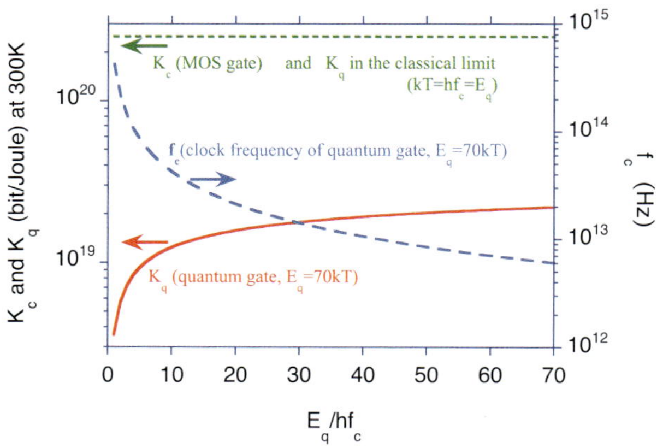

Figure 5. Bit/Joule performance of classical and quantum logic gates [5], see the derivation in the Appendix.

(bit). This aspect induces many new questions including the problem of computer architectures which are not sensitive to noise. In the Appendix, we show that the ultimate measure, Shannon's information channel capacity versus the energy dissipation, also gives better results in a classical rather than quantum gate, see Figure 5.

Conclusion

We all have been enjoying the fast growth of speed and memory size of computers during the last decades. Recently, the emerging fields of quantum computing and nanoelectronics have suggested that the future will be even more brilliant.

However, as soon as we confront physical laws and reality with expectations, the future seems to be less bright.

Acknowledgement

Discussions with Prof. Julio Gea-Banacloche and Prof. Minoru Fujishima are appreciated.

Appendix

A1. Accuracy versus power dissipation in CMOS and quantum computers

The lowest limit of the energy requirement, E_q, for a single operation of a quantum logic gate is given by Gea-Banacloche [2] as

$$E_q \approx \frac{\hbar}{\varepsilon_q \tau_q} = \frac{\hbar f_{c,q}}{\varepsilon_q} \tag{A1}$$

where ε_q and τ_q are the error probability and the time requirements of the quantum logic operation; and $1/\varepsilon_q$ is related to the accuracy of the gate operation. Then, for the case of sequential operation of the gate, the maximum clock frequency is given as $f_{c,q} = 1/\tau_q$. Though the required energy is not the dissipated energy but the energy requited to put into the quantum gate, we can consider this energy as a practically dissipated energy. The situation is very similar in CMOS technology, where the electric energy due to the gate capacitance charge is an ordered energy, it is most obvious that this energy is practically dissipated at each change of logic state.

For CMOS transistors, the mean frequency of bit errors [6] is given as:

$$\nu = \frac{2}{\sqrt{3}} \exp\left(\frac{-U_{th}^2}{2U_n^2}\right) f_c, \tag{A2}$$

where the U_{th} (logic threshold voltage) is the noise margin between the logic low (0) and high (1) levels, U_n is the effective thermal noise voltage on the CMOS transistors' resultant gate capacitance, and f_c is the RC time constant of the gate capacitance and its driving resistance, which is the maximal clock frequency. Then, if we drive the system at frequency f_c, the error probability is:

$$\varepsilon_c = \frac{\nu}{f_c} = \frac{2}{\sqrt{3}} \exp\left(\frac{-U_{th}^2}{2U_n^2}\right). \tag{A3}$$

Because U_{th}^2 and U_n^2 are related to the static and thermodynamical energies in the capacitor, respectively, this result can be generalized for any kind of available digital technology where, without exception, the logic threshold energy is dissipated during the change of logic state. Taking into account that the logic voltage change is the voltage change on the gate capacitance, the lower limit of energy dissipated during the change of logic state satisfies the relation:

$$E_c \geq \frac{1}{2} C U_{th}^2 \equiv E_{c,\min}, \tag{A4}$$

where C is the resultant capacitance at the gate electrode. Note, the actual value is $\frac{1}{2}C[(U_0 + U_{th})^2 - U_0^2] = \frac{1}{2}C(U_{th}^2 + 2U_0U_{th}) \geq \frac{1}{2}CU_{th}^2$, where U_0 is the low (0) logic voltage level, so Eq. (A4) indeed gives the lower limit (the $U_0 = 0$ case). The mean thermal noise energy in the capacitor is: $\frac{1}{2}CU_n^2 = \frac{kT}{2}$. Using Eq. (2), for room temperature, the minimum U_{th}/U_n ratio for error-free operation (<1 *false-bit-flip/year* due to thermal noise) was given 12 [1] which corresponds to

$$\frac{E_c}{kT} \geq 72 \approx 70 \tag{A5}$$

Thus, the error probability can be given as

$$\varepsilon_c = \frac{2}{\sqrt{3}} \exp\left(\frac{-U_{th}^2}{2U_n^2}\right) = \frac{2}{\sqrt{3}} \exp\left(\frac{-CU_{th}^2}{2CU_n}\right) = \frac{2}{\sqrt{3}} \exp\left(\frac{-E_{diss,min}}{kT}\right) \tag{A6}$$

which is a generalized result for digital gates, independent of technology. We can then express the minimum dissipated energy per bit-flip as follows:

$$E_{c,\min} = -kT\ln\left(\frac{\sqrt{3}}{2}\varepsilon_c\right) \approx kT\ln\left(\frac{1}{\varepsilon_c}\right). \quad \text{(A7)}$$

If, on the average, N transistors are changing their logical state in the processor during one clock period, the total dissipated power is

$$P_{total,min} = -Nf_c kT\ln\left(\frac{\sqrt{3}}{2}\varepsilon_c\right) \approx Nf_c kT\ln\left(\frac{1}{\varepsilon_c}\right). \quad \text{(A8)}$$

In Figure 3, the energy dissipation of one logical gate, for a single classical or quantum operation, versus the error ratio of the gate, is shown. The quantum case is given by the results of Gea-Banacloche [6]. At today's clock frequencies and those expected in the near future, the quantum gate dissipates more energy than the classical gate for error rates less than 10^{-6}. At the required error rate (10^{-25}) of today's classical gates, the quantum gate would need about 100 Joules for a single operation!

In Figure 4, the power dissipation of single logical gates, classical gates (CMOS) and quantum gates, driven with a given clock frequency, versus the error ratio of the gate, is shown. It is supposed that the classical gate changes its state in each clock frequency period. Already at a modest error rate of 10^{-16}, a *single* quantum gate would require more power than today's microprocessors. At the error rate of a classical logical gate ($\approx 10^{25}$), the *single* quantum gate would dissipate over 10^4 *MegaWatts*.

As an example, let us now estimate the classical power limit for today's microprocessors (2003). The number of transistors is $\approx 1.5 * 10^8$. Suppose that, at maximum load, all transistors are effectively changing their logic state at each clock period. Let the clock frequency be 3 GHz and the allowed total bit error frequency in the system of the $\approx 1.5 * 10^8$ transistors be 1/year [1]. Then

$$\varepsilon_c = 1/(3 * 10^9 * 3600 * 24 * 365 * 1.5 * 10^8) = 7.05 * 10^{-26} \quad \text{(A9)}$$

and from Eq. (A8) we get

$$P_{total,min} \approx 0.1W \quad \text{(A10)}$$

This calculation indicates that the energy efficiency of today's microprocessors with ≈100W power dissipation, at the given error probability, is around 0.1%. We will see in the next section that this low error rate, which is necessary for the error-free running of the programs and processor routines, is not necessary for the information content of data. In computers that are able to utilize the Shannon information, the required power is even less.

A2. Energy requirement of Shannon-information transfer in single classical (CMOS) and quantum gates

In this section, we evaluate the energy requirement of Shannon-information transfer. This measure, which cannot be improved by error correcting algorithms, is the ultimate one; the real characteristic of performance versus power dissipation. How to see this? Shannon's information channel capacity depends on two factors: the bandwidth and the signal to noise ratio (see below). A greater error probability can be compensated by a greater bandwidth in order to keep good performance. For example, the poor accuracy of quantum computers can, in principle, be compensated by a sufficiently higher speed of operation. Though the practical realization of such systems is not obvious, it is still interesting to explore the ultimate limits of performance for Shannon-information. If the signal to noise ratio (SNR), which is the ratio of the signal power, P_s, to the noise power, P_N, and the frequency bandwidth B is known, then Shannon's information channel capacity can be calculated by the Shannon formula. For the quantum gate, the SNR is $1/\varepsilon_q$, so using the best case given by Gea-Banacloche [2], we get:

$$C_q = B \log_2\left(1 + \frac{P_s}{P_N}\right) = \frac{1}{\tau_q}\log_2\left(1 + \frac{1}{\varepsilon_q}\right) = f_c \log_2\left(1 + \frac{E_q}{f_c \hbar}\right), \tag{A11}$$

where the maximum clock frequency is $f_c = \frac{1}{\tau_q}$. Let us introduce the normalized information channel capacity K (*bit/sec/W* or *bit/J*) which is the information channel capacity divided by the power dissipation. In the case of quantum gate, it is as follows:

$$K_q = \frac{C_q}{P_q} = \frac{C_q}{f_c E_q} = \frac{f_c}{f_c E_q}\log_2\left(1 + \frac{E_q}{\hbar f_c}\right) = \frac{1}{E_q}\log_2\left(1 + \frac{E_q}{\hbar f_c}\right), \tag{A12}$$

where P_q is the power dissipation of the quantum gate when driven by the maximum clock frequency. At non-zero temperature, the following natural limitations occur for the quantum limit:

$$\left.\begin{aligned} \hbar f_c &\gg kT \text{ (thermal (classical) decoherence constraint)}, \\ \hbar f_c &\leq E_q \text{ (quantum uncertainty error constraint,} \\ &\qquad \text{compare with Eq. 1) [2]}, \\ E_q &\gg kT \text{ (thermodinamic constraint for the quantum gate).} \end{aligned}\right\} \quad \text{(A13)}$$

The last condition is required to avoid flipping the logic state of the quantum gate by thermal activation and require the same kind of thermal noise considerations as Eqs. (3) and (5). As a practically motivated example, in Fig. 3, the maximum information transfer rate at 1 Watt power dissipation (bit/s/W) when the clock frequency is varied and E_q is set so that flipping the quantum gate's state by thermal noise can be neglected ($E_q = 70kT$); the dashed line shows how the clock frequency should be decreased to improve performance. The figure contains the whole range of meaningful the working range as expressed by Eqs. (A13). The right hand end of the x-axis corresponds to the classical thermodynamical (thermal decoherence) decoherence limit, and the left hand end to the case limited by quantum uncertainty, $\varepsilon_q = 1$ (see Eq. (A1)).

For a CMOS (classical) gate the error probability ε, when it is small, is

$$\varepsilon_c = \frac{\nu(E_c)}{f_c} = \tau_c \nu(E_c). \quad \text{(A14)}$$

Because the signal to noise ratio (SNR) is equal to the error probability, the information channel capacity can be given as

$$C_c = B \log_2 \left(1 + \frac{P_s}{P_N}\right) = f_c \log_2 \left(1 + \frac{f_c}{\nu}\right)$$
$$= f_c \log_2 \left[1 + \frac{\sqrt{3}}{2} \exp\left(\frac{U_{th}^2}{2U_n^2}\right)\right] = f_c \log_2 \left[1 + \frac{\sqrt{3}}{2} \exp\left(\frac{E_c}{kT}\right)\right] \quad \text{(A15)}$$

and for $E_c \gg kT$

$$C_c \approx f_c \frac{E_c}{kT}. \quad \text{(A16)}$$

Thus, in the case of a classical gate, the normalized information channel capacity is:

$$K_c = \frac{C_c}{P_c} = \frac{C_c}{f_c E_c} = \frac{f_c}{f_c E_c}\log_2\left[1 + \frac{\sqrt{3}}{2}\exp\left(\frac{E_c}{kT}\right)\right]$$
$$= \frac{1}{E_c}\log_2\left[1 + \frac{\sqrt{3}}{2}\exp\left(\frac{E_c}{kT}\right)\right] \approx \frac{1}{kT}. \quad \text{(A17)}$$

It is of interest to note that the dissipation is the same as Landauer's conjecture [8] about the energy dissipation during information erasure at reversible computing, which is about kT per bit. However, it is important to emphasize that our result is not for reversible computing, so it is not directly relevant for Landauer's case. The total energy in the CMOS capacitor is dissipated during discharge which is an irreversible process.

Equation (A17) indicates that the classical gate performs better by more than an order of magnitude than the thermal noise free quantum gate ($E_q = 70kT$), see Fig. 3. It is important to note that in the limit of

$$kT = E_q = \hbar f_c, \quad \text{(A18)}$$

the quantum gate would have the same performance as the classical, because then $K_q = 1/kT$. Though this case is excluded by Eq. (A13), it can be approached. Viewed from another angle, the conditions described by Eq. (A18) set the limit when the quantum system becomes classical. This finding, and the fact that then Eq. (18) yields the same results as Eq. (A17), confirms that both the classical and quantum theories are on the right track. However, in this limit the error probability ε_q approaches one (see Eq. (A1)) thus the quantum gate is useless for practical applications. To improve the accuracy to an acceptable level, we have to move toward the strongly quantum limit but then the value of K_q decreases. A simple estimation shows that for <0.1 error probability caused independently by both the thermal excitation and the quantum measurement, we would need $E_q \geq 2.4kT$ (Eq. (3)) and $E_q \geq 10\hbar f_c$ (Eq. (A1)). As, according to Eq. (A13), $\hbar f_c \gg kT$, the condition $E_q \geq 10\hbar f_c \gg 10kT$ satisfies both relations and then the quantum error would be dominant (0.1). Similar considerations would lead for $E_q \geq 70kT$ for an error probability requirement of 1.5% dominated by the quantum error again.

Finally, let us estimate how much power today's microprocessors would need in the most ideal case of a classical CMOS gate for processing of Shannon-information. The same conditions as at Eq. (A10) yield

$$P_{total,min} \approx 1.5 * 10^8 * 3 * 10^9 * 4 * 10^{-21} \approx 0.002\ W, \qquad \text{(A19)}$$

which is fifty times less than the minimal power required by the switching error rate estimation, see Eq. (A10). So, compared to the energy requirement of the processing of Shannon-information, the energy efficiency of today's microprocessors is around 0.002%.

The general conclusion of these derivations is that, at general purpose applications, where we have to have separate access to each memory elements in the computers, classical computers perform many orders of magnitude better than quantum computers when high accuracy of data is required. This is not really surprising. The phenomena of zero-point thermal noise [9,10] show that "quantum" can produce noise even when "classical" is certainly silent, which is at zero temperature.

References

[1] L.B. Kish, "End of Moore's law: thermal (noise) death of integration in micro and nanoelectronics", *Physics Letters* **305** (2002) 144–149.

[2] L.B. Kish, "Moore's law, Performance and Power Dissipation", *Encyclopedia of Nano Science*, Marcel Dekker, 2004, 2197–2200.

[3] Bing Huang, Jin Qin, Joerg Walters, Joseph B. Bernstein, "Development of Derating Guidelines for Semiconductor Devices", *Aerospace Vehicle Systems Institute Report*, (2003) unpublished.

[4] J. Kim, L.B. Kish, "Can single electronic microprocessors ever work at room temperature?", *Proc. SPIE Conference on Noise in Nanoelectronics, Sensors and Standards*, June 1–4, 2003, Santa Fe, NM (Editors: L.B. Kish, F. Green, G. Innaconne, J. Vig); and references therein (Likharev and Korotkov).

[5] L.B. Kish, "Moore's law is killed by classical physics; Can quantum information save it?", *Proc. SPIE Conference on Noise in Nanoelectronics, Sensors and Standards*, June 1–4, 2003, Santa Fe, NM (Editors: L.B. Kish, F. Green, G. Innaconne, J. Vig).

[6] Gea-Banacloche, J., "Minimum Energy Requirements for Quantum Computation" *Physical Review Letters* **89**, 217901 (2002).

[7] J. Gea-Banacloche, L.B. Kish, "Comparison of Energy Requirements for Classical and Quantum Information Processing, *Fluctuation and Noise Letters* **3**, C3–C7 (2003).

[8] R. Landauer, *IBM J. Res. Dev.* **5**, 183 (1961).
[9] A.E. Allahverdyan, Th.M., Nieuwenhuizen, "Testing the violation of the Clausius inequality in nanoscale electric circuits", *Phys. Rev. B* **66**, 115309 (2002).
[10] L.B. Kiss, "To the problem of zero-point energy and thermal noise", *Solid State Comm.* **67**, 749 (1988).

2.2. Who Wins the General-purpose Computer Race, Classical or Quantum?

Figure 2.2.1. Julio Gea-Banacloche. (Courtesy of J. Gea-Banacloche). More in the caption of Fig. 1.1.2.

2.2.1. The first SPIE symposium Fluctuations and Noise (FaN, including 5 separate conferences) in Santa Fe, New Mexico, in the June of 2003 ended with a public debate (published in FNL — *Fluctuation and Noise Letters*) about the future of quantum computing. Though I was the chair of FaN and fully immersed in setting it up and running it, and was totally exhausted, I decided to take part in the debate because of the following deeply puzzling question:

How can a quantum computer be better than the classical, if, at the same speed, it has more noise and errors?

(*Continued*)

(*Continued*)

The above question was answered via a collaboration (in 2004; see the paper in this Section) with an open-minded scientist, Julio Gea-Banacloche, who was not only in the "Con" debate team but also an Associate Editor for quantum informatics for *Physical Review A*; thus also representing an "official" authority on the subject. Via the comparison of his quantum dissipation and my classical dissipation formulas, and using his imagination of the essential elements of a quantum computer, we concluded (see paper below) that the lower limit of power requirement (dissipation) of a "general-purpose" quantum computer is about 1000 times greater than that of a classical computer at the same speed and temperature.

After all these, what remains of quantum computing is what we have today. The dreams of replacing our PCs and Macs with quantum computers are gone, and the only remaining possibility for quantum computers to ever play an important role is in the field of "special-purpose" computing, where for each computational task, a new computer must be built or rebuilt. That means a quantum computer should never be compared to a PC but to an analog computer.

The relevant example I had was the noise of a simple resistor in the classical and the quantum limits,[†] where the latter shows larger noise due to zero-point fluctuations, which is the manifestation of zero-point energy. While I was still in the learning phase of the related problems, the debate made a strong impression on me — in particular, seeing the strong contrast between the ongoing hype in leading scientific journals/magazines/media and the few humble, serious scientists who were trying to support the quantum dreams in the "Pro" debate team.

Note, even much of the special-purpose statements, for example about Shor algorithm, are just impractical dreams (see the open access article[‡] of Michel Dyakonov with thorough analysis in the proceedings of Hot Topics of Physical Informatics (HoTPI) conference).

[†]L.B. Kiss, "To the problem of zero-point energy and thermal noise", *Solid State Comm.* **67** (1988) 749.

[‡]M.I. Dyakonov, "Prospects for quantum computing: Extremely doubtful", *International Journal of Modern Physics: Conference Series* **33** (2014) 1460357 (9 pages). Link to the HoTPI papers is at the top of the page: http://www.ece.tamu.edu/~noise/HotPI_2013/HotPI_2013.html

2.2.2. Note: Similarly to the other papers in this book, the references in the following paper* will be listed at the end of the paper.

COMPARISON OF ENERGY REQUIREMENTS FOR CLASSICAL AND QUANTUM INFORMATION PROCESSING

JULIO GEA-BANACLOCHE
Department of Physics, University of Arkansas
Fayetteville, AR 72701, USA

LASZLO B. KISH
Texas A&M University, Department of Electrical Engineering
College Station, TX 77843-3128, USA

Received 5 August 2003
Revised 11 August 2003
Accepted 11 August 2003

By considering the energy requirements of quantum and classical computers we propose a criterion to separate the classical from the quantum regime and show that the classical scaling laws are much more favorable for conventional, general purpose computation.

Keywords: Energy efficiency; information versus energy dissipation; microprocessors; CMOS; quantum computing.

It is sometimes suggested, especially in the popular press, that quantum computers [1, 2] might be, in some sense, the natural successors of today's conventional digital computers, as the current trends in miniaturization reach the atomic level. While it is true that there are a few special tasks (notably, integer factoring [3, 4]) which a quantum computer could, exploiting some unique quantum mechanical effects, perform much faster than a classical computer, this does not address the question of whether it would actually make any sense to push conventional computers into the quantum domain, for anything other than these very special purpose tasks.

The energy dissipation and its relation to error-free operation has recently been identified as one of the most important problems in classical microprocessors [5]. Accordingly, in this note, we address the above issue by comparing the ultimate energy requirements of quantum and classical computers, based on our recent studies on classical [5] and quantum [6] systems, respectively. It was shown in [6] that the minimum error ϵ_q that is to be expected when an elementary logical operation is performed on a quantum computer satisfies

$$\epsilon_q > \frac{h}{E_T} . \tag{1}$$

*J. Gea-Banacloche, L.B. Kish, "Comparison of Energy requirements for Classical and Quantum Information Processing", *Fluct. Noise Lett.* **3** (2003) C3–C6.

Here h is Planck's constant, τ is the duration of the operation, and E is an amount of energy that needs to be put into the control system, and removed, after a time τ, in order to "switch" the state of the quantum bit, or qubit. If the control system is an oscillator (e.g. an electromagnetic pulse) of carrier frequency f, typically with $f \gg 1/\tau$, then f must be used in place of $1/\tau$ in Eq. (1). The error in (1) is ultimately due to zero-point ("vacuum") noise in the control system; this is of the order of hf for an oscillator, and a minimum of h/τ for a "static" field, switched on and off over the time τ. (See [7,8] for some relevant discussions on the consequences of zero-point noise in electric circuits.)

By contrast, the fundamental noise limit for classical computers is set by thermal noise [5], which has an associated energy $kT/2$ at a temperature T (k is Boltzmann's constant). To prevent thermal noise from randomly flipping a bit, the excitation energy E needs to be made sufficiently large. Then, under some reasonable asumptions [5], the error probability is given by

$$\epsilon_c > \frac{2}{\sqrt{3}} \exp\left(-\frac{E}{kT}\right) \tag{2}$$

where E is the minimum energy dissipated in the classical circuit in the course of the operation. A comparison of Eqs. (1) and (2) immediately reveals two fundamentally different scaling laws: in the quantum case the error decreases only inversely as the energy used, in the classical case it decreases *exponentially*.

We suggest that these different scalings can be used to effectively characterize a quantum and a classical regime for a digital computer. The quantum regime corresponds to

$$hf > kT \tag{3}$$

and is entered when the vacuum or zero-point noise in the control system exceeds the thermal noise at the applicable temperature. Here f is the frequency of the control oscillator, for an oscillating field, or the gate frequency $1/\tau$ otherwise.

Clearly, by Eqs. (1) and (2), the error rate in the quantum regime will always be greater than in the classical regime, for the same energy. Conversely, the exponential scaling shows that it is energetically much more advantageous, in order to achieve a low error rate, to stay in the classical regime.

Two points need further elaboration. In [6], it was stated that the energy E appearing in Eq. (1) does not necessarily need to be dissipated, only needs to be put into and removed from the control system to switch the desired evolution on and off. This is punctiliously true, but it must be added that there are currently no plans, in any proposed quantum computer architecture, to "recycle" that energy, which is, in fact, effectively lost; nor is it clear how one could actually attempt to recycle it. A static field with a certain energy, for instance, could be switched on and off by charging and discharging a capacitor, and it is hard to see how one could prevent the required energy from being dissipated in the discharging circuit. Similarly, one could think of keeping electromagnetic pulses around for a while, by bouncing them off of ultra-high reflectivity mirrors, but, in order to effectively reuse them, one would still need some kind of switch to send them off in the proper direction, at the right time, and that switch is again going to require an energy of the order given by Eq. (1) if it is to operate in a time τ with an error not exceeding ϵ_q.

In short, there are strong indications that at least an amount of energy of the order of $h/\epsilon_q \mathcal{T}$ must be dissipated per logical operation, if the switching mechanisms for the control fields in the quantum computer rely ultimately on conventional classical electronics.

In fact, the situation is similar in a classical microprocessor. The electric energy, which is stored in the MOSFET gate capacitors and is controlling the CMOS logic gate, could, in principle, be extracted by a proper LC circuitry and timed switches (the same technique that is applied in the "switching" power supplies of today's computers). The exception to this is a small part, of the order of kT, of the controlling electric energy. However, while energy dissipation problems are extremely serious in today's microprocessors [5], no initiative or realistic hope exists for extracting this energy in order to use it again. That would need an overhead circuitry of LC elements and switches which would be more complex than the original microprocessor, and its L elements would make it extremely bulky. Moreover, very similarly to the quantum case, because all the switches in this overhead circuitry would need similar controlling energy as the original CMOS gates, the efficiency of the energy-recycling operation would be strongly questionable. In conclusion, in both the classical and the quantum computers, it is only reasonable to consider the controlling energy pulses of gate operations as irretrievably lost, as we have done above.

The other point concerns quantum error correction [9], which can, in principle, reduce the overall failure probability for a logical operation on an encoded qubit, provided the individual logical operations can be carried out with an error probability smaller than a certain threshold ϵ_{th}. This reduction is superexponential for concatenated codes, n levels deep: $\epsilon_n = \epsilon_{th}(\epsilon_0/\epsilon_{th})^{2^n}$, whereas the energy cost of concatenation is only exponential in n: $E_n \geq (hf/\epsilon_0)N_c^n$, where N_c is the number of qubits in the base code (e.g. $N_c = 7$ for a popular code introduced by Steane [10]). This modifies the scaling (1), so that now one might have, for instance, for the 7-qubit code,

$$\epsilon_n = \epsilon_{th} \exp\left[-\left(\ln\frac{\epsilon_{th}}{\epsilon_0}\right)\left(\frac{\epsilon_0 E_n}{hf}\right)^{0.36}\right]. \tag{4}$$

This is typically better than Eq. (1), but it is still worse than Eq. (2) because of the large upfront energy cost: one must start out with $E \geq E_0 = hf/\epsilon_0$, which is very large compared to kT (since $hf > kT$ in the quantum regime and $\epsilon_0 < \epsilon_{th} \leq 10^{-4}$). A further disadvantage is the sublinear growth of the argument of the exponential with E; in the above example, the number 0.36 in Eq. (4) comes from $\ln 2/\ln 7$, where 2 is the number of total single-bit errors that cause the code to fail, and 7 the number of physical qubits used to encode a logical qubit. It must be noted, however, that more efficient codes exist, and concatenation may not even be the best strategy for special-purpose applications; see [11] and references therein for very thorough discussions of these points. Nonetheless, the general qualitative differences between quantum and classical cases remain, broadly speaking, as indicated above.

Yet, even this fails to tell the whole story. The quantum error correcting codes considered in [9–11] are necessarily complex because they are designed to prevent errors fatal to *quantum* algorithms; but if all one wants to do is to run a *classical* algorithm on quantum hardware, one only needs protection against one kind of error (namely, bit flips), and this can be provided comparatively easily by the quantum generalization of classical repetition codes. Such codes would use $2n+1$ qubits to encode a single logical qubit, and provide protection against n bit-flip errors; if ϵ is the probability of a single-bit flip, the total failure probability would then scale as $\epsilon^{n+1} \sim \exp((n+1)\ln\epsilon)$, where n is of the order of the total energy, so for these codes one again would have an exponential decrease of the error probability with the energy dissipated, just as in the classical regime.

Nonetheless, and even though the threshold for these codes is very moderate, one still has to pay an upfront energy penalty, relative to the classical case, and this would still be substantial. For instance, according to Eq. (2), to achieve the current error rates of about 10^{-25} of conventional computers, it would theoretically be sufficient to dissipate an energy of about $58kT$, in the classical regime. In contrast, in the quantum regime, we find it would be near-optimum in this case to start out with a single-qubit failure probability ϵ of the order of 0.1 (and therefore a minimum energy, per logical operation on a single physical qubit, of about 10 hf), which requires a scale-up in computer size of about $2n+1 = 107$ (this follows from a careful calculation, keeping track of combinatorial factors), and thus a total energy of at least $1070hf$ per logical operation on an encoded qubit. With $hf > kT$, this is easily seen to be at least two orders of magnitude larger than the classical result.

There are, of course, other inconveniences that would arise from having to rely so heavily on error correction, such as the scale-up in the physical size of the computer, already mentioned, and additional cycles (and energy) lost to error diagnostic and correction.

All in all, our conclusion is that, from an energy-efficiency point of view, there seems to be no reason why one might want to run a conventional computation on "quantum hardware". On the contrary, it appears to be energetically favorable to keep conventional, general-purpose computers away from the quantum regime whose characteristic scaling we have identified here. We should stress, however, that this logic need not apply to *special-purpose* quantum computers, such as those that might be built solely to, for instance, run the quantum factoring algorithm [3, 4]. We conjecture that whether such machines are ever built will probably depend, ultimately, on whether that specific application is still of strategic importance a decade or two from now.

Acknowledgements

Julio Gea-Banacloche's research has been supported by the Army Research Office. Laszlo Kish's research is supported by the Army Research Office, the Office of Naval Research and the TAMU Telecommunication Task Force. J. Gea-Banacloche is grateful to James P. Clemens for observations regarding quantum repetition codes. L. B. Kish appreciates quantum computing related discussions with H. Taylor, Ph. Hemmer, S. Sethuraman and J. Kim.

References

[1] C. P. Williams and S. H. Clearwater, *Explorations in Quantum Computing*, Springer-Verlag, New York (1998).

[2] M. A. Nielsen and I. L. Chuang, *Quantum Computation and Quantum Information*, Cambridge University Press (2000).

[3] P. W. Shor, *Algorithms for Quantum Computation: Discrete Logarithms and Factoring, Proceedings, 35th Annual Symposium on Foundations of Computer Science*, IEEE Press, Los Alamitos, CA (1994).

[4] A. Ekert and R. Jozsa, *Quantum computation and Shor factoring algorithm, Rev. Mod. Phys.* **68** (1996) 733–753.

[5] L. B. Kish, *End of Moore's law: thermal (noise) death of integration in micro and nano electronics, Phys. Lett. A* **305** (2002) 144–149.

[6] J. Gea-Banacloche, *Minimum energy requirements for quantum computation, Phys. Rev. Lett.* **89** (2002) 217901.

[7] A. E. Allahverdyan and Th. M. Nieuwenhuizen, *Testing the violation of the Clausius inequality in nanoscale electric circuits, Phys. Rev. B* **66** (2002) 115309.

[8] L. B. Kiss, *To the problem of zero-point energy and thermal noise, Solid State Comm.* **67** (1988) 749–751.

[9] See, e.g. J. Preskill, *Reliable quantum computers, Proc. R. Soc. Lond. A* **454** (1998) 385–410 for an introduction to quantum error-correcting codes.

[10] A. M. Steane, *Error correcting codes in quantum theory, Phys. Rev. Lett.* **77** (1996) 793–797.

[11] A. M. Steane, *Overhead and noise threshold of fault-tolerant quantum error correction*, preprint quant-ph/0207119.

2.3. The Kish–Sethuraman Cipher

As already mentioned, our main subject, the KLJN system is not a cipher. However, there was a software-based protocol created in 2004 which was correctly called a cipher: the Kish–Sethuraman (KS) cipher.[1]

2.3.1. My first real exposure to the secure communication problem started in 2004 as a motivation to attempt to challenge the dogma of the time, that only quantum informatics can offer unconditional security. The prologue to this motivation was the intensive debate on the feasibility of quantum computing mentioned in the former section. As a consequence of the publicity of the 2003 debate about the quantum computing dreams of the time and its negative

(*Continued*)

[1] L.B. Kish and S. Sethuraman, Non-breakable data encryption with classical information, *Fluct. Noise Lett.* **4** (2004) C1-C5.

(*Continued*)

conclusion, and probably of other factors, soon all defense-related funding for quantum computing dried out (while NSF was jump-starting funding for it at a much smaller scale). Thus, in 2004, only quantum communications research received significant support, in the range of half a billion dollars, according to the rumors of that time. Then, after this "killer-success", I started to explore the "impossible" to achieve absolutely secure communications by classical means, utilizing noise/randomness. The first result was the double-padlock protocol, where the first manuscript contained a mathematical error (assuming that permutation operation is commutative), which was corrected by Swaminathan Sethuraman (Swami) and that granted him his well-deserved co-authorship. Andreas Klappenecker named this protocol the "Kish–Sethuraman (KS) cipher".

The KS cipher was claimed to be an absolute secure data encryption, by using only classical information, *provided the required mathematical operators exist*. Swami identified a mechanical analogy of the method with a mail-carrier-box and two-padlocks, see Fig. 2.3.1.

The Sender uses a padlock to lock the box (operation A) before the mail is sent. When the Receiver receives the box, he uses another padlock to lock the box again (operation B). Then the double-locked box is sent back to the Sender where the first padlock is removed (operation A^{-1}). Then the Sender sends the box again, which is still locked by the Receiver's padlock, back to the Receiver where the box is opened (operation B^{-1}) thus the mail is delivered. In conclusion, the following operations are carried out on the

Figure 2.3.1. Mechanical analogy of the Kish-Sethuraman cipher. The operations represent locking (A and B) and opening (A^{-1} and B^{-1}) and they are to be carried out by the Sender (A and A^{-1}) and Receiver (B and B^{-1}), respectively.

Figure 2.3.2. Swaminathan (Swami) Sethuraman at the Palo Duro Canyon on his way to the first Fluctuation and Noise (FaN) Symposium in Santa Fe in 2003. Swami was a Master student studying stochastic resonance at *blue noise* under my supervision. Later on, he got fed up with engineering studies and switched to pursuing a PhD in math. After his PhD, he gained employment in the banking industry. When I told him about this "scrambling" operator based cipher idea and the paper I was writing on it, he lighted up and told me that the idea was exciting, but that the permutation operators I tried to use are not commutative, thus they are out of question. Moreover Swami identified the analogy for this cipher: a mailbox with two padlocks. For this important correction and the nice analogy, he deserved the second authorship.

t-th bit (or word, etc.) $m(t)$ of the message:

$$\begin{aligned} m(t) &\overset{\text{S}}{\Rightarrow} A(t)m(t) \overset{\text{R}}{\Rightarrow} B(t)A(t)m(t) \overset{\text{S}}{\Rightarrow} \\ &\overset{\text{S}}{\Rightarrow} A^{-1}(t)B(t)A(t)m(t) = B(t)m(t) \overset{\text{R}}{\Rightarrow} \\ &\overset{\text{R}}{\Rightarrow} B^{-1}(t)B(t)m(t) = m(t). \end{aligned} \tag{2.3.1}$$

The arrows with S and R mean that the resulting signal was generated by the Sender and the Receiver, respectively.

At the third step, the presumed mathematical condition is:

$$A^{-1}(t)B(t)A(t) = B(t), \tag{2.3.2}$$

which requires the commutativity of the operators:

$$A(t)B(t) = B(t)A(t). \tag{2.3.3}$$

The commutativity is necessary to get through the message to the Receiver.

Another mathematical condition concerns the security of the KS cipher:

(4) *It should be impossible to determine the message and the actual operator values by recording and comparing the publicly sent signals* $A(t)m(t)$, $B(t)A(t)m(t)$ and $B(t)m(t)$.

In other words, the comparison of a data sequent before and after the operators are used does not serve enough information to identify the data or the actual operator code.

However, one more condition is required that has been implicitly assumed but not explicitly stated:

(5) *The Sender and Receiver share no information and submit no public information about the actual operators* $A(t)$ *and* $B(t)$ *(except their class, which is public information).*

With conditions (4,5) the KS cipher would be absolutely secure against passive attacks, while condition (3) is needed for the basic functioning of this protocol.

However, it seems to be extremely difficult — or rather, impossible — to find *classical physical* operators satisfying all the three conditions. For example, the widely used RSA encryption operators do not satisfy condition (4) whenever prime factorization can be efficiently done, such as with a hypothetical quantum computer or noise-based logic engine; thus it is only conditionally secure. This was the point of Andreas Klappenecker in his comment-paper[2] about the KS cipher.

Using quantum physics, the situation is much easier. Three groups proposed solutions independently by using single photons; a Korean group,[3] Subhash Kak,[4] and Janos Bergou[5] with me. The information is the

---

[2]A. Klappenecker, "Remark on a "non-breakable data encryption" scheme by Kish and Sethuraman", *Fluct. Noise Lett.* **4** (2004) C25.

[3]W.H. Kye, C.M. Kim, M.S. Kim, Y.J. Park, "Quantum Key Distribution with Blind Polarization Bases", *Phys. Rev. Lett.* **95** (2005) 040501.

[4]S. Kak, "A three-stage quantum cryptography protocol", *Foundations of Physics Letters* **19** (2006) 293–296.

[5]J.A. Bergou, L.B. Kish, "An absolutely secure QKD scheme with no detection noise, entanglement and classical communication", http://arxiv.org/abs/quant-ph/0509097, (September 15, 2005).

polarization and the operators are random polarization rotations (locking the padlock) and their inversions (unlocking the padlock) at Alice's and Bob's ends. If Eve measures the photon, she destroys it and, from a single photon measurement, she cannot find out the polarization angle. Though these quantum realizations are interesting, they lack the practicality of the classical physical system and its robustness against various types of invasive (active) attacks.

2.3.2. Janos Bergou and I were sitting at the beachfront of a hotel in Gallipoli on the sunny morning of June 10 after the 2005 conference on Unsolved Problems of Noise, *UPoN*; the conference series I founded in 1996. "Let us try to do something together", I proposed, and Janos answered, "Let's do that". In ten minutes we had the full double padlock quantum scheme developed, and later, wrote it up. Only after submitting it to a journal did we learn about the fact that Kye, *et al.* had published a scheme with the same hardware but different protocol in *Physical Review Letters*, a few months earlier. Their protocol was not analogous to the KS system because it corresponded to an impossible situation of placing the message into the box when it is already locked by Alice. However, it was the same hardware; thus Janos wanted to withdraw the paper, and so I withdrew it. That was a good decision. Much later, I learned from Subhash Kak about his paper which — though was published only later, in 2006 — had been submitted for publication about a month earlier than the date Janos and I created the idea. Kak's system was exactly the same system we had in our manuscript in arxiv. Neither Kye, *et al.* nor Subhash said anything about the classical physical analogy, the KS cipher, which had been published much earlier, in 2004; otherwise they could have used it as a nice analogical picture. This kind of "telepathy" in scientific ideas can often be seen in the history of physics; even a similar form of Einstein's special relativity was published some 10 years before Einstein by Poincare with a somewhat different philosophy; while its mathematics was created by Lorentz a year before Einstein.

The most advanced geometrical (rotational) operators (Clifford's geometric algebra) so far were explored by James Chappell,[6] *et al.*, while

[6]J.M. Chappell, L.J. Gunn, D. Abbott, "The double-padlock problem: Is secure classical information transmission possible without key exchange?", *Internat. J. Mod. Physics: Conf. Ser.* **33** (2014) 1460355.

Figure 2.3.3. (From left to right): Lachlan Gunn, the Author and James Chappell at the Mao monument during the HoTPI-2013 conference, in Changsha, China. Both Lachlan and James brought interesting ideas on how to realize the operators for the KS cipher. That was also the occasion on which I first heard Lachlan's new cracking idea against the KLJN. However it was immediately obvious that the attack cannot work (see Sections 4.6.1–4.6.3 and 5.8.5).

Lachlan Gunn,[7] *et al.* explored the utilization of random transit times during internet communications. Both schemes offer only conditional security.

2.3.3. A related personal story about the KS cipher shows that interesting ideas may not require any higher education — just unbiased thinking. When my son, Lazar, learned about matrices in the 6th grade, he came up with a new idea on how to realize the KS operators with classical physics. He said that, if Alice is doing permutations on the locations of the columns of a matrix and Bob

(*Continued*)

[7] L.J. Gunn, J.M. Chappell, A. Allison, D. Abbott, "Physical-layer encryption on the public internet: A stochastic approach to the Kish-Sethuraman cipher", *Internat. J. Mod. Physics: Conf. Ser.* **33** (2014) 1460361.

(*Continued*)

is doing permutations on the location of the rows, the permutations of Alice and Bob are commutative; thus Swami's objection about non-commutativity is nullified. I was amazed about the simplicity of the idea. Unfortunately the analysis showed that, while the KS system became functioning with this type of permutations, it was not secure because the columns and rows with random numbers are like fingerprints which Eve can locate based on their patterns and thus follow the operations.

2.3.4. Many years later, Derek Abbott sent me an email with the very same idea. I quickly cooled his enthusiasm with the analysis's results.

Chapter 3

The Birth of KLJN (2005–2008)

In this chapter, the papers leading to the discovery of KLJN and the first KLJN papers are shown with notes to clarify the science and to indicate some of the complex history behind it in the areas of science, politics or personal matters.

Also one can observe how the notion and deeper understanding of *unconditional security* (see also Sections 1.6 or 5.1 for advanced definitions) have progressed — an evolutionary process that can be followed through this book.

And of course, all those debates...

Figure 3.1. Janos Hajdu with his wife, Anita (2012). He is currently Emeritus Professor of Theoretical Physics at the University of Cologne, Germany, and a real deep thinker. I was visiting scientist there for short periods in 1989 and 1990. With him and Janos Kertesz in 1990, we showed that the famous claim that the mesoscopic universal conductance fluctuations are the source of the 1/f noise in solids is a myth and that the related celebrated theoretical models violate basic rules of theoretical stochastics.[1] His personality and critical attitude to scientific hypes and fashionable mistakes have been a very rewarding experience.

3.1. Birth Pains: Noise-free, Random Kirchhoff-loop (RKL) Cipher

3.1.1. In May 2005, I decided that, from then on, I would not waste my time writing grant applications (unless I was invited to join such an effort), and instead, focus on doing something really original. The results came soon. In that summer, we had seminars and interviews for Head of Department Candidates; and during two subsequent interview-seminars, two independent ideas popped into my head. The first one was the stealth (zero-signal-power) communicator

(*Continued*)

[1] L.B. Kiss, J. Kertesz and J. Hajdu, "Conductance noise spectrum of mesoscopic systems", *Z. Phys. B*, **81** (1990) 299.

(*Continued*)

(see Sections 3.2–3.3). The second was the cipher outlined below. The facts about these:

(i) The former invention works but it is only conditionally secure.
(ii) The second one is unconditionally secure but it does not work: Eve cannot extract the information but Alice and Bob cannot exchange information either.

Similarly, during my other explorations with classical physical security, I typically got either (i) or (ii). The KLJN was to be discovered next (see Section 3.4); it has been the only exception by offering a functioning unconditionally secure system.

The reasons for showing the non-working system are that its security and the fact that it does not function are important aspects for a deeper understanding of KLJN and its security.

Let us call it the Random-Kirchhoff-loop (RKL) cipher, which is indeed a cipher with its intended function. It looks the same as the KLJN key exchanger: two resistors and two voltage generators (see Figure 3.1.1).

For Alice and Bob, only the two values (voltage and resistance) at the other end are unknown. Everybody can measure two quantities: voltage and current, and can set up two equations with them, including four parameters. Eve would need to solve four independent equations to extract the four secret

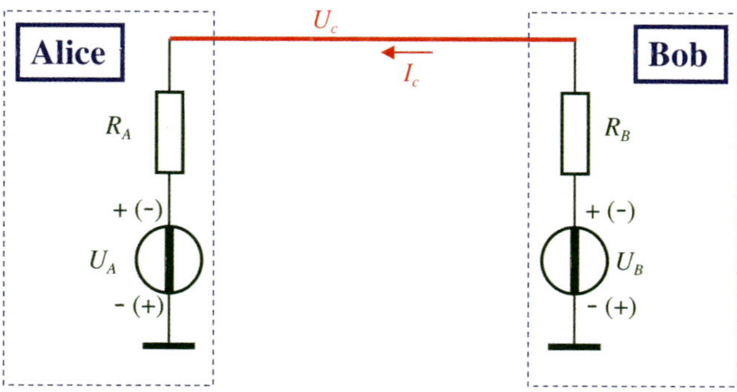

Figure 3.1.1. The Random-Kirchhoff-loop (RKL) cypher, which was created in the summer of 2005, a few weeks before the KLJN. The generators of Alice and Bob serve random but static (DC) voltages with zero expected value. Their resistors are also random. For Eve, all four loop parameter-values are secret; for Alice and Bob only two are unknown. Everybody can measure current and voltage and set up two equations about them.

parameters. However, there are only two equations: one for the current and one for the voltage. For example, Alice can write:

$$I_c = \frac{U_B - U_A}{R_A + R_B} \tag{3.1.1}$$

$$U_c = U_A + R_A I_C \tag{3.1.2}$$

The idea was that, because Alice knows U_A and R_A, and by measurements I_C and U_c, she solves this system of equations and gets Bob's U_B and R_B. However the idea was wrong because the current and voltage equations are not independent. The easiest to see is by checking the solution of Bob's voltage. According to Eqs. (3.1.1) and (3.1.2) or alternatively by Ohms law:

$$U_B = U_c - R_B I_c. \tag{3.1.3}$$

This can be rearranged as:

$$U_B + R_B I_c = U_c. \tag{3.1.4}$$

Similarly, Bob's solution to extract the information from Alice's side is:

$$U_A + R_A I_c = U_c. \tag{3.1.5}$$

It is obvious that for a given pair of measured I_c and U_c, there is an infinite number of different U_B and R_B pairs as well as U_A and R_A pairs that can satisfy Eqs. (3.1.4) and (3.1.5).

In conclusion: while Eve cannot not extract Alice's and Bob's voltages and currents from the static measurement data, Alice is also unable to extract Bob's voltage and resistance, and Bob cannot extract Alice's voltage and resistance either. The cipher therefore does not function; one may say it is "too secure" because even Alice cannot decrypt Bob's data and Bob, likewise, cannot decrypt Alice's data.

To make a functioning system, some new mathematical relation between the voltages of the generators and the resistance of the resistors is needed; this will allow Alice and Bob to extract the data. However, this mathematical relation should not offer Eve a means to crack the cipher.

We will see in Section 3.4 that, in the functioning KLJN scheme, the voltage randomness is dynamic: it is the thermal noise voltage of the resistor. The missing relation to make the cipher work is the Johnson–Nyquist formula, which interrelates the resistance and its *mean-square* thermal noise voltage.

We will also see that it is more efficient to use the KLJN system as a key exchanger than a cipher because, after exchanging a secret key, both Alice and Bob can use that key in a high-speed secure communication with software-based ciphers via optical, wire or radio channels.

3.1.2. The first people to whom I told the RKL cypher idea during the summer of 2005 were quantum optics experts Suhail Zubairy and Janos Bergou, at a Golden Corral lunch during Janos's visit. They looked skeptical (with good reasons; see above segment) but could not provide any concrete objection. Thus, after I wrote up and submitted the stealth (zero-signal-power) communicator paper (see next section) on August 18, I quickly started writing the paper on the RKL cipher. I soon got to the point where it was obvious that it could not function; it secured Alice's and Bob's information so well that neither could communicate. It was then that an intensive struggle started; one that would go on for several weeks. I tried many ideas; different signal times, timings, circuit modifications, but the resulting system was either non-secure or non-communicating. The idea of KLJN came when I was giving up the fight and a final question appeared: what if the noise voltage generators were the thermal noises of Alice's and Bob's resistors? The result is shown in the paper in Section 3.4. But first let us see the other invention of that summer, the Johnson noise based "stealth" cipher, which functions but has only conditional security.

3.2. Pre-KLJN: Zero-energy (Stealth) Communication

3.2.1. I created this scheme about a month earlier than the KLJN and submitted it for publication in the August of 2005. It is almost the KLJN; however it did not (yet) utilize the Second Law of Thermodynamics — thus it was not yet unconditionally secure. Its fundamentally interesting aspect is its ability to communicate without emitting signal energy into the information channel. The parameters of the channel are modulated but no signal energy is emitted. It is similar to the mythical usage of mirror flashes in ancient Greek and Roman communications, and to the corresponding *heliograph* invented by Gauss and used by the Germans and the Brits in the 19th century.

(*Continued*)

(Continued)

Like all papers in this book, the following paper has its own independent numbering system for its references, equations, etc.

Paper: Stealth Communication: Zero-power classical communication, zero-quantum quantum communication and environmental-noise communication†

Abstract. An alternative physical way of communication, *communication by the inherent background noise*, is proposed which does not need net energy transfer in the information channel. The communicator devices do dissipate energy; however, they do not emit net energy into the channel. Instead, they modulate the parameters of inherent spontaneous fluctuations in the channel. The method can use two different mechanisms: thermal noise (Johnson-Nyquist noise) for classical communication, and vacuum fluctuations/zero point energy (quantum uncertainty noise) for quantum communication. The strongest advantage of this method is that this is apparently the *most hidden* (*stealth*) way of communication, because it uses the inherent background noise for communication, therefore making it extremely difficult or impossible to discover its presence. With proper wave-based arrangements and specific conditions, the sender and the receiver can easily detect eavesdropper activities, so that the eavesdropper is detected as soon as she extracts a single bit of information; thus the security of this method is comparable to the security of quantum communication/quantum key distribution schemes. Finally, concerning practical applications, environmental noise, out of the fundamental/inherent fluctuations, can also be used for this kind of communication provided that is sufficiently stationary.

Keywords: Stealth communication; eavesdropper detection; thermal noise; zero-point energy; vacuum fluctuations; communication by noise.

Below, it is shown that it is possible to communicate via an information channel without putting *net energy* into the channel. This surprising result is achieved by the modulation of the parameters of *equilibrium-thermal* or *zero-point-quantum* fluctuations in the channel. Thus this communication is achieved using the inherent energy fluctuations, which is called *channel noise* in the literature, to carry the information.

First of all, an important fact is emphasized; namely that *this method is very different from the so-called energy-free communication.* There

†L.B. Kish, *App. Phys. Lett.* **87** (2005), Art. No. 234109.

have been some discussion and debate by prominent scientists about the possibility of energy-free communication [Porod 1988, Landauer 1996]. However it is important to note that the *methods and issues presented in this paper are very different* because:

- The method presented in this paper utilizes and detects the *equilibrium-thermal* and/or *zero-point-quantum* fluctuations in the information

Figure 3.2.1. M. Suhail Zubairy (Courtesy of M.S. Zubairy). He has an interesting new "photonless" quantum communications scheme,[‡] which is similar in nature to the classical zero-energy scheme described in this section. Suhail is a Texas A&M University Distinguished Professor (Holder of the Munnerlyn–Heep Chair in Quantum Optics), in the Department of Physics. He is an open-minded scientist and my interactions with him have always been enlightening. His interests range from quantum optical applications to quantum computing and quantum informatics. He has also been interested in quantum state measurement of the radiation field and sub-wavelength atom localization. His other interests include coherent atomic effects and quantum thermodynamics. He has received many honors including the Willis E. Lamb Award for Laser Science and Quantum Optics; the Alexander von Humboldt Research Prize; the Outstanding Physicist Award from the Organization of Islamic Countries; the Abdus Salam Prize in Physics; the International Khwarizmi Award from the President of Iran; the Orders of Hilal-e-Imtiaz and Sitara-e-Imtiaz from the President of Pakistan; and the George H. W. Bush Award for Excellence in International Research. He is an elected member of the Pakistan Academy of Sciences and a Fellow of the American Physical Society, and the Optical Society.

[‡]H. Salih, Z.H. Li, M. Al-Amri, and M.S. Zubairy, Protocol for Direct Counterfactual Quantum Communication *Phys. Rev. Lett.* **110** (2013) 170502.

channel. All the debated arrangements mentioned above make strong efforts to *suppress* these fluctuations *below the detection limit* where they do not disturb the information processing any more than the detection limit.

- Even if there is *no net energy* communicated in the channel, there is a *huge energy*, much more than *kT/bit*, dissipated in the communicator devices at the sender and receiver's side; unlike in the debated arrangements mentioned above, where there is an effort to regain the energy spent. We do *not* make any effort to regain any type of energy used to run the system.

In conclusion, the debate [1, 2] about energy-free communication, though intriguing, is irrelevant here.

After realizing the fact that the inherent fluctuations can be used for communication, the realization and the various solutions are self-evident and straightforward. From the many possibilities, we show only a few examples in this paper. Without the restriction of generality, we demonstrate the idea by voltage fluctuations described by the *fluctuation-dissipation theorem* (FDT) in the classical physical and the quantum physical limits, respectively. According to the FDT [3–8], the generalized Johnson–Nyquist formula of equilibrium thermal noise in the voltage can be written as:

$$S_u(f) = 4[N(f, T) + 0.5]hf\,\mathrm{Re}[Z(f)] \qquad (1)$$

where $S_u(f)$ is the power density spectrum of the voltage noise on the open-ended impedance $Z(f)$; $\mathrm{Re}[Z(f)]$ is the real part of the impedance; and h is the Planck constant. The Planck number $N(f, T)$ is the mean number of hf energy quanta in a linear harmonic oscillator with resonance frequency f, at temperature T:

$$N(f, T) = [\exp(hf/kT) - 1]^{-1}, \qquad (2)$$

which is $N(f, T) = kT/(hf)$ for the classical physical range; $kT \gg hf$ so that the mean energy in the oscillator is $kT + 0.5hf$, where the second term is the zero-point energy. The familiar thermal noise voltage formula is valid in this classical limit, that is for the low frequency range $f \ll kT/h$, where the first energy term dominates:

$$S_{u,class}(f) = 4kT\,\mathrm{Re}[Z(f)] \tag{3}$$

In the quantum limit, which is the high frequency limit $f \gg kT/h$, the zero-point fluctuations can be measured with a linear voltage amplifier:

$$S_{u,quan}(f) = 2hf\,\mathrm{Re}[Z(f)]. \tag{4}$$

Although the proper physical interpretation of Eq. (4) is somewhat debated [5, 7, 8] (whether the observed noise is an explicit zero-point fluctuation or if it is the manifestation of the uncertainty principle, etc.), the validity of Eq. (4) for linear voltage amplifiers in the quantum limit is commonly accepted, confirmed by experiments, and applied in engineering [4, 6, 7]. Therefore, because the practical existence of this quantum noise phenomenon *in the voltage* is a fact, the debate mentioned above is irrelevant for the present paper, and here we can take Eq. (4) as the voltage noise of impedances in the quantum range. It is however important to mention [6] that this noise is zero in a photocell or in other detectors which are detecting particle number instead of field amplitude, and that measure only the Planck radiation. Thus, we always envision a *linear voltage amplifier* input at the receiver's side when we discuss *quantum noise* examples in the present paper.

The *general outline* of the zero-signal-power communication system *for classical information* is shown in Fig. 1. The $f \ll kT/h$ classical physical condition holds and the impedance generating the thermal noise of the channel (see Eq. (3)) is modulated by the sender. The receiver detects the signal, which is the modulation of the statistical properties such as spectrum, hidden correlations, etc. in the thermal noise. Though controlling the communication devices requires energy, the signal energy pumped into the channel and the net energy transfer between the sender and receiver is zero. Note that with alternative, wave-based solutions, the sender can modulate the reflection coefficient of the channel at his end and introduce time-correlations in the noise (see below).

An example of a simple realization of the *zero-signal-power classical communicator* can be seen in Fig. 2. The information-bit is represented

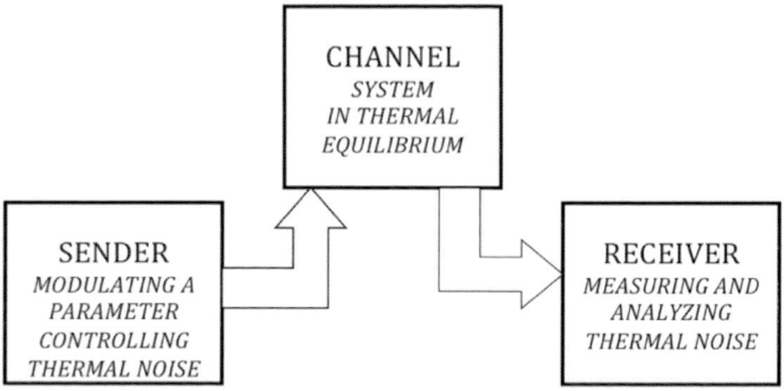

Figure 1. Generic outline of the zero-signal-power communication (ZPC) scheme for classical information.

by the choice of two different impedance-bandwidths and the two corresponding thermal noise spectra (see Eq. (3)) provided by the two different capacitors at the sender's side of the channel, which is made of two wires of short length $L \ll c_p/f$, where c_p is the propagation velocity of electromagnetic fields in the cable. The noise analyzer can measure power density spectra, zero-crossing frequency autocorrelation or other related quantities. The advantage of the method shown in Fig. 2 is that for everybody, except for an educated eavesdropper, it looks like there is no communication in the channel because only the usual background noise (thermal noise) is present and the signal power is zero. However the educated eavesdropper with proper equipment and settings can observe that the high-cut-off frequency of the thermal noise is switching between two distinct values.

The *general outline* of the zero-signal-power communication scheme *for quantum information* is also shown in Fig. 3, with the condition $f \gg kT/h$.

The impedance determining the spectral properties of vacuum fluctuations in the channel (see Eq. (4)) is modulated by the sender. Though controlling the communication devices require many energy quanta, no quanta are sent through the information channel and the energy transfer between the sender and receiver is zero. A similar solution can also be achieved by two coupled quantum systems; one at the sender's side and

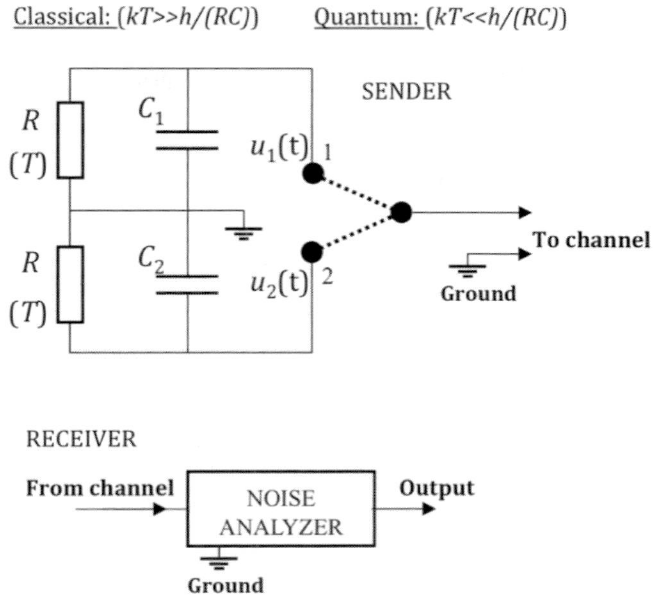

Figure 2. Simple realization of the zero-signal-power classical or quantum communicator. The information-bit is represented by the choice of two different impedance-bandwidths and the two corresponding thermal noise spectra. The noise analyzer can measure power density spectra, zero-crossing frequency autocorrelation or other related quantity.

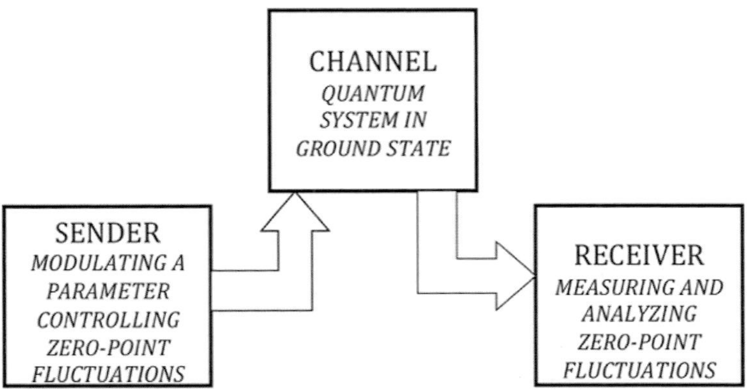

Figure 3. Zero-quantum quantum communication. General outline of the zero-signal-power communication scheme in the quantum limit.

one at the receiver's side, respectively. Note that concerning *electronic* solutions. there is an apparent disadvantage compared to the classical schemes. The $kT \ll hf_c$ condition requires amplifying voltages at very high frequencies or being in a very low temperature environment. Therefore this kind of *electronic quantum communication* seems to be more relevant for communication in space where the background thermal noise temperature is around 2.7 K. However heterodyne techniques such as TerraHertz or optical phase sensitive techniques may still be relevant.

The example outlined in Fig. 3 can also be used as a simple realization of the *zero-quantum quantum communicator* provided that the $kT \ll hf_c$ condition holds, and f_c is the upper cut-off frequency. The information-bit is represented by the choice of two impedance-bandwidths and the two corresponding uncertainty-noise spectra, provided by the two different capacitors at the sender's side of the wire-channel. The advantages and disadvantages of this system are the same as those of the classical system described above. An extra problem is the need for a low temperature environment and/or high frequency.

At considering *wave-based realizations*, for the sake of simplicity and practical relevance, we restrict ourselves to the case of classical thermal noise, though considerations about the quantum case would be similar. Wave-based arrangements need longer cables than half of the wavelength corresponding to the bandwidth, $L \gg c_p/f$. They can also be realized with the bandwidth modulation systems shown in Fig. 2, however modulation of the reflection can be a more advanced option. Figure 3.2.5 shows a wave-based arrangement with a coaxial cable and matching resistor at the receiver's end, and the modulation of the reflection at the sender's end. The detection of the information at the receiver's side needs a correlator (a multiplier and an averaging circuit) and a delay line to determine the value of the autocorrelation function at time equal to the time of return flight of the wave along the cable. The three stages of the switch at the sender's side represent a three-level information system: (1) no reflection (no correlation); (2) positive reflection (positive correlation); (3) negative reflection (anti correlation). Observing the reflection properties via the autocorrelation function provides extraordinary sensitivity to detect

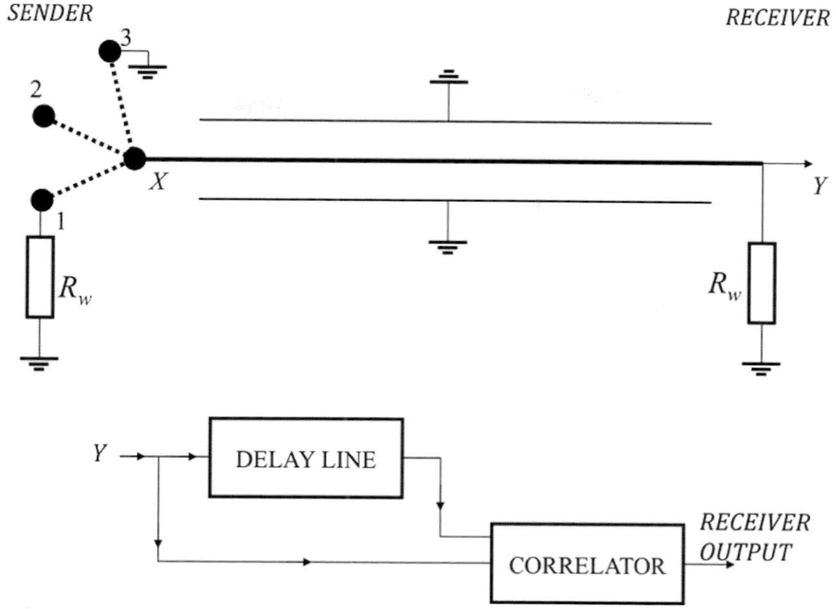

Figure 4. Wave-based arrangements with a coaxial cable and matching resistor at the receiver's and the modulation of the reflection at the sender's end.

possible eavesdropper activities because any activity along the line will cause reflections at different time delays.

The presence of an eavesdropper can be detected in a wave-based arrangement with two parabolic antennas. The arrangements at the sender's and receiver's sides are the same as in Fig. 4, however this time the signal propagates in the space, see Fig. 5. When used as a classical communicator, in *thermal equilibrium*, this arrangement has the highest sensitivity to detect an eavesdropper. Eve, by using a tilted reflector, can couple out a part of the *signal-carrying-noise* from the communication channel between the two antennas; however she will always induce excess noise when she uses a matched resistor. Otherwise, if she leaves the antenna electrodes open or shorted, she causes excess reflection maxima/minima in the correlation function at the receiver's device. Because the out-coupled *signal-carrying-noise* has the same wave intensity as the *useless thermal noise* of the external space, out-coupling has to be significantly

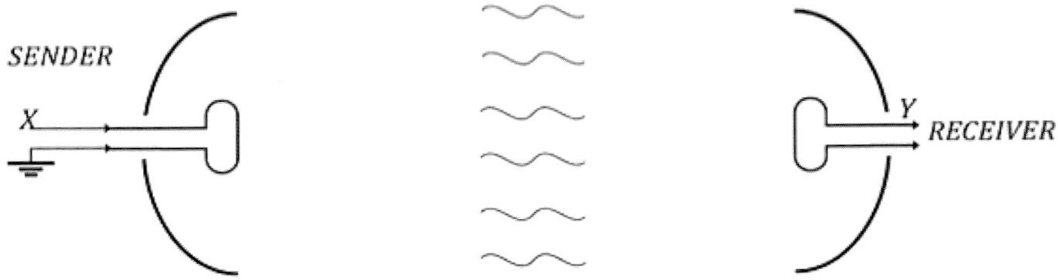

Figure 5. Wave-based arrangements with microwaves, two parabolic antennas and the reflection modulation technique [for full understanding, see Fig. 3.4 too].

strong to have a reasonably good statistic of the detection. What is most important is that the receiver will observe at least as large a change on the autocorrelation function as the signal detected by the eavesdropper (Eve). Thus, as soon as Eve detects a bit with a certain probability, the receiver will detect Eve's "invasive presence" with the same probability. Therefore while the Eve detects a single bit, the receiver detects the presence of the eavesdropper.

However, these nice security characteristics of the classical communicator holds *only in thermal equilibrium* between Eve and the system — thus this communicator is only *conditionally secure*. It has been shown by the late Henry Taylor (see history below) that an eavesdropper using sufficiently cooled nonreciprocal devices (e.g. the Faraday isolator) would be able to extract information while staying hidden. Thus, *at least* in the case of a *classical communicator*, the absolute security holds only in situations where the eavesdropper is unable to use devices with a lower temperature than the temperature of the communicator system.

If this communicator is operated in the *quantum limit*, the temperature has no role in determining the working conditions because the communication takes place via parametric modulation of the vacuum fluctuations of the ground state. Therefore, the study of the visibility of the eavesdropper needs a quantum treatment of the situation. Taking into the account the properties of the quantum noise of resistors due to zero-point fluctuations [3–5, 8] (and the fact that there is no lower noise energy than zero-point energy), we suspect that the visibility of the eavesdropper will be similar in the case of thermal equilibrium to that of the classical communicator: that is the eavesdropper can probably be discovered after extracting a single bit of information. However, this conclusion about the quantum-case is not yet proven.

Finally if, instead of the thermal background, stationary environmental noises dominate the channel and the sender can modulate the bandwidth, reflection or some statistical property of noise, the zero-signal-power communication can be executed in the same ways as described above. The properties will be similar but the noise will be stronger than the fundamental limits.

References

[1] W. Porod, *Appl. Phys. Lett.*, **52** (1988), 2191; and references therein; W. Porod, R.O. Grondin, D.K. Ferry, *Phys. Rev. Lett.*, **52** (1984), 232–235; W. Porod, R.O. Grondin, D.K. Ferry, G. Porod, *Phys. Rev. Lett.*, **52** (1984), 1206; and references therein.
[2] R. Landauer, *Science*, **272** (1996), 1914–1918.
[3] H. B. Callen and T. A. Welton, *Phys. Rev.*, **83** (1951), 34.
[4] Ch. T. Stelzried, *IEEE Trans. Microwave Theor. Techn. MIT*, **16** (1968), 646–655; and references therein.
[5] W. Kleen, *Solid-State Electron.*, **30** (1987), 1303–1304; and references therein.
[6] H. Heffner, *Proc. IRE*, **50** (1962), 1604.
[7] A. E. Allahverdyan, Th. M. Nieuwenhuizen, *Phys. Rev. B*, **66** (2002), 115309.
[8] L.B. Kiss, *Solid State Comm.*, **67** (1988), 749.
[9] H. Taylor (TAMU), personal communication.

To conclude this segment (Section 3.2), the interesting property of the ZPC method described above, as compared to other kinds of classical and quantum communication (quantum key distribution), is that it is stealth because only thermal or zero-point (quantum) background noise can be detected in the channel. However, it is only conditionally secure. For further details about the history of this issue, see the next section.

3.3. Henry Taylor's Notes about the Stealth Communicator

This section is different in nature from the others; it is dedicated to the memory of Henry Taylor, the Director of Texas A&M University's Solid State Electronics Division at the time when the KLJN was created.

3.3.1. The fact that the system (see Section 3.2) is not secure in wave-based scenarios was not obvious to me in the August of 2005 when I submitted the first version of the paper for publication. When I bumped into Henry Taylor in the stairway in Zachry building at Texas A&M later that year, instead of the usual "*How are you — Fine thanks*" small talk, I outlined the pre-KLJN idea (in Section 3.2) to him. He became very enthusiastic about it. Thus I sent him the manuscript and, unexpectedly, Henry soon produced the evaluation below that, for the first time, pointed out the lack of unconditional security.

Henry Taylor's unpublished evaluation puts this system in its place by comparing it with other stealth schemes and shows that not even the wave-based version is unconditionally secure because Eve can use lower temperature devices and a Faraday insulator arrangement to tap into the communication. However, the wire-based version can be made secure, when it is expanded to become the KLJN system (as we will see in the next section). Below is a reproduction of Henry's final note.

Comments on Laszlo Kish's scheme for zero-power communication (ZPC)

The idea is to use a noise waveform (e.g. electronically generated Johnson noise) as a carrier for communication. The noise carrier would be modulated (e.g., spectrally modulated) for transmitting information. When properly implemented, the communication would be covert since an eavesdropper would have a difficult time extracting the signal from the noise.

A question is: what are comparable schemes in conventional communication? In particular, let's consider free space RF or microwave communication.

Firstly, ZPC brings to mind the idea of a self-jammed communication link. Normally, jamming is used by an adversary to make it difficult for the receiver to extract the information signal from a high-power, broadband jamming waveform. Here, we envision the jamming waveform being generated and transmitted from the same antenna as the signal.

Spread spectrum communications (SSC) is a conventional, covert scheme which is proposed as a baseline for comparison. In SSC, multiple single-frequency carriers are used. In the simplest case, **N** *fixed frequencies can be generated and only one frequency is transmitted at a time. It is also possible to have more than one of the* **N** *frequencies transmitted simultaneously. The most versatile "frequency hopping" scheme features spectrally agile carriers, rather than set fixed frequencies.*

ZPC would replace fixed frequency or variable frequency carriers and on-off modulators or switches with a single broadband noise carrier and single-frequency filters (e.g., notch filters) or tunable spectral filters. Switches might also be utilized in ZPC.

Two limiting factors in SSC are the number of carrier frequencies which can be generated and the speed at which these frequencies can be switched and/or tuned. In ZPC, the comparable measure would be the number of independent spectral filters and the speed at which they can be switched and/or tuned. If, in a particular application, spectral filters were cheaper and/or demonstrated greater agility than single frequency sources, a case might be made for the use of ZPC.

In ZPC, as in other covert systems, a tradeoff will exist between the transmitted signal power to noise power ratio (TSNR), and the achievable bit rate.

Unlike SSC, transmitter power is "wasted" in ZPC through having a substantial part of the transmitter power allocated to noise. This is particularly an issue if maximizing the range to the receiver is important.

In quantum encryption (QE) schemes, the eavesdropper has no chance as long as the transmitter generates only one photon of known state (e.g., polarization, wavelength) at a time. The eavesdropper can succeed if the transmitter generates multiple photons at a time. From a practical standpoint, QE suffers from short range, low bit rate, and high cost, but it can be made eavesdropper-proof.

In the case of free-space RF or microwave ZPC, an eavesdropper can intercept the modulated carrier without affecting the transmitter by inserting a tilted reflector into the wave path. The only observable effect would be the attenuation of the received carrier. In the real world, the worst case scenario involves having an eavesdropper near the transmitter location. In typical systems, the received power is several orders of magnitude below the transmitted power. This means that the eavesdropper can enjoy the same signal level as the receiver while attenuating the average power at the receiver by a small fraction of the modulated carrier power.

In summary, ZPC is an interesting new approach to covert communication which will no doubt stimulate a lot of discussion in the years ahead. As always with a new concept, the big question is practicality. Ideally, a "killer application" will be found. Cost and performance of the components, which differ from those needed for conventional communication, will likely be a key factor in determining ultimate success.

Henry Taylor
Aug. 21, 2005

Figure 3.3.1. Henry F. Taylor (1940–2006) circa in year 2000. He was Distinguished Professor of Electrical and Computer Engineering at Texas A&M University, USA. Born in Ft. Worth, Texas, he attended the Rice University in Houston, Texas and received his B.A., M.A., and PhD degrees in 1962, 1965, and 1967, respectively, all in Physics. He was employed as a Research Physicist at the Naval Ocean Systems Center (formerly the Naval Electronics Laboratory Center) in San Diego, California, USA, from 1967 to 1978. From 1978 to 1980 he was employed by Rockwell International in Thousand Oaks, California, where he was Principal Scientist of the Optoelectronics Department of the Microelectronics Research and Development Center. From 1980 to 1985 he was Head of the Optical Techniques Branch of the Naval Research Laboratory in Washington, DC.

Henry's prediction that this scheme "*will no doubt stimulate a lot of discussion in the years ahead*" has not been realized so far. KLJN, which can also work in stealth (zero-power-communication) mode, uniquely dominated the discussions even though the zero signal energy aspect has not been part of the debates. Henry pointed out in a personal discussion that Eve can use a *cooled* Faraday insulator in the wave-mode for eavesdropping. As a consequence, the published version of the paper claims security that is limited only to thermal equilibrium between Eve and the rest of the system, which makes it insignificant for practical security applications. Note, the KLJN system is fully defended even against these types of attacks via its defense tools against active (invasive) attack.

Equity, fairness, and plain human decency were dear to Henry's heart. He was a deep thinker with unwavering convictions, and a formidable and respected scientist. His diligence was a key factor in his success and an inspiration to his students and colleagues. His legacy lives on.

3.3.2. At that time (August 2005), I did not understand why Henry wrote this official-report-looking text and signed/dated it. However dramatic events soon after this may serve as a probable explanation. A few weeks later, I found a thick book by Richard Rhodes in my mailbox, *The Making of the Atomic Bomb*, with a note from Henry that I should read it because it contained very interesting facts about Leo Szilard, the Hungarian physicist whose discoveries in physical informatics really got recognized only after his death. It was about the same time that we learned that Henry had esophagus cancer. He died next March. The same period was heavily turbulent for me because my mother was dying in Hungary while the KLJN issue simultaneously exploded with so many emails requiring responses, thus I did not have time/motive to read Henry's book then. When I met him, we did not talk about science; rather I asked him if he was afraid of dying. He said he was not afraid of death because of his faith in God and Heaven. I read the Rhodes book only after Henry's death. Then I found a computer tomography invoice in it giving higher significance to that particular period that also correlated with Henry's actions; of giving a signed/dated report and mentioning Szilard. When I read the book, the relevance to Szilard was obvious especially in the context of the dramatic events about the KLJN system. It is a great pity that we did not have the chance to discuss the KLJN paper (completed a month later, in September 2005) with Henry.

3.4. Enter the Dragon: The First KLJN Paper

3.4.1. After struggling for several weeks, right at the point when I was about to give up, the new idea of using thermal noise voltages in the RKL cypher (see Section 3.1) resulted in a surprise: after measuring the noise current and voltage in the wire channel, Eve could calculate the resistor values; however, she did not know where they were located. On the other hand, Alice and Bob knew their own resistors. Whenever their resistances were different, they knew that the other resistor value belonged to the resistance at the other end. I had a secure communicator!

3.4.2. It was obvious from simple circuit and noise considerations that, even if the system is weakly non-ideal — such as non-zero wire resistance, inductance

(*Continued*)

(*Continued*)

capacitance, or inaccurate resistor or temperature values — the system will still be reasonably secure against *passive* (listening) attacks, and no infinite measurement speed, accuracy nor infinite computational speed can help Eve to extract more information than a miniscule amount. In other words, even the non-ideal system is unconditionally secure. Unfortunately, I did not think it was important to mention these facts in the first paper, which had to be brief because *Physics Letters A* (PLA) was a letter-journal. That was an error on my side because this negligence resulted in several attacks by others; however, it was also beneficial because the attacks not only brought attention to the "show" but often helped to identify and solve other problems, too.

3.4.3. The big question for me was not the passive (listening) attack but the *active* (invasive) one. What if Eve modifies the system by injecting a probing current or adding a shunt resistor? I devoted space to this analysis in this first paper and the results were impressive: active attacks are useless. More effective active attacks, such as the *man-in-the-middle-attack* were dealt with by the subsequent paper in Section 3.5.

3.4.4. I submitted the paper to PLA on September 15, 2005, and its arxiv version got featured by *Science*[+] two weeks later on September 30. In a month it got accepted for publication by Charlie Doering the PLA Editor. However he held the publication back because Koby Scheuer and Amnon Yariv had submitted a comment featuring their wire resistance attack, and Doering wanted to publish them together. That plan did not work out because of problems with Koby's paper, yet the publication of my paper was delayed for 6 months. Thus, by all means, *Science* was the first journal to show it. My only result published in *Science* — but I am not the author of the publication!

Paper: Totally Secure Classical Communication Utilizing Johnson (-like) Noise and Kirchoff's Law[1]

Abstract. An absolutely secure, fast, inexpensive, robust, maintenance-free and low-power-consumption communication is proposed. The states of the

[1]L.B. Kish, "Totally secure classical communication utilizing Johnson (-like) noise and Kirchoff's law", *Physics Letters A*, **352** (2006), 178–182; also at http://arxiv.org/physics/0509136.

[+]Adrian Cho, "Simple noise may stymie spies without quantum weirdness", *Science* **309** (2005, September 30), 2148.

information bit are represented by two resistance values. The sender and the receiver have such resistors available and they randomly select and connect one of them to the channel at the beginning of each clock period. The thermal noise voltage and current can be observed but Kirchoff's law provides only a second-order equation. A secure bit is communicated when the actual resistance values at the sender's side and the receiver's side differ. Then the second order equation yields the two resistance values but the eavesdropper is unable to determine the actual locations of the resistors and to find out the state of the sender's bit. The receiver knows that the sender has the inverse of his bit, similarly to quantum entanglement. The eavesdropper can decode the message if, for each bits, she injects current into the wire and measures the voltage and current changes in the two directions. However, in this way she gets discovered by the very first bit she decodes. Instead of thermal noise, proper external noise generators should be used when the communication is *not* intended to be stealth.

Keywords: Secure communication; stealth communication; classical information; eavesdropper detection; classical bit entanglement.

1. Introduction

Much effort has been spent to find a data communication method which is absolutely secure. Quantum key distribution offers methods where the eavesdropper gets discovered as soon as she extracts a certain number of (noisy) bits (the number depends on the specific method), because of the extra noise she is introducing into the communication channel. Thus quantum encryption is considered to be absolutely secure. However they are expensive, slow, maintenance-needy and work only for short distances without a repeater.

Concerning classical physical information, there is a widespread opinion that classical information channels cannot be made secure. Such an opinion sounds reasonable because neither the *collapse of wavefunction during measurement* nor the *quantum no-cloning theorem*, which make quantum encryption secure, hold for classical information.

> **3.4.5.** Challenging this *unfounded* quantum-dogma, that is, only quantum physics can offer unconditional security, was originally the main motive of my efforts. At that time, I did not know that something potentially more important — such as securing communications in instruments and computers, or in the Smart Grid — would come out of KLJN, and I did not even care. Dogmas have no place in science, even though some scientists love to play the role of "high priests" or

(*Continued*)

(Continued)

"magicians". I consider myself a ("heretic", *non*-"churchian") Christian, and regard the teachings about seeking the Truth as part of the most important directive — without which one remains in the dark, and brainwashing prevails. In my opinion, the same rule should hold for anybody who calls himself/herself a scientist. We simply cannot serve two masters; we have to choose between Truth and Mammon (Money/Pride/Popularity). When we yield to one we tend to suppress the other...

Concerning today's classical encryptions, the weak point of the mathematical methods used generally is that the sender and receiver have to somehow share the key they use to encode and decode the message, while the eavesdropper is recording the communication between them. To share the key, they start with a bit of common public information manipulated mathematically in a successive way, sending it back and forth several times until they are able to extract the key. The problem with this approach is that the eavesdropper may record everything and can also extract the key even though she is only a passive observer with less information — but with a sufficient computational power or efficient algorithm at hand. Therefore those encryption methods, such as RSA encryption systems [1, 2], where the sender and receiver *share some public information about the key*, are not absolutely secure.

3.4.6. The last sentence above is not fully correct. In such cases, the eavesdropper can deterministically deduce the key if enough time is available; thus these software-based systems offer ***zero*** information-theoretic security. Thus they are not future-proof, either (see Section 1.6).

The cipher to be described in this paper provides absolute security with classical information and is based on the simultaneous physical encrypting of the information by both the sender and the receiver.

3.4.7. I also fell into the same trap as others initially: calling KLJN a cipher and actually, this paper is proposing a cipher-style use of the scheme. However this system and its protocol is a key generation and distribution system — at least, when it is binary, which is the form it got popular. With a continuum (analog) system, it could operate as a cipher but that would be extremely slow.

It is relevant to mention that we have made an earlier mathematical attempt to double encrypt the message while bouncing it between the sender and receiver, the mathematical operator version of a *double-padlock* method [3]. However, the finding of proper encryption (padlock) operators satisfying the necessary mathematical conditions of working/security is hopeless [2] with classical information, unless we use operators like those of RSA encryption systems [1], where there is a common public knowledge about the key — that however terminates the absolute security of the cipher. On the other hand, quantum operators can be applied easily and the result is a quantum communication scheme less noisy and more secure than earlier quantum key distribution systems [4]. However, even in this improved quantum case, the eavesdropper can extract many noisy bits until she is discovered with a reasonable probability [5].

Another attempt [6] with classical (and quantum) information is *stealth communication*.[2] This is a zero-signal-power communication method utilizing parametric modulation of classical or quantum noise in the information channel that can provide absolute security for classical information with radio waves in thermal equilibrium only [6]. If the eavesdropper uses cooled devices, she can hide for a much longer time period than one bit.

In this paper a new classical physical method is shown to make absolutely secure data transfer with classical information where the information is encrypted by both the sender and the receiver — an approach similar to the double-padlock systems [3, 4]. However the way of communication is very different from the double-padlock system, and it is very special because these encryptions take place simultaneously, and because the information is carried by *two physical quantities* with *zero crosscorrelation* between them: *voltage* and *current*. The method succeeds to achieve the difficult goal that there is no public knowledge about the keys, which are independent random sequences of the numbers 0 and 1 and independent Gaussian voltage noises generated locally at the two sides. They are neither shared nor transferred anywhere, and they are not saved either. These properties make the communication absolutely secure.

---

[2] See Sections 3.2 and 3.3.

While this communication is similar to the absolute security of quantum communication, it has some major advantages: this communication is inexpensive to set up; very robust; maintenance-free; recognizes the eavesdropper after she extracts a single bit; works for much longer distances without the need of repeaters; or if it is used for the same distance as the range of quantum communication, is much faster.

> **3.4.8.** This statement is too vague. The fact is that quantum communication has an exponential cut-off of speed versus range due to the absorption of photons. The speed of the KLJN key exchange has a power function (1/range or the square or it) scaling. That means, there is always a critical range beyond which KLJN will perform progressively faster than Quantum Key Distribution (QKD). This crossover may happen somewhere in the range of a few kilometers, to a few hundred kilometers — depending on many factors, and on issues debated within the quantum field. Such debate is, for example, Horace Yuen's claim that, for unconditional security, QKD should not go beyond the bit rate of 0.1 bit/second.

Because the eavesdropper has zero information, *computational power does not help to break the code*; thus even an *infinitely advanced quantum computer would not be able to break this code*.

2. Thermal noise voltage and current of two parallel resistors[3]

For the sake of readers unfamiliar with thermal noise in resistors, here we present some simple considerations and facts [7]. In thermal equilibrium at temperature T, in the low-frequency limit, the power density spectrum of the thermal noise voltage on an open-ended resistor R is given by the Johnson formula:

$$S_{u,th}(f) = 4kTR \tag{1}$$

[3]See Section 1.4.

If the two ends of the resistor are shorted with a wire then the current in this loop will be

$$S_{i,th}(f) = \frac{4kT}{R} \tag{2}$$

due to Ohm's law. Let us suppose that we have two resistors, R_0 and R_1, which are *connected in parallel*. Then from Eq. (1) and Kirchoff's laws it follows that the power density spectrum of the resulting noise voltage on the two parallel resistors will be:

$$S_{u,R||}(f) = 4kT\frac{R_0 R_1}{R_0 + R_1}. \tag{3}$$

From Eq. (2), it follows that the power density spectrum of the noise current flowing through the loop consisting of the two resistors is:

$$S_{i,R||}(f) = \frac{4kT}{R_0 + R_1} \tag{4}$$

It is an important consequence of thermal equilibrium that the mean power flow between the parallel resistors is zero, and that the crosscorrelation between the voltage on these parallel resistors and the current flowing in this Kirchoff loop is zero.

3.4.9. This claim is the direct consequence of the Second Law of Thermodynamics, which is the impossibility to construct a perpetual motion machine of the second kind. Such a machine would be able to induce a heat-flow between two separate sections of an isolated system and build a heat engine utilizing the resulting temperature difference to execute work. This is a termodynamical version of the energy conservation law. Violating this law would allow, for example, creating work (energy) from *nothing*, which also means creating mass from *nothing*, as Einstein's energy-mass equivalence states that both mass and energy are direct measures of each other. All the KLJN systems that work at a single temperature are governed by this law. Two new schemes, see Section 5.8, the VMG-KLJN and the RRRT-KLJN systems operate at different temperatures and they are governed by the more general Fluctuation-Dissipation Theorem (in the classical physical limit).

3. Secure, threefold-encrypted Kirchoff-loop-Johnson-noise (KLJN) cipher

3.1. *Physical base*

Let us suppose that both the sender and the receiver have a resistor each and these resistors are connected in parallel by a pair of wires, which represents the information channel between the sender and the receiver. The eavesdropper has no information about the resistance of these resistors. The key questions are:

(i) Can the eavesdropper find out the resistance values from voltage and current measurements?
(ii) If yes, can she find out the location of the resistances; and which resistor is at the sender and which one is at the receiver?

The answer to question (i) is *yes*. By measuring the current in the wire and the voltage between the wires, the eavesdropper can determine $S_{u,R||}(f)$ and $S_{i,R||}(f)$. Then using Eqs. (3) and (4) she can set up a second order equation and the two solutions provide the two resistance values. However, the answer to question (ii) is *no* and this answer guarantees the absolute security of the KLJN cipher. There is no way to determine the exact location of these resistances. The specific scaling between the resistance and the thermal noise voltage and current (see Eqs. (1)–(2)) makes it sure that the net energy flow between the sender and the receiver is zero; thus even the Pointing vector or similar cross-correlation measures provide zero information about the location.

3.2. *The absolutely secure communication scheme*

Accordingly, the absolutely secure classical communication scheme utilizing Kirchoff laws and Johnson noise is shown in Fig. 1. The information channel is a wire. The sent information bit is carried by the sender's choice of resistor R_s selected from two significantly different values, $R_0 \ll R_1$, where the two values represent the two states of the information bit (0 and 1). The sender encrypts this message with the random generator voltage $U_S(t)$ of the actual resistor. The receiver double-encrypts the message by using his randomly chosen resistor R_R, which (for simplicity but not necessarily) can

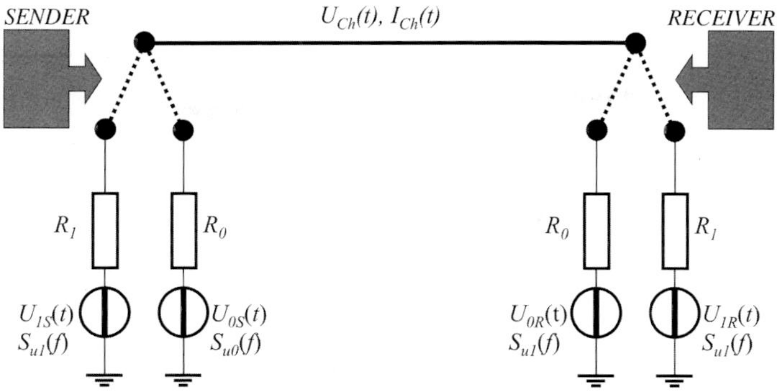

Figure 1. Absolutely secure classical communication scheme utilizing Kirchoff laws and a threefold encryption. The information channel is a wire. The message is carried by the sender's choice of resistor value. The sender encrypts this message by the random generator voltage U_S. The receiver double-encrypts the message by using his randomly chosen resistor R_R and the random generator voltage U_R. The random voltage generators are either the Johnson noises of the resistors or an artificial noise generator with much larger noise voltage but with the same scaling relation between the resistance and the noise voltage spectrum as that of Johnson noise. The eavesdropper may have access to the measurement of the voltage and current in the channel; however, this information is not enough to break the code. (*Note, this time, following the cipher idea, Alice and Bob was called Sender and Receiver.*)

be selected from the same set of two resistance values and the corresponding random generator voltage $U_R(t)$. At each clock period, a new information bit is sent and, at the same time, the receiver's resistance is changing randomly between the two values. The clock period is selected so that sufficiently good statistics can be made on the measured noise voltage or current in the channel. The random voltage generators are either the Johnson noises of the resistors or are products of an artificial noise generator with a much larger noise voltage, but with the same linear scaling relation between the resistance and the noise voltage spectrum as that of Johnson noise. We call this a *threefold encryption scheme* because the data are encrypted by three independent stochastic processes: the two noise voltage generators, and the resistance at the receiver. Because the sender and receiver know their own actual resistor value they can both determine the resistor value at the other side by measuring either the voltage or the current using Eq. (3) or (4), respectively. For the identification of *secure bits* (see below) it is *necessary* that the sender also determines the resistance value at the other side and that can be used for a simultaneous two-way communication.

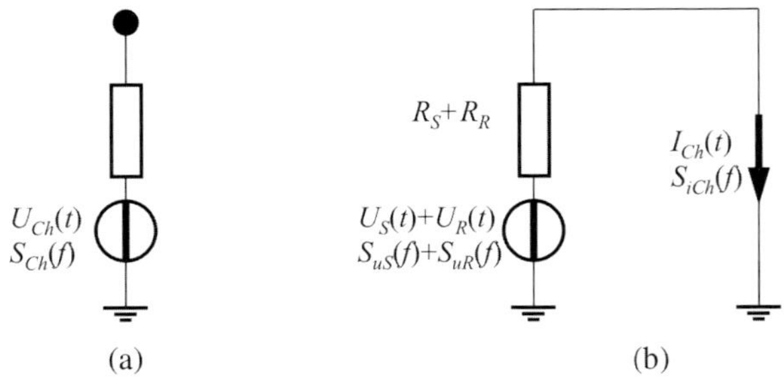

Figure 2. Circuit models for the calculation of the channel voltage and current via the Kirchoff's law and the Johnson noise equation. The eavesdropper can measure the voltage and the current; and after determining the spectra, she can derive the resistor values from Eqs. (3) and (4), but not their location.

The eavesdropper may also have access to the measurement of the voltage and current in the channel (see Fig. 2); however, this information is not enough to break the code. As we have shown above, by measuring both the voltage and current in the channel, she can extract only the two momentary resistance values, and remains unable to locate the resistors when their resistances are different. Consequently, she cannot decide if the sender is communicating 0 or 1.

3.3. *Non-secure bits and simultaneous two-way communication*

If both the sender and receiver use the same resistance value, the eavesdropper can recognize that fact because the resultant resistance is either the lowest or the highest of the three possibilities; thus she can find out the resistor value and the status of the information bit at the sender. Thus such a bit is not secure and cannot be used for secure data transfer. On the average, 50% of the sender and receiver bits will be different thus we can use only that 50% for secure communication. Similar situations are well known in quantum communication and it is easy to make use of the successfully communicated secure bits because the sender is aware of the security or non-security of the communicated bit. For example, after a whole random sequence is sent, it is possible to announce through a public channel which secure bits forms the message and what is their proper order.

3.4.10. This here is an attempt to use the KLJN as a combined key exchanger and cipher, as a one-time-pad. Though such a scheme can be made to work, this complicated scheme is unnecessary if we simply use KLJN as a key exchanger, and separate the cipher function.

In this way, both the sender and the receiver can simultaneously communicate a random message and the random message will be the inverse of each other at the two sides — similar to *quantum-entanglement* in quantum information processing. Thus the receiver can also send information simultaneously; therefore *both sides can simultaneously act as sender and receiver* by randomly choosing between the two resistor values at each clock period. In this way the 50% bit loss due to the security requirement is made up by the simultaneous operation mode which is a speed doubler.

3.4. *External noise generators*

We have already mentioned above that proper external noise generators can and should also be used because the same considerations hold for resistances and corresponding external noise voltage generators that follow the same scaling law between the resistance and the mean-square voltage noise as that of thermal noise:

$$S_{u,th}(f) = VR, \tag{5}$$

where V is the scaling coefficient and its value has to be equal at the sender's and the receiver's ends. Then all considerations and conclusions described above remain valid including the absolute security. The advantage of using external noise generators is that V can be arbitrarily large instead of the $4kT$ value that results in the case of thermal noise.

3.4.11. Using artificial noise generators have many advantages. It allows strong signals (noise temperatures corresponding beyond a trillion Kelvin) and control of the temperature with astronomically high accuracy — which makes the Hao attack meaningless (see Sections 3.7 and 3.9 of this book). It also allows a number of enhanced features for security and speed (see Section 5.3).

(*Continued*)

(*Continued*)

It also allows the strongest defense tools against active (invasive) attacks because Alice and Bob can have and monitor a full and exact system model and compare that with their measurements.

Thus, using an external noise generator of large V makes the communication free of background noise effects, which is a strong advantage compared to quantum communication where the background noise of the quantum source, the channel and the detectors significantly reduces performance, compromising security. The only advantage of the thermal noise driving, which has originally been shown for the sake of simplicity, is that it is a *stealthy communication* method that only uses the background noise in the channel, see [6]. In this respect, Fig. 1 also shows the way to provide *absolute security for the stealth communication* arrangement using wire connections that was not achieved in the originally proposed secure communication [6].

4. Breaking the code and the immediate detection of the eavesdropper[4]

It is possible to physically break the KLJN cipher (see Fig. 3) if the eavesdropper injects a small stochastic current $\Delta I_E(t)$ into the channel and determines the following crosscorrelation functions:

$$C_{ES} = \langle \Delta I_E(t) \Delta I_{ES}(t) \rangle \quad \text{and} \quad C_{ER} = \langle \Delta I_E(t) \Delta I_{ER}(t) \rangle, \tag{6}$$

where $\Delta I_{ES}(t)$ and $\Delta I_{ER}(t)$ are the induced current changes by the induced voltage change $\Delta U_{E,Ch}(t)$ and the linear response.

Obviously, the following rule due to Kirchoff's law allows the determination of the information bit:

$$R_R/R_S = C_{ES}/C_{ER}, \tag{7}$$

where R_S and R_R are the actual resistances at the sender and the receiver ends, respectively. The easiest way to uncover the eavesdropper is to use a public channel to exchange and compare the measured current values at the sender and the receiver. In this case, the eavesdropper is *immediately*

[4]This is the very first invasive (active) attack treatment of KLJN.

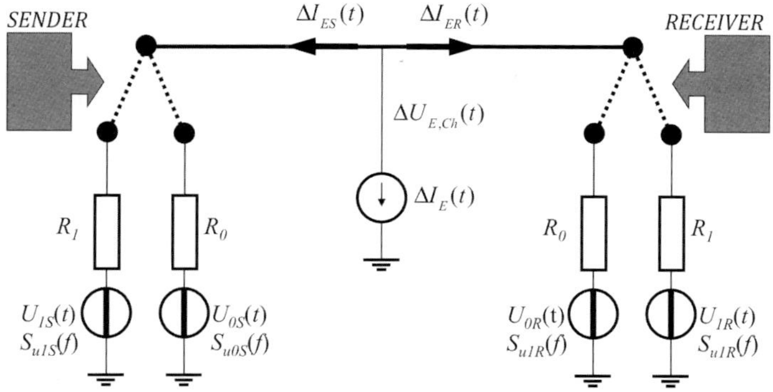

Figure 3. Breaking into the system requiress injecting a current into the channel. The current has to have dynamical and statistical characteristics that differ from the voltage noises used by the sender and receiver. However, as soon as the eavesdropper extracts a single bit her action gets discovered because the sender and the receiver observe the same deviations in the voltage, the current and their statistical properties.

discovered, before she can make any statistics, because she needs to cause a measurable current difference while the entangled bits are communicated. Thus the eavesdropper has only one choice: instead using a small current and making a statistics, to inject a fast and large current spike in the channel and use the relation

$$R_R/R_S = \Delta I_{ES}/\Delta I_{ER}, \tag{8}$$

to extract the bit values. Thus the eavesdropper can extract a single bit but she gets immediately uncovered. Therefore the communication is absolutely secure. It is more secure than quantum communication where a certain number of noisy bits can always be extracted by a quantum amplifier before the eavesdropper is uncovered.

5. Practical questions: speed, range and cost

There are two factors limiting speed and range. Firstly, Kirchoff's laws should hold. That means that

$$L \ll \lambda_{\max} = c/f_{\max} \quad \text{or} \quad f_{\max} L \ll c, \tag{9}$$

where L, $f_{\max}$ and c are the range; the maximal frequency bandwidth and the velocity of EM waves in the wire, respectively. Secondly, the *clock frequency* (*effective bandwidth*) f_c should be low-enough for reasonably good noise statistics to distinguish the resistors $R_0 \ll R_1$ securely, thus $f_c \ll f_{\max}$. For this reason, it is much faster to use the measurement of the mean-square noise voltage and current which are the integrals of the corresponding power density spectra [7] and provide a good statistics in a much shorter time.

For a practical estimation, let us suppose that $c = 2 * 10^8$ meter/s, $f_{\max} L = 0.1 * c$, and $f_c = 0.1 * f_{\max}$. Then the effective *bandwidth-distance product* $f_c L = 2 * 10^6$ meterHz. This is slightly (factor of 2–3) better than present quantum communicator arrangements [8].

> **3.4.12.** At sufficiently long distances, cable capacitances can cause a square-law reduction of speed versus range. While this is obvious from simple calculations, such analysis has never been published due to the lack of time. If the resistances were chosen to be less than the wave-impedance of the cable (this is not practical due to wire resistance issues), the cable inductance would take over the speed-limiting role with the same, square-law scaling. For the cable capacitance attack, see Section 5.8.

However, there is a simple and inexpensive way to improve speed radically. Because this communicator is extremely inexpensive (100–1000 times cheaper than quantum communicators of the similar performance) it is natural to build parallel channels to enhance speed and security. Using chip technology and a multi-wire cable with 100 wires, the speed of the system can be increased by a factor of 100 and the price would still be much lower than the price of a quantum communicator. Moreover, the security can further be enhanced because the information can be cross-encrypted among the wires by a public code; thus, if the eavesdropper would read out a single bit from a single wire, that would still be zero bit of extracted information without reading the other 99 wires.

Finally it should be mentioned that this technology is not only inexpensive but also very robust and virtually maintenance-free compared to quantum communicators, moreover it consumes much lower power. It is very robust: vibrations, dust, temperature gradients and material ageing, which are serious problems in quantum communication, cause no problem here.

3.4.13. The factor that is as expensive in this technology as in quantum communication is the research and development (R&D). The system looks simple (QKD schemes are simple too) but, when it gets to approaching perfect security, it requires as large an interdisciplinary team as QKD and as much work. There is no free lunch. However, as soon as KLJN is developed and put on a chip, billions of copies can be printed via silicon technology. It is difficult to imagine that QKD will ever reach such an advanced phase of development.

3.4.14. After KLJN was published on arxiv on September 18, 2005 and picked up by the media, inventor Davide Antilli contacted me and claimed he had already patented this idea, (see patent number EP1952573A2, "System and method for the propagation of deterministic and non-deterministic values by means of electrical conductors") and called "Orlando system". Its claimed (but not confirmed) priority date was July 2005, while the official patent filing date was July 2006, patent publication date August 2008, and withdrawal date July 2011. Charles Bennett and Jess Riedel came up with the same idea 8 years later (see our analysis of it in Section 4.4, and the 10 different ways of fully cracking its security). A quick check in 2005 immediately showed that the "Orlando system" of that time had nothing to do with KLJN because there were no resistors and noise in it — just DC voltage or DC current generators and wire. When I told that to Davide, he claimed that it was still the same system, and quickly added resistances and noise to the system — but that was still not thermal noise. Though, the final, published patent years later also contained thermal noise, it still did not contain any Johnson formula or security analysis. I was perplexed as to how Davide managed to file for a patent on the idea even though the first KLJN paper and my response to the Scheuer–Yariv argument predated the patent, and had both been cited by the parent examiner. How is this possible? I learned from our patent officer, Nick Chremos, that according to new rules patents are not necessarily held by/awarded to the verified originator of the idea. Instead, the patent holder is the one who patents it. Thus a publication of an idea by Alice can be later patented by Bob even if Bob had nothing to do with creating the idea. It is *hard to believe that it is so, but the facts seem to confirm this rule.*

3.4.15. Another strong evidence that this new crazy rule is in effect: *Siemens patented KLJN* in 2011! (See patent number: DE102010025639A1.) Its title is incorrect; it calls the system "reliable" instead of "secure": Johannes Kammerer, *et al.*, "Absolutely reliable signal transmission by means of thermal noise". The

(*Continued*)

(*Continued*)

KLJN principle is shown and the first KLJN paper is cited. The patent was abandoned recently in March 2014 — which is good news because who can fight Siemens when he/she wants to use KLJN? Anyway, KLJN was intended to be public property.

3.4.16. Final note: by now, all our patent disclosures have been made public except one. The only patented KLJN scheme is the unconditionally secure Smart Grid scheme (see Section 5.5).

6. Summary

In this Letter, we outlined a new way of communication through a wire. The main advantages of this method are absolute security and the use of classical information in a form that is fast, inexpensive, robust (it is vibration, dust, temperature and aging resistant), maintenance-free, and that has a low power consumption rate. Even quantum computers are unable to break this code. Moreover, it is more secure than quantum communication. If it is applied for the same range as a quantum communicator, its speed can easily be made 100 times greater; or alternatively, at the same speed it can go for a 100 times longer distance without the need for repeater stations. The communication can also use the stealth mode when the noise generators are the thermal noises of the resistors.

Acknowledgements

This paper is the subject of a TAMU patent disclosure submitted on September 15, 2005. The manuscript was uploaded on the arxiv preprint server arxiv.org/abs/physics/0509136 (on September 15). Subsequent discussions with Janos Bergou, Julio Gea-Banacloche, Bob Biard, Adrian Cho, Imre Naggyorgy, Christian Vandenbroeck, Frank Moss, Derek Abbott, Peter Heszler and Suhail Zubairy are appreciated.

References

[1] R. L. Rivest, A. Shamir and L. Adleman, "A method for obtaining digital signatures and public-key cryptosystems", *Commun. ACM* **21** (1978), 120–126.

[2] A. Klappenecker, "Remark on a "Non-breakable data encryption" scheme by Kish and Sethuraman", *Fluctuation and Noise Letters* **4** (2004), C25–C26; 5. L.B. Kish, "Response to Klappenecker's Remark on a Non-Breakable Data Encryption Scheme by Kish and Sethuraman" *Fluctuation and Noise Letters* **4** (2004), C27–C29.

[3] L.B. Kish and S. Sethuraman, "Non-breakable data encryption with classical information" *Fluctuation and Noise Letters* **4** (2004), C1–C5; L.B. Kish, S. Sethuraman and P. Heszler, "Non-Breakable Data Encryption with Classical Information?" Proc. Fourth International Conference on Unsolved Problems of Noise, Lecce/Gallipolli, Italy, June 6–10, 2005, American Institute of Physics Press (2005).

[4] J.A. Bergou and L.B. Kish, "An absolutely secure QKD scheme with no detection noise, entanglement and classical communication", http://arxiv.org/abs/quant-ph/0509097, (September 14, 2005).

[5] A. Geresdi, J.A. Bergou and L.B. Kish, unpublished.

[6] L.B. Kish, "Zero-Signal-Power Communication: Zero-Power Classical Communication, Zero-Quantum Quantum Communication and Environmental-Noise Communication", *Applied Physics Letters* **87** (2005) Art. No. 234109. http://arxiv.org/abs/physics/0508135, (November 1, 2005).

[7] C.D. Motchenbacher and J.A. Connelly, *Low-Noise Electronic System Design*, John Wiley & Sons, (New York, 1993).

[8] I. Marcikic, H. de Riedmatten, W. Tittel, H. Zbinden, and N. Gisin, Long distance teleportation of qubits at telecom wavelength, *Nature* **421** (2003), 509.

3.5. Terminating the Man-in-the-middle and Other Hackers

3.5.1. Soon after the time when the KLJN feature came out in *Science* (October 2005), my ailing mother in Mako, Hungary, was taken to the hospital. Then she was moved between various hospitals in the area. I managed to visit her for a week, and had planned to do it again during the Christmas break. However, she died on December 9 right after my wife visited her. At the hour she died, I was in the midst of an intense email debate about the security of KLJN with *Bruce Schneier* and his software-security experts.

Figure 3.5.1. Left: my mother (Belane Kiss) in the summer of 2005; about the same time when the pre-KLJN idea was formed. Although she died in the middle of the first KLJN debates, I did not receive any "parapsychological" sign that my mom had just died. I only received an email an hour later while I was still in the middle of the debate in my office. It was then that I quit the debate and went out into the hall of Zachry building to find somebody to speak to. That was when I bumped into Henry Taylor, whom I told about my mother's demise. He soon followed the same fate (see Section 3.3. — particularly Note 3.3.2). When I left the office, the weather was cold — the remains of an ice storm from the night before. I had wished then that I could be with my mom instead of defending KLJN. Right: My father, Bela Kish, in 1989, at the cannons of the Uppsala Castle, when he visited me during my first visiting scientist stay at Uppsala University. We went to the Registrar to check the family legend about a remote relative being an Honorary Doctor at Uppsala. It turned out to be untrue. My dad passed away eight years later. However, 21 years later, when I received my Honorary Doctor title at the very same place we had visited together, I felt that his spirit was also there. My mother came from a very rich family that had been robbed by the Communists after 1945. As a result, she was kicked out of high school, for being an "enemy of the proletariat". So she could not become more than a typist. Though it had then seemed as if fate was cruel, it was due to those very circumstances that my dad (who was a poor chemical shop manager) and mom could meet and eventually marry.

3.5.2. Bruce Schneier is a famous security specialist/entertainer called a "security guru" by the *Economist*, and who has written at least 12 books on security. Bruce was part of the first cyber KLJN debate. The debate was very educational. He had close contacts with software-security experts, including people who had discovered (and named) the RSA protocol. During these debates they did not succeed with challenging the security of KLJN; however, I learned a lot and found KLJN more secure than I had originally thought.

(*Continued*)

(*Continued*)

3.5.3. When Bruce or one of his associates asked me how I would defend KLJN against a man-in-the-middle-attack, my first answer was that there is no unconditionally secure communications that could be defended against such. I was wrong. At that time, even I did not know my own system deeply enough.

3.5.4. At the middle of December 2005, I visited Janusz Smulko and his department in Gdansk, Poland. In the evenings at the hotel, I felt heavy waves of mourning kicking in. At the same time, I also felt that I had been given the opportunity to explore new defense methods based on statistical physics and Kirchhoff's theory. The idea of measuring and comparing both the currents and voltages at the two ends completely nullified the man-in-the-middle-attack. I quickly wrote the resulting paper and submitted it to both arXiv and FNL from the hotel on December 19th.

3.5.5. In *quantum systems*, similar continuous monitoring methods *would fail* because quantum information lacks the robustness of classical information. *Due to the fragility of the quantum bit, the signal cannot be monitored without destroying it.* Thus, the security of KLJN against (invasive) active attacks is clearly due to the classical physical nature of current and voltage in the system. Even a single shared raw key bit is protected against the man-in-the-middle attack, moreover against any active (invasive) attacks. No similarly efficient defense method is known in quantum encryption.

Paper: Protection against the man-in-the-middle-attack for the Kirchoff-loop-Johnson(-like)-noise cipher and expansion by voltage-based security†

Abstract. It is shown that the original Kirchoff-loop-Johnson(-like)-noise (KLJN) cipher is naturally protected against the man-in-the-middle (MITM) attack, if the eavesdropper is using resistors and noise voltage generators just like the sender and the receiver. The eavesdropper can extract zero bits of information before she is discovered. However, when the eavesdropper is using noise current generators, though the cipher is protected, the eavesdropper may still be able to

†L.B. Kish, *Fluct. Noise Lett.* **6** (2006) L57–L63.

extract one bit of information while she is discovered. For enhanced security, we expand the KLJN cipher with the comparison of the instantaneous voltages via the public channel. In this way, the sender and receiver have full control over the security of measurable physical quantities in the Kirchoff-loop. We show that, when the sender and receiver compare not only their instantaneous current data but also their instantaneous voltage data, then the zero-bit security holds even for the noise current generator case. We show that the original KLJN scheme is also zero-bit protected against that type of MITM attack when the eavesdropper is using voltage noise generators only. Thus, man-in-the-middle-attack does not provide any advantage within the mathematical model scheme. The remaining possibility is the short, large current pulse described in the original paper, which uncovers Eve during the exchange of a single bit.

Keywords: Unconditionally secure communication without quantum; man-in-the-middle attack; stealth communication; noise.

1. Introduction: totally secure communication without quantum information with Kirchoff-loop and Johnson(-like) noise

Recently, a totally secure classical communication scheme was introduced [1, 2] utilizing two pairs of resistors and noise voltage generators, the physical properties of an idealized Kirchoff-loop and the statistical physical properties of thermal noise. In the idealized scheme of the Kirchoff-loop-Johnson-(like)-noise (KLJN) cipher, the passively observing eavesdropper can extract zero bit of information. The *intrusive eavesdropper*, who emits a large and short current pulse in the channel, can extract *only one bit* of information while she is getting discovered [1]. The issue of the man-in-the-middle (MITM) attack was not studied in the original paper because of the author's earlier belief that no core physical secure layer is protected against such an attack and security against the MITM attack has to be provided by additional tools, signatures, a trusted third party, etc.

In this paper, it will be shown that the originally proposed KLJN scheme [1] is naturally protected against the MITM attack. Moreover, we propose an easy enhancement of the security that is also the exchange and comparison of the instantaneous voltage amplitudes. This step further enhances the security against MITM attacks, and can also be beneficial against other types of attacks of any practical realizations.

2. General clarifications about the issue of security of physical secure layers

In secure communication, any one of the following cases implies absolute security; however the only interesting cases for physical secure layers are points 3 and 4:

1. The eavesdropper cannot physically access the information channel.
2. The sender and the receiver have a shared secret key for the communication.
3. *The eavesdropper has access and can execute measurements on the channel but the laws of physics do not allow extracting the communicated information from the measurement data.*
4. *The eavesdropper can extract the communicated information*; *however, when that happens; it disturbs the channel so that the sender and receiver discover the eavesdropping activity.*

Keeping in mind that the relevant points are 3 and 4, we can classify the subject of research as follows:

(i) *Absolute security of the idealized situation.* This is the most fundamental scientific part of the research and the mathematical model of the idealized physical system is tested. The basic question is: how much information can be extracted from the data served by the physical measurements allowed in the idealized situation? The original paper and the present study investigate this question.

(ii) *Absolute security of the practical situation.* This part of the research requires an interdisciplinary effort including the fields of physics, engineering and data security. Because no real system can totally match the physical properties of the ideal mathematical model system — this kind of absolute security *does not exist in reality*; it is, rather, approached with a security that may be *only practically absolute*. For example, in quantum communication, we have no ideal single photon source, no noise-free channel, and no noise-free detectors, and any of these deficiencies compromise absolute security.

(iii) *If the code is broken, how many bits can the eavesdropper extract before she is discovered due to the disturbance of the channel*? This question can also be treated at the idealized-fundamental level and at the practical one. Some answers for the idealized case: RSA: infinite number of bits; Quantum: 20–10000 bits; KLJN cipher: 1 bit.

In the rest of the paper, we strictly focus on question (*i*) while we discuss the MITM attack in terms of point *4* above (that the eavesdropper can extract communicated information, but at the cost of disturbing the channel and being discovered).

3. Security of the KLJN cipher against the man-in-the-middle attack and expanding the system with voltage based security

In this section we show that the original KLJN cipher arrangement [1] is secure against the MITM attack. The goal is to study the idealized model and show that the eavesdropper is discovered when she executes the attack. Finally, we propose a simple expansion of the scheme by comparing the voltage data, too, for enhanced security.

Figure 1 shows the MITM attack by using resistors with the same values, and noise voltage generators with the same parameters, as those of

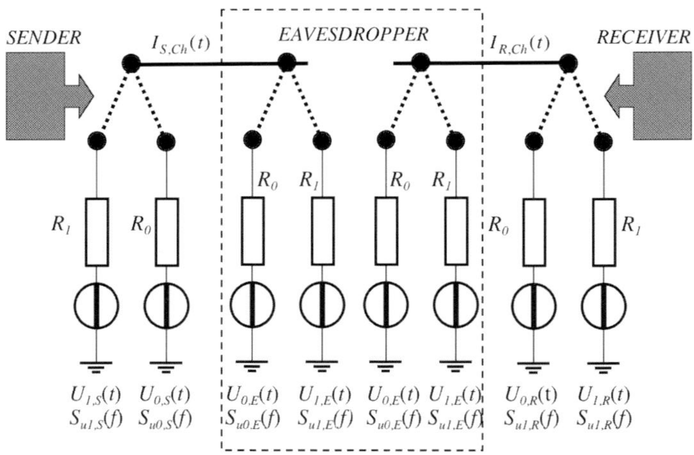

Figure 1. Man-in-the-middle-attack by using resistors with the same values and noise voltage generators with the same parameters as those of the sender and the receiver. (*Note, this time, following the cipher idea, Alice and Bob was still called Sender and Receiver.*)

the sender and the receiver. The eavesdropper breaks the line at the middle and installs two KLJN communicators, one for the sender and another one for the receiver. According to the original scheme, the instantaneous current amplitudes, at the sender's end and at the receiver's end, are compared via a public channel. Because the eavesdropper's noise generators are different representations of the corresponding stochastic processes, their instantaneous amplitudes are different from those of the sender's and the receiver's noise generators. Therefore, the current amplitudes are different in the two loops and the eavesdropper is discovered within the reciprocal of the bandwidth, within the time resolution of the communicated noise, before extracting a single bit of information. We remind the reader that the extraction of information from the noise needs to make (short range) statistics through an averaging time window, which is the clock period, thus a single time (noise) sample provides zero information.

Mathematically, we can write as follows. Let $i \in 0, 1$; $m \in 0, 1$; $k \in 0, 1$; $p \in 0, 1$. Then all the following voltages $U_{i,S}(t)$, $U_{m,E}(t)$, $U_{k,R}(t)$ and $U_{p,E}(t)$ (see Fig. 1) are statistically independent Gaussian stochastic processes with zero mean. The current at the sender's end and the receiver's end can be written as

$$I_{S,Ch}(t) = \frac{U_{i,S}(t) - U_{m,E}(t)}{R_i + R_m} \quad \text{and} \quad I_{R,Ch}(t) = \frac{U_{k,R}(t) - U_{p,E}(t)}{R_k + R_p}, \tag{1}$$

respectively. As the most pessimistic case, let us suppose that the denominators are equal and the nominators also contain the same type of noise generators; for example, $i = k$ and $m = p$. Then the RMS value of $I_{S,Ch}(t)$ and that of $I_{R,Ch}(t)$ are equal. Then the probability P_0 that $I_{S,Ch}(t_0) = I_{R,Ch}(t_0)$ at a given time moment t_0 is roughly equal to the ratio of the amplitude resolution Δ of the measurement system and the RMS value of these currents:

$$P_0 \approx \frac{\Delta}{I_{RMS}} \tag{2}$$

Let us suppose 7 bits resolution of the measurement (a pessimistic value), then $P_0 = 1/128$, which is a less than 1% chance of staying hidden. On the other hand, P_0 is the probability that the eavesdropper can stay hidden during the correlation time τ of the noise, where τ is roughly the inverse of the noise bandwidth. Because the KLJN cipher works with statistics made on the noise, the actual clock period T is $N \gg 1$ times longer than the correlation time of the noise used [1]. Thus, during the clock period, the probability of staying hidden is:

$$P_{clock} = P_0^N \tag{3}$$

Supposing a practical $T = 10\tau$ (see [1]) the probability at the other example is $P < 10^{-20}$. This is the estimated probability that, in the given system, the eavesdropper can extract a single bit without getting discovered. The probability that she can stay hidden while extracting 2 bits is $P < 10^{-40}$; for 3 bits it is $P < 10^{-60}$, and so on. In conclusion, we can safely say that the eavesdropper is discovered immediately before she can extract a single bit of information. The probability of staying hidden can be estimated in the same manner in the rest of this paper, however we skip these calculations because of their trivial nature, and the obviousness of the considerations below.

Because the defense described above is based on the difference of the instantaneous current amplitudes at the two sides, a natural question arises. Can we provide security when the eavesdropper executes the MITM attack by using *two noise current generators* with the same instantaneous current amplitudes and RMS values imitating a reasonable noise current in the channel? Figure 2 shows the MITM attack by using twin noise current

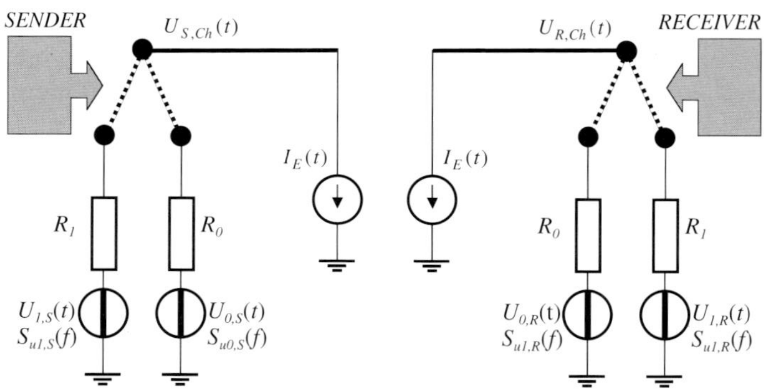

Figure 2. Man-in-the-middle-attack by using twin noise current generators with the same instantaneous current amplitudes.

generators with the same instantaneous current amplitudes. In this case, the equality of the current amplitudes at the two sides is guaranteed by the twin noise current generators, therefore comparing the current amplitudes cannot be used for protection. For the best protection, we can expand the original KLJN cipher by a voltage-based security measure. The sender and the receiver can compare the instantaneous voltage amplitudes at their end via a public channel. Because the sender's noise generator and receiver's one have either different parameters or they are different representations of the same stochastic processes, their instantaneous amplitudes differ most of the time. Therefore, the voltage amplitudes are different at the two ends and the eavesdropper is discovered practically immediately, before extracting a single bit of information.

However, it is important to note that the original scheme is also secure against this last kind of attack though not at the zero-bit security level but at the one-bit level. Normally, the power density spectrum $S_{u,Ch}(f)$ of the voltage noise in the channel is *smaller* than that of the actual noise voltage generator $S_u(f)$ of the sender and the receiver, respectively, because it is proportional to the parallel resultant of those resistances:

$$S_{u,Ch}(f) < S_u(f) \tag{4}$$

However, during the MITM attack shown in Figure 2, the following relations hold:

$$S_{u,Ch}(f) = S_u(f) + R^2 S_{i,E}(f) \quad \text{therefore } S_{u,Ch}(f) > S_u(f). \tag{5}$$

That means, the MITM attack will be discovered after sufficient statistics have been made on the spectrum during the clock period. During the communication of the secure bit, the side using the larger resistor will discover the attack faster. At the same time, the eavesdropper, while getting discovered, can extract one bit of information. Thus, the expansion by the voltage-based security measure described above provides an enhanced security.

Because the defense described in the context of Figure 2 is based on the difference of the voltages at the two sides, a natural question arises. Can we provide security when the eavesdropper executes the MITM attack by using *two noise voltage generators* with the same instantaneous voltage amplitudes and RMS values imitating a reasonable noise voltage in the

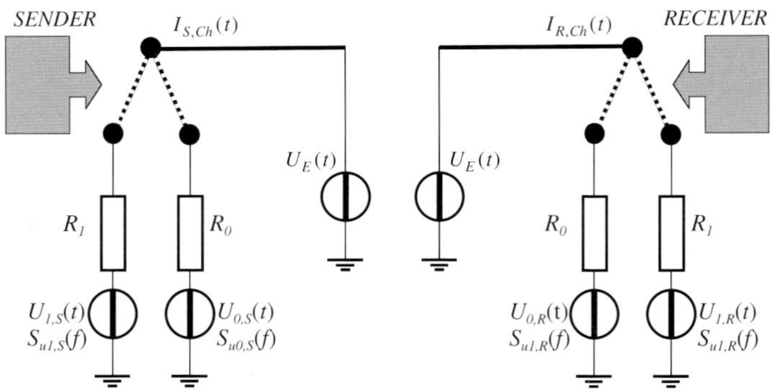

Figure 3. Man-in-the-middle-attack by using twin noise voltage generators with the same instantaneous current amplitudes.

channel? Figure 3 shows the man-in-the-middle-attack using twin noise voltage generators with the same instantaneous current amplitudes. In this case, the twin noise voltage generators guarantee the equality of the instantaneous voltage amplitudes at the two sides. Therefore comparing the voltage amplitudes cannot be used for protection. For protection, the sender and the receiver can compare the instantaneous current amplitudes, via a public channel. Because the sender's noise generator and receiver's have either different parameters or they are different representations of the same stochastic processes, their instantaneous amplitudes are different. Therefore, the current amplitudes are different in the two loops and the eavesdropper is discovered immediately, before extracting a single bit of information.

3.5.6. This paper does not discuss KLJN emulation with active devices. Pao-Lo Liu introduced a *circulator-based* emulation in 2009. Following that (see the paper in Section 4.1), I showed a circulator-based man-in-the-middle-attack and the means of defending against it.

4. Conclusion

The KLJN cipher with a public channel for comparing currents is naturally protected against the MITM attack and the eavesdropper is discovered with a very high probability while or before she can extract a single bit of information. Enhanced security can be reached by comparing the voltages.

Acknowledgments

Discussions with a number of security experts and scientists/engineers challenging the security of the KLJN scheme and emphasizing the importance of the MITM attack are appreciated. Without the ability to provide a complete list of emailers, I list here, a few most significant commenters: Jonathan Wolfe, Bruce Schneier, Steven M. Bellovin, David Wagner, Charlie Strauss, Adrian Cho, Matthew Skala, Greg Kochanski, Christian Vandenbroeck, Frank Moss, and Derek Abbott.

References

[1] L. B. Kish, "Totally secure classical communication utilizing Johnson (-like) noise and Kirchoff's law", *Physics Letters A* **352** (2006) 178–182.

[2] Adrian Cho, "Simple noise may stymie spies without quantum weirdness", *Science* **309** (2005) 2148.

3.5.7. Soon after submitting this paper for publication, I had my first perplexing encounter with the National Science Foundation (NSF). Let me share this personal experience. When the *Science* magazine feature about KLJN came out at the end of September 2005, I sent a copy to the relevant Division Director of NSF. Sometime in January 2006, I received an email from him inviting me to apply to a call for proposals on emerging technologies (*quantum, bio, genetic, and other systems*) in computation and communications. It was a short deadline but I happily agreed. I applied for support for the experimental demonstration of the KLJN idea. I knew that my only chance to get serious consideration was if I exposed all the weak sides of quantum encryption. Thus I detailed out its security problems due to the impossibility of generating single photon pulses and noise; its bulkiness and high maintenance-need; and the extraordinary price of US$100k for a single communicator pair (at that time). This action massively backfired. Within 24 hours of submission, I received a phone message from the Program Manager saying that I should withdraw my application because it is irrelevant. When I called her to ask for an explanation, she said that the program was for quantum computing, genetic computing, etc., and that the KLJN scheme was irrelevant. I pointed out to her that in the announcement there was the phrase "and other systems" at the end, and that I didn't mind getting turned down if I could see the reviews for them. She said

(*Continued*)

(*Continued*)

that I should apply to some other call about software-based security, which will probably come out soon. I had already spent 2 weeks writing the application, and it was obvious that software-based security was a very wrong place for the KLJN. I insisted on preferring to see the formal (if applicable, rejection) reviews and decision, and mentioned that I had been invited by her Division Director to prepare and submit the proposal. The next day, I received another phone message saying my proposal had been returned without being reviewed by the board. I could not believe to my ears then. The Texas A&M administrator who handled the submission was also shocked by the treatment we had received. That encounter caused me to develop an aversion towards writing further applications to the NSF. I did not do so again until 2011, when I prepared an application for *noise-based logic*, because I had been personally encouraged to do so. The Program Manager for this 2011 submission was excellent and strongly supportive of our initiative, but the review board's response was less than favorable — perhaps because they had wanted to defend the *status quo* of the mainstream of that time (basically Moore's law of miniaturization).

3.5.8. Why had the Program Manager in 2005 acted in the way that she did? There were many possibilities. My personal guesses are various. She might have acted on her own initiative, or had been commanded to do so. Alternatively, it was also possible that exposing the practical weaknesses of quantum informatics in my proposal could have hit the sensitive political balance within the review board, and or boosted other classical physical schemes (genetic computing, etc.) too much. Anyway, staying without a grant helped me to focus on more interesting and deeper aspects and, by the end of the same year, the demonstration of KLJN was done independently, in collaboration with friends at the University of Szeged, Hungary (see Section 3.9). *It is a shame that the only technology that has ever been able to compete with quantum encryption got experimentally verified in Hungary, while the concept had been invented in the United States*. It seemed to me that Leo Szilard's idea to create the NSF to support "unidentified ideas" had not worked out as he may have envisioned. Szilard (ironically) may have foreseen this subversion. In his sci-fi satire "*The Voice of the Dolphins*", Szilard presented a research foundation where the democratic voting process in a committee containing a genius and two mediocre scientists reduced support for breakthrough-science. See also, the excellent humorous analysis by *J.G. Cramer, "The Retarding of Science"* (published in the *Analog Science Fiction & Fact Magazine*, 1984), http://www.npl.washington.edu/AV/altvw04.html.

3.6. The (Bergou–) Scheuer–Yariv Attack and Its Correct Derivation by Kish–Scheuer

3.6.1. After I submitted the first KLJN paper for publication, I held a seminar about it in Andreas Klappenecker's quantum computing seminar series on October 18, 2005. I showed a slide about practical (non-ideality-related) information leaks that do not, however, limit unconditional security (see Fig. 3.6.1.). One of them was the wire resistance. After that, I learned the hard lesson: that one should at least mention in the foundation paper any aspects, even trivial ones, that impact security. Mentioning them at seminars do not protect against attacks later. I had not shown these in the paper owing to the paper length constraint of *Physics Letters A*.

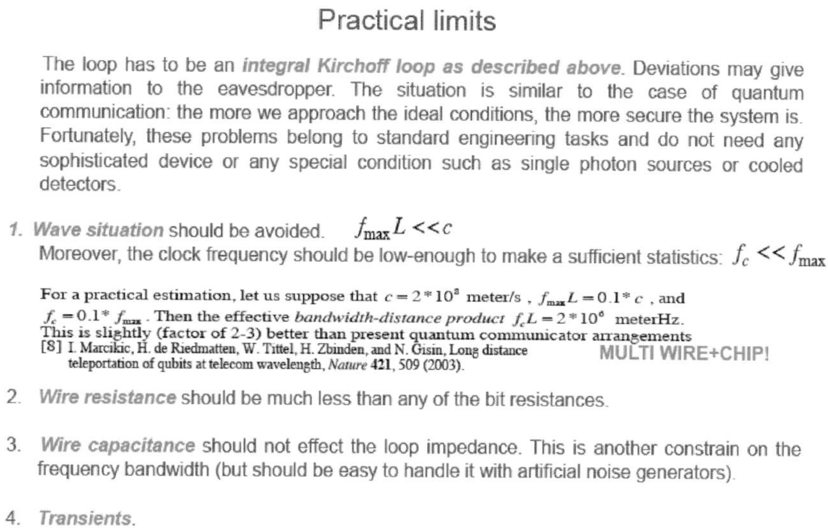

Figure 3.6.1. The slide about the major practical information leak sources, as presented at the very first KLJN seminar arranged by Andreas Klappenecker in his "Quantum Computing Seminar Series", on October 18, 2005. Due to the limited space in *Physics Letters A*, only point 1, the requirement for the steady state (non-wave) requirement was discussed in the first KLJN paper. Twelve days later, Janos Bergou mentioned point 2 (the wire resistance issue) to *Science* magazine reporter Adrian Cho. The following year, Jacob (Koby) Scheuer and Amnon Yariv published a derivation of this effect with incorrect results (see the corrected treatment below).

So Jacob (Koby) Scheuer and Amnon Yariv, independently from me, wrote a paper[1] where they claimed that KLJN was not perfectly secure in various situations. Only one of these situations, the band-limited noise with non-zero wire resistance was valid because the other situations (wide-band noise, switching transients and waves) were excluded from the permitted operation of the KLJN; thus their assumptions were illegal. However, their noise-calculations were flawed.

3.6.2. In 2010, Koby visited me. I welcomed him with a manuscript containing the improved derivation, and pointed out to him that the Scheuer–Yariv derivation was conceptually wrong, and that some of their results did not even have the correct physical unit. I invited him to become the second author of my manuscript to conclude the case smoothly. He agreed and, to my surprise, he even found a minor calculation error; thus really deserved the co-authorship. By his correction, Eve's information further decreased. This paper is shown below.

3.6.3. When I was assigned to review the Scheuer–Yariv paper, I told to Charlie Doering, the acting Editor of PLA at that time, that there was a fundamental error in their derivation, but that I was neither interested in correcting their error, nor in making the correct derivation at the reviewer level.; but that I rather preferred to write a separate paper about it later on. Charlie did not agree with my decision, and said that it was in everybody's interest that the paper was published free from errors. I felt differently because reviewers in PLA are anonymous and cannot be credited if they make major contributions to a paper. Thus their paper was published with the errors still present in it, which later justified publishing the paper in this section.

3.6.4. In my published first response [*Phys. Lett. A* **359** (2006) 741–744; http://arxiv.org/abs/physics/0602013], I kept silent about their fundamental error because I planned to publish a detailed analysis later; instead I pointed out that the about 1% voltage drop difference Scheuer–Yariv came up with was acceptable, because that level of information leak should be the typical

(*Continued*)

[1] J. Scheuer, A. Yariv, "A classical key-distribution system based on Johnson (like) noise — How secure?", *Phys. Lett. A* **359** (2006) 737–740.

(Continued)

practical goal. And this small leak can be easily eliminated (for example, through privacy amplification or the filtering out of high-risk bits; see Section 5.1).

3.6.5. From this lesson I learned that, when one knows ways of attacking his new scheme, it is very important to mention this fact in publication, even if only briefly.

Note: the final word on the (Bergou–) Scheuer–Yariv attack can be found in Section 5.2 where this information leak is totally eliminated by properly

Figure 3.6.2. Jacob (Koby) Scheuer with his wife Michal (senior computer scientist researcher at IBM) at Yosemite Park in 2013 (Courtesy of J. Scheuer). Scheuer is an Associate Professor with a background in photonics at Tel Aviv University. When Koby visited me in 2010, we published the corrected derivation[2] (see below) resulting in 1000 times less signal-to-noise ratio for Eve than their earlier results indicated. The attack was recently completely eliminated by a new defense technique; see Section 5.2. In 2011, on Koby's second visit, we published a joint work (lead by Tamas Horvath) on a simple *privacy amplification* scheme (see Section 1.6). Koby's most interesting related contribution is an optical, giant fiber laser key exchanger[3] inspired by the KLJN scheme. Instead of asymmetric resistances, narrow-band filters with frequency offsets are used. Since their system is wave-based and the security is not provided by the laws of physics, it is only conditionally secure. Koby agrees that his system is not unconditionally secure but he says it is technically difficult to crack.

[2]L.B. Kish, J. Scheuer, "Noise in the wire: the real impact of wire resistance for the Johnson (-like) noise based secure communicator", *Phys. Lett. A* **374** (2010) 2140; http://arxiv.org/abs/1002.0087.

[3]J. Scheuer and A. Yariv, "Giant Fiber Lasers (GFL): A New Paradigm for Secure Key Distribution", *Phys. Rev. Lett.* **97** (2006) 140502.

offsetting the temperatures. Thus this attack now has only an educational and historical importance. In the same paper, we also show a more efficient attack (*Second Law attack*). This new attack is also nullified by the same defense method.

3.6.6. The treatment in the paper below follows the algebraic rules of linear operations on noise (see Section 1.2; Noise essentials). Had Scheuer–Yariv checked the physical units of their result, they would have seen that something was very wrong. But then, if that had happened, Koby and I would have missed the opportunity to meet and co-publish the joint paper that follows.

Paper: Noise in the wire: the real impact of wire resistance for the Johnson (-like) noise based secure communicator[2]

Abstract. We re-evaluate the impact of wire resistance on the noise voltage and current in the Johnson-(like)-noise based secure communicator, correcting the result presented in [*Physics Letters A* **359** (2006) 737]. The analysis shown here is based on the fluctuation-dissipation and the linear response theorems. The results indicate that the impact of wire resistance in practical communicators is significantly lower than the previous estimation.

1. Introduction

Following the introduction of the Kirchhoff Law and Johnson (-like) Noise (KLJN) cipher [1], it was suggested that the system was vulnerable to time domain analysis of the currents and voltages evolution in the system [2, 3] as well as to the voltage drop on the non-ideal wire [2, 4]. The time domain claims and related statistical issues were answered in [5, 6]. In the present paper, we revisit the analysis of the impact of the wire's resistance, showing that in fact the expression derived in [2] was inaccurate. Here we present the correct derivation of the problem of wire resistance.

Before the analysis we would like to emphasize the following remarks:

(a) In the experimental realization of the KJLN communicator by Mingesz, *et al.* [7], the wire resistance was 2% of the sum of Alice and Bob's driving resistances. Based on the measured voltage distributions, the raw information leak accessible by Eve in this case (before privacy amplification) was estimated to be 0.19% of the key bits, a value which is substantially better than in any typical quantum key distribution. This value indicates a much higher security than practical quantum mechanical based systems where Eve has utilized proper non-idealities for attack [8–10].

(b) We emphasize that the fully protected version of the KLJN key exchanger [7, 11] is less susceptible to any attack based on detecting differences in the voltage or current along the wire. Alice and Bob are continuously measuring the voltage and current data at the two ends of the line, and are broadcasting and comparing these data via public channels [7, 11]. Thus they know exactly all the information accessible to Eve and can act accordingly [6] by discarding key bits where security has been compromised beyond a certain threshold. Just for comparison, QKD systems also offer some relevant protection but at a lower efficiency level because it necessitates the evaluation of the *error statistics* over a sequence of bits, an operation that requires a long record. However, in the KLJN protocol, Alice and Bob have full knowledge of Eve's best guess *for each bit* [6], thus knowing if this guess is correct or not. This is in contrast to QKD (where Alice and Bob do not know Eve's measurement outcome), and originates from the classical physical nature and robustness of the information transfer [6].

2. The results in [2] and their implications

According to Eq. (12) in [2], the voltage and current power density spectra measured by Eve at an asymmetric point in the wire [2] with wire resistances R_{W1} and R_{W2} toward Alice and Bob, respectively ($R_{W1} + R_{W2} = R_W$) are:

$$S_{ch}(f) = \frac{4kT[R_A(R_B + R_{W2}) + R_B(R_A + R_{W1})]}{R_A + R_B + R_{W2} + R_{W1}} \qquad (1)$$

$$S_i(f) = \frac{4kT(R_A + R_B)}{R_A + R_B + R_{W2} + R_{W1}}, \qquad (2)$$

where $S_{ch}(f)$ is the power density spectrum of the channel voltage at the given point in the wire, $S_i(f)$ is the power density spectrum of the current in the wire (loop current), R_A is Alice's resistance, R_B is Bob's resistance, k is the Boltzmann constant, and T is the effective absolute temperature of the noise generators of Alice and Bob. The Johnson noise of the wire is neglected because T is orders of magnitudes greater than the ambient temperature, corresponding to an external voltage noise generator.

To observe the highest voltage drop on the wire, that is, to extract the greatest amount of information, Eve must measure at the two ends of the wire and compare the voltages.

Thus, for the sake of simplicity, we suppose: $R_{W1} = 0$ and $R_{W2} = R_W$. Eq. (1) then yields that the noise voltage at Alice's end is:

$$S_{ch,A}(f) = \frac{4kT[R_A(R_B + R_W) + R_B R_A]}{R_A + R_B + R_W} \tag{3}$$

Applying Eq. (3) for Bob's end of the channel we use $R_{W1} = R_W$ and $R_{W2} = 0$ thus:

$$S_{ch,B}(f) = \frac{4kT[R_A R_B + R_B(R_A + R_W)]}{R_A + R_B + R_W}, \tag{4}$$

Finally, Eve extracts the most information when she measures at both ends and compares the voltages. Note, that in the practical run, Eve does not even have to measure anything because the current and voltage values measured at the two ends are broadcasted (and compared) via public channels by Alice and Bob for a full protection against invasive attacks [6, 11]. This also means, as we have already mentioned, that Alice and Bob are well aware of the actual voltage drop and the best guess of the bits by Eve [6].

Eve can extract significant information only when (the voltage-based alarm [11] is off and) the difference between the mean-square of the noise voltages (spectra integrated over the whole frequency range of operation) at the two ends is approaching or greater than the statistical inaccuracies of them during the finite sampling which is limited by the duration of the clock period. The difference between the spectra is:

$$S_{u,A-B}(f) = \frac{4kT[R_W(R_A - R_B)]}{R_A + R_B + R_W} \tag{5}$$

3. Analysis of the results of [2] shown above

Referring to Eqs. (3, 4), we note that in the case of zero wire resistance $R_W = 0$, the voltage noise spectrum is twice as large than what can be expected of a system comprising a set of two resistors R_A and R_B connected in parallel. This fact indicates that the results in [2] are inaccurate. The correct result, shown below, indicates significantly less information for Eve than the relevant result in [2], as shown in Section 5.

4. The wire resistance problem

In Fig. 1, the model circuits for the straightforward calculation of the voltage noise spectra at Alice's end (left panel) and Bob's ends (right panel) are shown. Alice's voltage generator has $U_A(t)$ voltage amplitude and $S_A(f) = 4kTR_A$ spectrum, where T is the effective temperature. Similarly, Bob's voltage generator has $U_B(t)$ voltage amplitude and $S_B(f) = 4kTR_B$ spectrum. Similarly to [2], the effective temperature T of the voltage generators is assumed to be substantially higher than the thermal noise of the wire, which is, therefore, neglected.

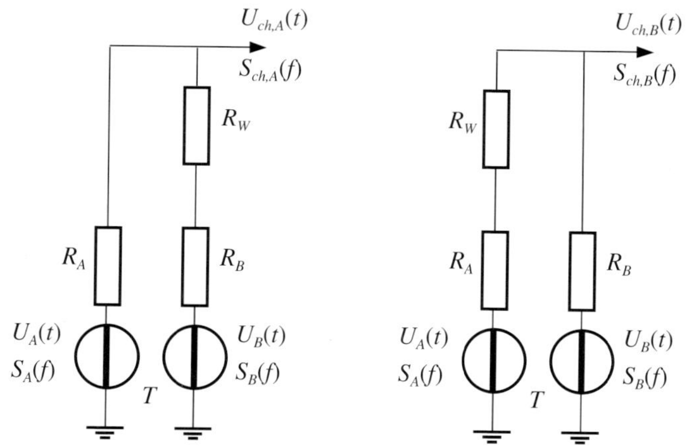

Figure 1. Circuit model to calculate the power density spectra of the channel voltage at Alice's and Bob's ends (left and right circuits). Similarly to the assumptions of [2], the effective temperature T of the voltage generators is supposed to be so high that the thermal noise of the wire is negligible.

The instantaneous channel voltage amplitude at Alice's end of the wire is the linear superposition of Alice's and Bob's generator voltages after reduced by appropriate voltage divider terms:

$$U_{ch,A}(t) = U_A(t)\frac{R_W + R_B}{R_A + R_W + R_B} + U_B(t)\frac{R_A}{R_A + R_W + R_B} \quad (6)$$

Similarly, the channel voltage at Bob's end:

$$U_{ch,B}(t) = U_B(t)\frac{R_W + R_A}{R_A + R_W + R_B} + U_A(t)\frac{R_B}{R_A + R_W + R_B} \quad (7)$$

Because the two noise voltages are statistically independent, the rule of linear operations on noises can be applied to get the separate noise spectra, and no cross-term appears:

$$\begin{aligned} S_{ch,A}(f) &= S_A(f)\left(\frac{R_W + R_B}{R_A + R_W + R_B}\right)^2 + S_B(f)\left(\frac{R_A}{R_A + R_W + R_B}\right)^2 \\ &= 4kT_{eff}\left[R_A\left(\frac{R_W + R_B}{R_A + R_W + R_B}\right)^2 + R_B\left(\frac{R_A}{R_A + R_W + R_B}\right)^2\right] \end{aligned} \quad (8)$$

and

$$\begin{aligned} S_{ch,B}(f) &= S_B(f)\left(\frac{R_W + R_A}{R_A + R_W + R_B}\right)^2 + S_A(f)\left(\frac{R_B}{R_A + R_W + R_B}\right)^2 \\ &= 4kT_{eff}\left[R_B\left(\frac{R_W + R_A}{R_A + R_W + R_B}\right)^2 + R_A\left(\frac{R_B}{R_A + R_W + R_B}\right)^2\right] \end{aligned} \quad (9)$$

where T_{eff} is the effective temperature of the noise sources ($T_{eff} \gg T$). We note that for the zero wire resistance case, Eqs. (8, 9) do provide the correct expression for the Johnson noise: $S_{ch}(f) = 4kT_{eff} R_A R_B/(R_A + R_B)$.

The difference between the measured voltage noise spectra at the two ends is the difference between (8) and (9):

$$S_{ch,A-B}(f) = 4kT_{eff}\left[\frac{R_W^2(R_A - R_B)}{(R_A + R_W + R_B)^2}\right] \quad (10)$$

5. Comparison of the results of [2] and that of the current derivation

Let us suppose, that at a given secure bit exchange, Alice uses the larger resistor and Bob uses the smaller one. For simplicity, we assume the following relations between the resistances:

$$R_W \ll R_B \ll R_A \tag{11}$$

Then, for the difference of the spectra at the two ends, according to Eq. (5), yields:

$$S^{[2]}_{ch,A-B}(f) \approx 4kT_{eff} R_W, \tag{12}$$

while (10) yields:

$$S_{ch,A-B}(f) \approx 4kT_{eff} \frac{R_W^2}{R_A} \tag{13}$$

For the relative difference of the spectra compared to the channel voltage spectrum, $\approx\ 4kT_{eff} R_B$, we get the following estimations by Eq. (12 and 13):

Based on Ref. [2]:

$$\frac{S^{[2]}_{ch,A-B}(f)}{S_{ch}(f)} \approx \frac{R_W}{R_B} \tag{14}$$

Based on Eq. (13):

$$\frac{S_{ch,A-B}(f)}{S_{ch}(f)} \approx \frac{R_W^2}{R_A R_B} \tag{15}$$

For example, with $R_A = 100\,k\Omega$, $R_B = 1\,k\Omega$, $R_W = 100\,\Omega$, Eq. (14) yields 0.1 while Eq. (15) yields 10^{-4} which is 3 orders of magnitude lower. This means that, according to this analysis, Eve needs 10^3 times longer measurement window to resolve the bit situation compared to the results of [2] because the uncertainty is mean-square estimation is inversely proportional with the measurement time window for band-limited white noise in the long-time limit.

3.6.7. How do Eve's measurement results, shown by Equation 15, impact the quantitative security measures of the key? That was evaluated only years later following the advice by Horace Yuen and a relevant remark by Renato Renner. These quantitative security measures are Eve's probability p of successfully guessing a key bit and the statistical distance (variational distance) between the actual key and a perfectly secure key of the same length (see the analysis in Sections 4.4 and 5.1). Both measures indicate that KLJN is unconditionally (information-theoretically) secure even if the defense method to nullify the (Bergou–) Scheier–Yariv attack (Section 5.2) is not used. The perfect security situation can be approached by investing more resources (thicker wire), and/or discarding the risky bits, and/or by privacy amplification (see Section 5.1).

Conclusions

The research of secure communication is progressing by a continuous process of proposing new schemes, attempting to break/attacking them, followed by the development of defense mechanisms and/or modified schemes. We revisited the impact of finite resistance on the KLJN cipher and found that the predicted information leak due to parasitic wire resistance is significantly less than it was previously estimated. In addition, the scheme can be improved to counteract such attacks because Alice and Bob can determine Eve's best guess based on the voltage difference at the two ends [6]. If these security measures are still insufficient, the same path can be followed as with quantum communicators: applying privacy amplification to the exchanged key.

3.6.8. While the attack described above became the most famous one (probably due to its easy conceptual understanding), its practical importance is negligible — like the Hao attack described in Section 3.7. The information leak is minor and, if that poses a problem, it can be totally removed (see section 5.2). In practical applications, the cable capacitance will be a much greater issue at longer ranges, see Section 5.8 for the cable capacitance attack. For KLJN security applications within computers, games, instruments or *physically unclonable function* (PUF) hardware keys, none of these security issues exist due to the short distance.

Acknowledgements

LBK is grateful to V. Makarov for information on quantum hacking.

References

[1] L.B. Kish, "Totally secure classical communication utilizing Johnson (-like) noise and Kirchhoff's law", *Physics Letters A*, **352** (2006), 178–182.

[2] J. Scheuer, A. Yariv, "A classical key-distribution system based on Johnson (like) noise — How secure?", *Physics Letters A*, **359** (2006), 737–740.

[3] Pao-Lo Liu, "A new look at the classical key exchange system based on amplified Johnson noise", *Physics Letters A*, **373** (2009), 901–904.

[4] J. Bergou in: Adrian Cho, "Simple noise may stymie spies without quantum weirdness", *Science*, **309** (2005), 2148.

[5] L.B. Kish, "Response to Scheuer-Yariv: 'A Classical Key-Distribution System based on Johnson (like) noise — How Secure?'", *Physics Letters A*, **359** (2006), 741–744.

[6] L.B. Kish, T. Horvath, "Notes on Recent Approaches Concerning the Kirchhoff-Law-Johnson-Noise-based Secure Key Exchange", *Physics Letters A*, **373** (2009), 2858–2868.

[7] R. Mingesz, Z. Gingl, L.B. Kish, "Johnson(-like)–Noise–Kirchhoff-loop based secure classical communicator characteristics, for ranges of two to two thousand kilometers, via model-line", *Physics Letters A*, **372** (2008), 978–984.

[8] V. Makarov, *New Journal of Physics*, **11** (2009), 065003.

[9] David Robson, "Laser cracks 'unbreakable' quantum communications", *New Scientist* (October 3, 2008); http://www.newscientist.com/article/dn14866-laser-cracks-unbreakable-quantum-communications.html

[10] V. Makarov, Q. Liu, L. Lydersen, S. Sauge, C. Weichers, I. Gerhardt, Ch. Kurtsiefer, A. Lamas-Linares, and A. Anisimov, manuscript in preparation (2010).

[11] L.B. Kish, "Protection against the man-in-the-middle-attack for the Kirchhoff-loop-Johnson(-like)-noise cipher and expansion by voltage-based security", *Fluctuation and Noise Letters*, **6** (2006). L57–L63.

3.7. The Hao Attack and the Response to It

3.7.1. While the attack described below is practically insignificant because of the extraordinary accuracy of the noise temperature that can be set with digital noise generation, the attack is very interesting from academic and historical points of views.

(*a*) This is the first attack where the direct goal is *not* the determination of who has the high and low resistances. The goal is simply to *separate* the noise levels of the 01 and 10 bit situations of Alice and Bob. Then, for example, Eve sets up two hypotheses about the bit situations: (*i*) 01 is when the noise level is higher; (*ii*) 10 is when the noise level is the lower. The two hypotheses provide two keys that are the inverse of each other. Then Eve simply tries both keys and one of them will work.

(*b*) Accordingly, the same attack works for any asymmetry that separates the 01 and 10 noise levels. (In my response below, I showed that this is true also for the inaccuracy of the resistors of Alice and Bob, which is a much more practical issue.)

(*c*) This genuine idea was developed by a (then) PhD student, Feng Hao, at Cambridge, UK. We got into a heated debate on the internet, which was mainly my fault. I saw that the scheme was practically unimportant however I should have appreciated the originality of Feng. The title of his paper (saying that KLJN is insecure) was incorrect because unconditional (information theoretic) security was not challenged at all, and by that time I was fed up with the large number of attacks — most of them extremely naive. Feng's was not naive at all; it had simply came at a personally bad time for me and his journal did not let me to write a comment bounded to his paper, which was the journal's fault, not Feng's. Thus I answered in *Fluctuation Noise Letters* (FNL). Years later, we exchanged some emails and both of us apologized, while I also congratulated him for this invention.

(*d*) The Hao attack is *unique*. So far (as of January 2015), it is the *only* attack *created by an independent author*, where the *derivation is correct*! There is *no error* in it *except in the title* of the paper, which is completely wrong. Correctly, it *should have been* something like: "*Information leak in the KLJN scheme due to temperature inaccuracies*".

Figure 3.7.1. Feng Hao in 2014. (Courtesy of F. Hao). He created the Hao attack against the KLJN scheme in 2006, while he was a PhD student in a software-based security group in Cambridge. Currently, Feng is a Reader and security researcher with an impressive publication list at the School of Computing Science, Newcastle University, UK.

Paper: Response to Feng Hao's paper "Kish's key exchange scheme is insecure"[1,2]

Abstract. In a [then] forthcoming paper in *IEE Proceedings: Information Security*, Feng Hao claims that temperature inaccuracies make the key exchange scheme based on *Kirchhoff Loop with Johnson-like Noise* insecure. First we point out that this claim is irrelevant for the security of the idealized/mathematical communicator because it violates basic assumptions. Furthermore, in general practical applications, as it has been pointed out in the very first paper, the use of thermal noise is disadvantageous; therefore the issue of temperature and its accuracy is unimportant even for the security of common practical realizations. It is important to emphasize that any deviation from the idealized scheme can lead to a finite level of security. Thus, if the above-mentioned author had stressed the inaccuracy of the resistor values; his claim would have been practically valid.

[1] L.B. Kish, Response to Feng Hao's paper "Kish's Key Exchange Scheme is Insecure", *Fluct. Noise Lett.*, **6** (2006), C37–C41; http://arxiv.org/abs/physics/0612193

[2] See also Section 5.8 for the new Vadai-Mingesz-Gingl system and the Random-Resistance-Random-Temperature KLJN scheme. They are both immune against the Hao attack. The last schee is immune against all the currently known attacks (January 2016).

However the impact of such systematic errors can be kept under control. Here we cite our relevant analysis (*Phys. Lett. A*, **359** (2006), 741–744) of 1% voltage inaccuracy along the line due to wire resistance effects, which can be considered as a general result for inaccuracies of this order. The necessity to build statistics to extract information during the properly chosen clock time leads to extremely poor eavesdropper statistics, and this situation provides a good practical raw bit security that can be orders of magnitude beyond the raw bit security of idealized quantum communicators.

Keywords: Secure classical communication with Johnson-like noise and Kirchhoff loop.

1. Feng Hao's comments

In [1], Feng Hao claims that temperature inaccuracy/difference at the resistors in the *Kirchhoff Loop Johnson-like Noise* (KLJN) communicators [2–8] compromises its security. Indeed, if we run the communicator with itself, the Johnson noise of the resistors (stealth communication [9]) and the temperature of one of the resistors deviates from the temperature of the other resistors, and there will be a net power flow between the communicators (Alice and Bob) (see Fig. 1). As pointed out in the first paper [2], non-zero net power flow between Alice and Bob compromises security. In the present case, if the clock period were infinite, the situation when the resistor

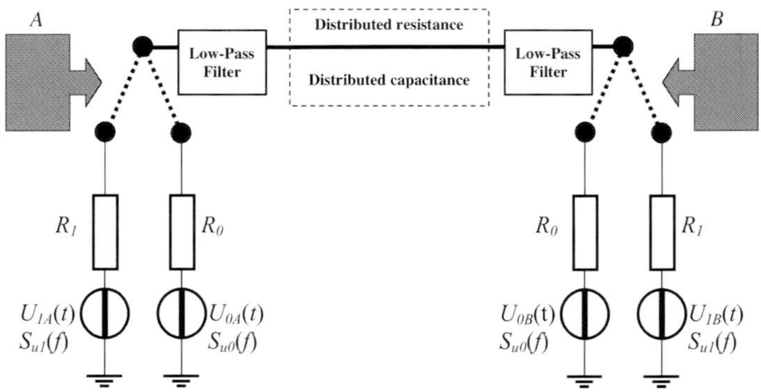

Figure 1. The Kirchhoff Loop Johnson-like Noise (KLJN) setup. Feng Hao's comment is based on the assumption that the voltage noise spectra $S_u(f)$ are different because the temperatures of the resistors are different at practical (non-ideal) conditions. However, in practical cases, the voltage noise generators are not thermal but external generators [2] with 1–100 million times greater voltages than the thermal noise; therefore the temperature has no relevance.

with a different temperature is connected to the line could be identified by the measurement of the net power flow and its direction. Though the eavesdropper (Eve) cannot identify the bit, and whether it is high or low, she can identify this situation whenever it happens, provided the temperature is fixed and the clock period is long enough. That means 50% of the shared key or its inverse could be extracted. Similarly, with proper differences between the temperatures of the other resistances, the whole secure key or its inverse could be extracted by Eve provided the clock period is sufficiently long. After getting the key or its inverse, Eve can test it and its inverse on the message and can decrypt the data.

In Section 2, we show several reasons why Feng Hao's comments [1] are though valid, practically irrelevant. In Section 3, we show how these comments can be modified and be made partially relevant for practical situations within the class of non-ideality problems of security. Finally, we show the solution: how to keep the practical security at any required level, beyond that of quantum security. This is the very same method that was already described in the first paper [2] and security data were published in [6].

2. Why are Feng Hao's comments irrelevant?

2.1. *Irrelevance for the case of idealized KLJN communicators*

It is important to emphasize that the comments [1] are irrelevant for the idealized/mathematical KLJN key exchange scheme [2] because the comments [1] violate the mathematical conditions assumed for total security. In the KLJN scheme it is assumed that the temperatures or the spectra of the voltage noise generators are equal, and violating that key assumption naturally leads to compromised security. It is like assuming that quantum communicators requiring single photons for security cannot produce single photons but only larger packages of photons, which is a valid practical problem in quantum communication. Though all the practical quantum communicators suffer from this deficiency compromising practical security, we do not say that the idealized/mathematical quantum communicator schemes are insecure. Therefore, we conclude that *Feng Hao's comments are irrelevant for the conceptual/theoretical KLJN scheme which remains totally secure.*

2.2. *Irrelevance for the case of practical KLJN communicators*

From the very beginning [2], *Johnson (-like) noise* was used for the KLJN scheme and the Johnson noise based considerations were used only for educational purposes, for the sake of simplicity, because the well known statistical physical characteristics of Johnson noise make the total security of the idealized KLJN cipher obvious.

3.7.2. The issue that I was stressing here and in similar, non-idealities-based information leak type attacks is the following (when expressed with the language of unconditional information-theoretic security):

(*a*) These papers (Scheuer–Yariv: Section 3.6; Hao: this section; Pao-Lo Liu: Section 4.1; Bennett–Riedel: Section 4.4) explicitly or implicitly suggested that the security of KLJN was cracked by these attacks. Instead (see Sections 4.4 and 5.1), these are only limited information leaks, which can (practically) be arbitrarily reduced by approaching the ideal situation or by privacy amplification software. The minimum price for these (when investments into infrastructure improvements are out of question) is the slowing down of the key exchange.

(*b*) Quantum key exchange (QKD) has had many types of information leaks described in a vast body of literature. Privacy amplification is used to fix these. However, in general, QKD has a nasty problem: its error probability is huge. Because privacy amplification is an error enhancer not only for Alice and Bob but also for Eve, QKD needs to run error correction algorithms, which unavoidably give out extra information to Eve. This fact is the object of one of the harshest criticisms by Horace Yuen from Northwestern University, who claims the security of QKD is unproven; but his deep analysis is largely disregarded by mainstream quantum informatics. (See, H.P. Yuen, "Security Issues Associated With Error Correction And Privacy Amplification In Quantum Key Distribution", http://arxiv.org/abs/1411.2310 (2014)).

(*Continued*)

(*Continued*)

(*c*) KLJN can reach astronomically small error probabilities (down to 10^{-20} see Section 4.3) without error correction algorithms by discarding suspicious bits without giving out information to Eve. Thus privacy amplification is very safe for KLJN, and Horace's arguments do not hold here. See our paper in Section 5.1, which Horace elected to invited paper rank in his special issue with Osamu Hirota in the *Journal of Quantum Information Processing*. (According to the editorial, only our paper and another one by Shapiro, *et al.* made it to the rank of invited paper; O. Hirota, H.P. Yuen, "Introduction to special issue on secure quantum communication", *Quant. Inf. Proc.* **13** (2014) 2153–2154. DOI: 10.1007/s11128-014-0813-z

(*d*) In conclusion, while information leaks due to non-idealities have been well-accepted facts of quantum communicators, these KLJN-attack papers suggested that similar leaks were killers for KLJN. However, in fact, the situation of KLJN is much more suitable for privacy amplifications; thus KLJN's unconditional security went unchallenged by these attacks. These issues indicated that the *essence of unconditional (information theoretic) security of physical key exchangers has not been widely understood.* Note, the general security proof in Section 5.1 also describes other, *non-privacy-amplification-type* methods to approach perfect security by using the robustness of classical physical information to filter out risky bits.

As it has been pointed out in the very first paper [2], in practical applications, the use of thermal noise is disadvantageous (except in stealth communication [9]) therefore the issue of temperature and its accuracy is unimportant even for the security of common practical realizations. In practical applications, the voltage noise generators are not thermal but external voltage noise generators [2] with 1–100 million times greater RMS noise voltages than Johnson noise. *Therefore actual the temperature of the resistors has no relevance.*

3. A valid practical case: inaccuracies as non-idealities and how to treat them

It is important to emphasize that any deviation from the idealized scheme of a physical secure communication scheme can lead to a finite level of security. Relevant examples for that are practical quantum communicators with the impossibility of generating strictly ideal single photon packages or with the impossibility to avoid detector noise which also compromises quantum security. Thus, if the above-mentioned author had stressed the inaccuracy of the resistor values, his claim would have been practically valid. However the impact of such systematic errors can be kept under control in practical KLJN realizations. Here we cite our relevant analysis [6] of 1% voltage inaccuracy (voltage drop) along the line due to wire resistance effects. The results can be considered as general results for the security leak at inaccuracies of this order (1%). The necessity to build statistics to extract information during the properly chosen clock time leads to extremely poor eavesdropper statistics and this situation provides an excellent practical *raw bit security* that can easily exceed the *raw bit security* of idealized quantum communicators.

Figure 2 shows the results of an analysis of the information leak due to systematic errors/inaccuracies [6], with 1% relative effect; which is a significantly worse situation than the 0.5% systematic temperature error supposed by Feng Hao [1]. In the case of Feng Hao's supposed situation, Eve would need a four-time longer clock period to extract the same amount of information as the analysis indicates below. Practical parameters and noise

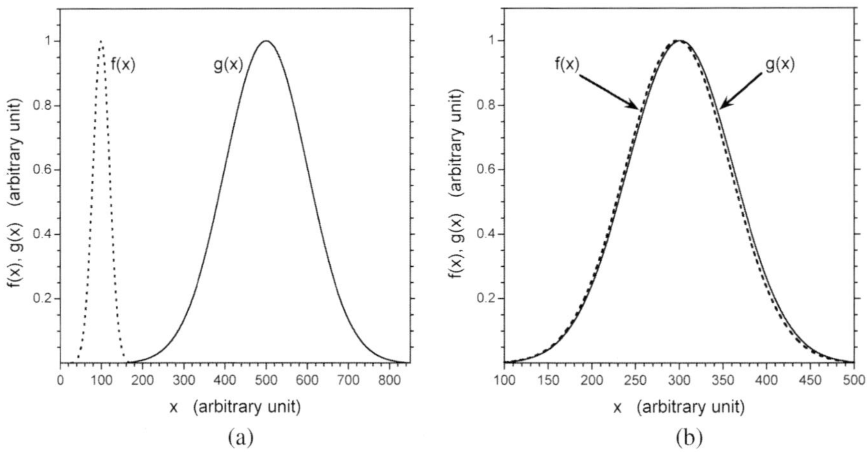

Figure 2. Analysis of the information leak due to systematic errors/inaccuracies [6], with 1% relative effect, which is a significantly worse situation than the 0.5% systematic error supposed by Hao. In the case of Hao's situation, Eve would need a 4-time longer clock time period to reach the same situation as indicated by these amplitude density functions.

rectifiers have been assumed (see [6]). The noise bandwidth (correlation time) and clock time period are selected so that Alice, Bob and Eve are able to extract only a few independent samples during the clock period. Alice and Bob must decide between the amplitude distributions f and g on Fig. 2(a) and Eve must decide between the amplitude distributions f and g on Fig. 2(b) while they have a few random hits at the x coordinates. It is obvious that Alice and Bob have an easy decision and Eve has a very poor case. Still, the application of Shannon's channel code theorem indicates that, if Eve finds out the best possible decoding method, she may be able to extract about 0.7% of the exchanged bits. However this raw bit leak is still better than raw bit leaks in idealized quantum communicators (see [6]).

Finally we note that a 1% inaccuracy is an "economical" one and, if higher raw bit security is needed, the accuracy can be improved with sufficient resources. For example, a 10-time increase in wire diameter decreases the 1% inaccuracy (voltage drop) in [6] to 0.01%, which would make the curves f and g in Fig. 2(b) indistinguishable to the naked eye and decrease the bit leak to 0.00007%, which is a *raw bit leak*/security much beyond the reach of any quantum communicator.

3.7.3. While these quantitative data effectively illustrate the nature of the matter, more elegant tools are the statistical (variational) distance measure and the simpler measure of Eve's successful bit guessing probability p (see Section 5.1 or 4.4). The perfect security is indicated by $p = 0.5$, which is equivalent with Eve's use of a perfectly random coin instead of eavesdropping. At the experimental demo (Section 3.9) the value was $p = 0.525$ with the (Bergou–) Scheuer–Yariv attack.

3.7.4. It was shown [T. Horvath, L.B. Kish, Y. Scheuer, "Effective Privacy Amplification for Secure Classical Communications", *EPL*, **94** (2011). 28002] that, by using a simple privacy amplification technique, which does *not* induce correlations between key bits (and security risks by that), a large $p = 0.55$ value can be reduced to $p = 0.50005$ (a value sufficient for any application), at the price of a 4-times slowdown. An astronomically large information leak of $p = 0.85$ can be reduced to near-to perfect level, $p = 5.0001$, at the price of a 32-times slowdown. The negligible error probability of KLJN makes these improvements possible (see Section 4.3).

References

[1] Feng Hao, "Kish's key exchange scheme is insecure", to appear in IEE Information Security; manuscript is at http://www.cl.cam.ac.uk/~fh240/pdf/total.pdf and discussions are at: http://www.lightbluetouchpaper.org/2006/10/08/kishs-totally-secure-system-is-insecure (version downloaded at October 29, 2006).

[2] L.B. Kish, "Totally secure classical communication utilizing Johnson (-like) noise and Kirchhoff's law", *Physics Lett. A*, **352** (2006), 178–182; also at http://arxiv.org/physics/0509136.

[3] Adrian Cho, "Simple noise may stymie spies without quantum weirdness", *Science*, **309** (2005), 2148.

[4] L.B. Kish, "Protection against the man-in-the-middle-attack for the Kirchhoff-loop-Johnson(-like)-noise cipher and expansion by voltage-based security", *Fluctuation and Noise Letters*, **6** (2006), L57–L63; also at http://arxiv.org/physics/0512177.

[5] L.B. Kish and R. Mingesz, "Totally secure classical networks with multipoint telecloning (teleportation) of classical bits through loops with Johnson-like noise", *Fluctuation and Noise Letters*, **6** (2006), C9–C21.

[6] L.B. Kish, "Response to Scheuer-Yariv: "A Classical Key-Distribution System based on Johnson (like) noise — How Secure?", *Physics Letters A*, **359** (2006), 741–744; doi: 10.1016/j.physleta.2006.07.037; http://arxiv.org/physics/0602013.

[7] L.B. Kish, («YEAR?«) "Response to Bollinger's 'On the Impossibility of Keeping Out Eves Using Only Classical Physics' and to Some Other Comments at Bruce Schneier's Blog Site", [online] (updated on 17 February 2006). Available at: http://www.ece.tamu.edu/~noise/research_files/Response_Bollinger.pdf

[8] L.B. Kish, "How Secure, How to Design, and How to Run Practical Kirchhoff-loop-Johnson-like-Noise Ciphers", unpublished.

[9] L.B. Kish, "Stealth communication: Zero-power classical communication, zero-quantum quantum communication and environmental-noise communication", *Applied Physics Lett.*, **87** (2005), 234109.

3.8. Telecloning and the First KLJN Network Application

During the intensive debates about KLJN on the web, I was contacted by a researcher from the National Security Agency of Germany, who was asking about a current injection attack. KLJN's immunity against it and the defense scheme used are outlined in the first paper (see Section 3.4). It has also been fully tested recently, at practical conditions (see Section 5.8).

To my surprise one morning, I received a phone call from the Pentagon. An expert who had been following the debates, told me that while he was sympathetic towards my efforts, it was hopeless to try to obtain funding for KLJN. He was right. He said the (then) new buzzword was "quantum", and that people at high levels had already invested in that direction. However, history shows that "quantum" security has not happened either (at least, not to the time of writing this book). I was told that NSA will not approve any security scheme that is either controversial, or not fully understood — a comment made in regard to governmental infrastructure.

As for civilian security — in our present time, with international terrorism stretching out globally, having an inexpensive unconditionally secure system, such as the KLJN, available for personal use, is unlikely to be supported. The very recent public debate between the FBI and Apple, and related plans for new laws are direct indications of that. A governmental contractor who was interested in KLJN once told me that in such a case, a backdoor must be installed for Big Brother before it can be marketed.

3.8.1. At the spring of 2006 I created a KLJN network and coauthored the paper in this section (Section 3.8.) with Robert Mingesz because, in his capacity as a reviewer for the paper, he had found a security flaw in its original protocol. The resulting protocol is the very first network application. It may look simple but it is actually quite tricky when it gets to security. It still remains to be seen if it will ever be used but, from a scientific point of view, it is an exciting advancement.

(*Continued*)

(Continued)

3.8.2. While writing this book, I came to realize that KLJN represents a classical physical telecloning of information during secure bit exchange. The definition of telecloning is that we copy the information to a remote place so that, in the space between the two places, this information does not exist.

3.8.3. Example: Alice generates a random bit and teleclones that to Bob. Suppose it is the bit value 1.

(*a*) She sets in the 1 resistor and they start the operation. Bob will randomly set a resistor value of 1 or 0. If Bob sets 1 then the communication is not secure, and Alice and Bob have failed in the telecloning, and must discard the bits and start again. This happens 50% of the time.

(*b*) They start again. Say, Alice generates 0 and Bob 1. Then a KLJN secure key exchange is executed. The information is copied to Bob, while this information does not exist in the wire line of the ideal KLJN system. Thus the telecloning of a bit took place.

(*c*) The situation is even better with *analog KLJN* (continuum resistor values). Then the probability of failure is theoretically zero. Then Alice and Bob can choose any resistor value from the continuum. After the KLJN operation, they will extract the resistor values at the other end by solving the current and voltage equations and choosing the solution that is not identical with their own resistor value. Its success means that both Alice and Bob have telecloned their continuum resistor value to the other end.

(*d*) Note: the situation (*c*) is a cypher, and this two-way communication was described in the first paper for discrete bits (see Section 3.4). So, for this special, non-practical situation, the "cypher" name indeed holds.

(*e*) It is not teleportation (even if that word has also been used there) because, in teleportation, the information must disappear at the origin; otherwise it is an incident of telecloning rather than teleportation. An interesting question: is there a way to achieve teleportation? Can we make a KLJN system in which the information at the point of origin is *naturally* erased (not artificially erased) by the telecloning process?

Paper: Totally secure classical networks with multipoint telecloning (teleportation) of classical bits through loops with Johnson-like noise[1]

Abstract. First, we show a new inexpensive defense against intruders and the man-in-the-middle attack in the Kirchhoff's-loop-Johnson-like-noise (KLJN) cipher. Then instead of point-to-point communication, we propose a high efficiency, secure network. The (in the idealistic case, totally secure) classical network is based on an improved version of the KLJN cipher. The network consists of two parallel networks: (*i*) a chain-like network of securely communicating electrically isolated Kirchhoff-loops with Johnson-like noise, driven by a specific switching process of the resistances; (*ii*) and a regular non-secure data network with a Coordinator-server. If the classical network is fast enough, the chain-like network of N communicators can generate and share an N-bit long secret key within a *single clock period* of the ciphers; and that implies a significant speed-up compared to the point-to-point key exchanges used by quantum communication or RSA-like key exchange methods. This is a teleportation-type, multiple telecloning of the classical information bit because the information transfer can take place without the actual presence of the information bit at intermediate points of the network. At the point of telecloning, the clone is created by the product of a bit coming from the regular network and a secure bit from the local KLJN ciphers. The same idea could be implemented with quantum communicator pairs producing entangled secure bits at the two ends. This is the telecloning of classical bits via quantum communicator networks without telecloning the quantum states.

Keywords: Unconditionally secure; network key distribution; classical telecloning; telecloning of classical bits via quantum channel; stealth communication.

1. Introduction: totally secure statistical physical communicator with Kirchoff-loop and Johnson(-like) noise

Recently, a totally secure classical communication scheme was introduced [1, 2] utilizing two identical pairs of resistors and noise voltage generators, the physical properties of an idealized Kirchoff-loop and the statistical physical properties thermal noise (see Figs. 1 and 2). The resistors (low bit = small resistor R_S, high bit = large resistor R_L) and their

[1]L.B. Kish, R. Mingesz, *Fluct. Noise Lett.*, **6** (2006), C9–C21; http://arxiv.org/abs/physics/0603041.

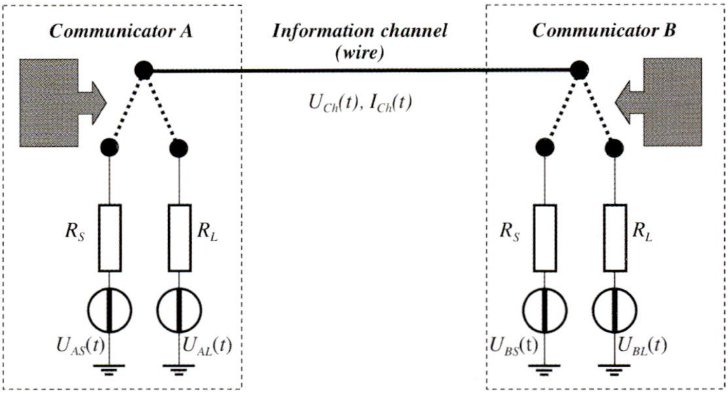

Figure 1. Secure key generation and distribution via the Kirchhoff-loop-Johnson(-like)-noise (KLJN) cipher (idealized picture). The joint information of the two end-bits is carried by the statistics of the *rms* noise voltage and current in the channel. Secure bit exchange takes place whenever the end-bits are different (intermediate value of the *rms* noise voltage and current). To extract the secure information bit of one end, the knowledge of this joint information and that of the information bit at the other end is needed. The passive eavesdropper can access only the joint information, so she knows that the bits are opposite, but she has zero information about the actual end bit.

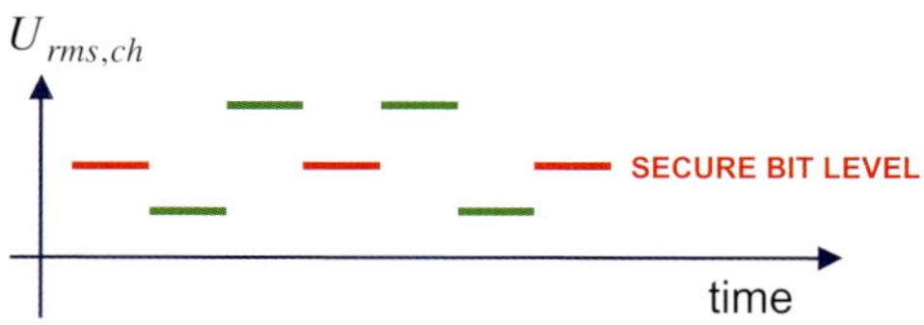

Figure 2. Channel noise *rms* voltage level due to the random bits at the two ends of the line; and secure bit exchange in the KLJN cipher. The *rms* value is obtained through a single clock period. On average, secure bits are only exchanged during 50% of the clock periods. The eavesdropper does not know which end of the line has the low bit and which one has the high bit.

noise voltage generators are randomly chosen and connected at each clock period at the two sides of the information channel. A *secure bit exchange* takes place when the states at the two ends are different, which is indicated by an intermediate level of the *rms* noise voltage on the line, or that of the *rms* current noise in the wire. The most attractive properties of the KLJN cipher are related to its *security* [1, 3–5] and the *robustness* of classical information. In the *idealized* scheme of the

Kirchoff-loop-Johnson(-like)-noise (KLJN) cipher, the passively observing eavesdropper can extract *zero bit* of information. The total security of the idealized system is based on Statistical Physics, and inherently connected to the Second Law of Thermodynamics (Fluctuation-Dissipation Theorem), the Energy Conservation Law (Kirchhoff's Loop-Law) and the rules of algebraic linear equation systems (N independent equations are needed to determine N unknown variables).

2. On the security of the KLJN cipher

Though there have been many comments about the fact that the practical cipher is not totally secure (see their refusals [4, 5] and a correct point [6]), the *total security of the idealized (mathematical model) system remains unchallenged.* The ideality of the KLJN cipher means that the physical/circuit model [1] shown in Fig. 1 *must be* the correct description of the situation. This requirement implies an ultimate high-frequency limit [1] for any normal or parasite frequency components in the channel to avoid wave affects (delay, propagation, reflection, etc.). This "no-wave" bandwidth limit scales with the reciprocal of the wire length [1]. Moreover, any other deviations from the model in Fig. 1, such as parasite elements or other modification, pose the risk of information leak [3–6]. The most significant, practical, non-ideal features are the cable resistance (pointed out by many commenters; first by Janos Bergou [6]) and the cable capacitance [3]. However, by using sufficiently thick wires, these problems can be eliminated and the idealized situation and total security can *arbitrarily* be *approached* depending on the *available resources*. The situation is similar to the case of quantum communication, where theoretical security can only be approached, but never reached. due to the lack of an ideal single-photon source, the noise in the channel and the detectors.

3. The earlier and new, economical solutions for defense against intruders and man-in-the-middle attacks

The *intrusive eavesdropper*, who emits a large and short current pulse in the channel, can extract *only one bit* of information while she is getting discovered [1]. The KLJN cipher, with a public channel for comparing currents, is naturally protected against the *man-in-the-middle* (MITM) attack [3] and the eavesdropper is discovered with a very high probability while or before she can extract a single bit. Enhanced security can be reached by comparing the voltages and then the eavesdropper is discovered with a very high probability before she can extract a single bit of information [3]. The fastest key distribution with the highest security needs these defenses (see Sections 7 and 8). A more economical solution is to detect the eavesdropping using only the statistical analysis of the noise (similar to quantum communication); however that method would allow the eavesdropper to extract a number of bits before she gets discovered, *just like in quantum communication*. On the other hand, in the case of the man-in-the middle attack, an average security of 4-bits can be reached if the parties only communicate the fact of a secure bit exchange via the public channel, which is a satisfactory and inexpensive solution. It is so because the man-at-the-middle cannot predict the time when the end bits and its own bits will favor a secure bit pair alignment. Thus, on the average, after 4 bits, one communicator will see a secure transfer while the other does not; consequently their exchanged public information will contradict and the "eavesdropping alarm" goes off. This economical solution requires exchanging the same number of bits via the secure and public channels. However (see Sections 7 and 8), the unconditional security will require extra clock steps which will slow down the key generation and distribution by a factor of 2 or 3.

4. The enhanced KLJN communicator with complementary loops

The original KLJN cipher exchanges secure bits only 50% of the time. This property can cause an exponential slow-down versus distance if the information is measured and further transmitted in a network of coupled KLJN ciphers. In the rest of this paper, first we show an enhanced KLJN communicator which exchanges bits at 100% of the time. Then we show a network solution where the intermediate network elements do not have to decode the information in order to forward it. Because the information "gets through" a chain of transmitters without even being there, the mechanism is a *classical teleportation* [7] effect, more precisely, *telecloning* because the information will be present at both the source and the receiver units. In this paper, we will apply the following convenient mathematical fashion. We will represent the "high" information bit with $+1$ and the "low" information bit by -1. Thus if A is the logic state of a port driving one end of a KLJN cipher then the inverse satisfies $\bar{A} = -A$.

The enhanced KLJN communicator makes use of each clock period by using two KLJN ciphers driven in anticorrelated way (see Fig. 3). This lossless KLJN communicator consists of two *electrically isolated* parallel KLJN loops with ports driven in the same fashion at one end and in a complementary way at the other end. Then one of the loops will always communicate a secure bit because the logical state will be the opposite at the ends of one of the loops.

An important weakness is that the communicator shown in Fig. 3 is not secure if the parties must agree through a regular network about who is running the correlated ports and who does the complementary ports. As soon as the eavesdropper learns the port arrangement, she can apparently decode the bit status at the two ends.

However, by using the totally secure nature of KLJN communication, it is possible to arrange the port distribution in an unconditionally secure way; let us call this the Secure Complementary (SC) cipher. The secure sharing of the parallel KLJN ports of Communicators (A) and (B) is

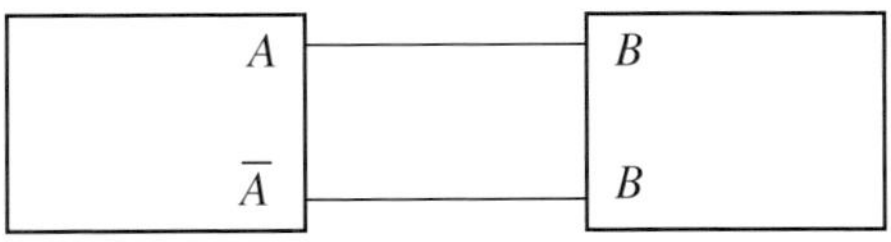

Figure 3. Lossless unsecure KLJN communication with two parallel ports. Each clock period transfers a bit. However the data are not secure as soon as the eavesdropper learns the port arrangement.

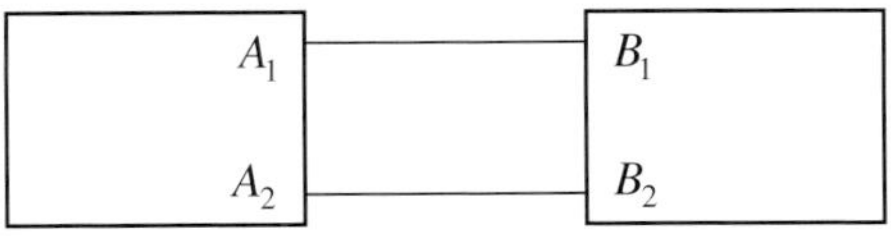

Figure 4. The Secure Complementary (SC) cipher and the secure distribution of the port control.

Table 1. Truth table of sharing the port controls at the SC cipher.

Situation of secure bits	Low (−1) at B_1 High (1) at A_1	Low (−1) at A_1 High (1) at B_1
Resulting port control	$A_1 = A_2$ $B_1 = \bar{B}_2 \equiv -B_2$	$A_1 = \bar{A}_2 \equiv -A_2$ $B_1 = B_2$

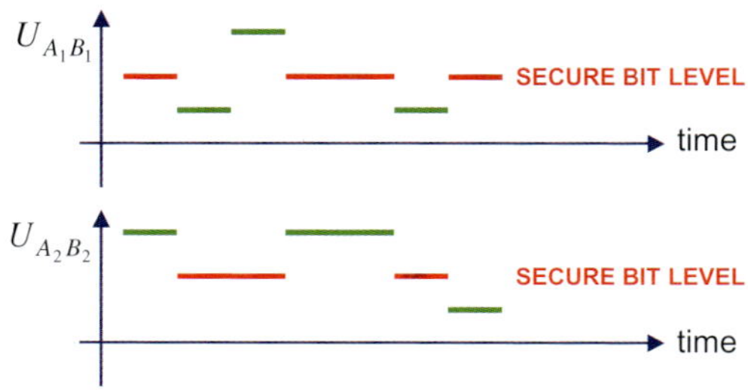

Figure 5. Example of channel noise *rms* voltage levels at lossless KLJN communication. All clock periods provide a secure exchange of (inverse) bits via one of the lines. Because the eavesdropper does not know the port control rules, she cannot access the information.

done once only, at the moment of initialization. The ports A_1 and B_1 are randomly driven (connecting one of their resistors) in each clock period until a secure bit (an intermediate *rms* noise voltage level) is exchanged between them. Then, for example, they can use the following pre-agreed table for sharing:

The SC communicator is also lossless, and communicates a secure bit at each clock period with total security in the idealized case (see Fig. 5).

5. Networks: initialization and high-speed bit-exchange via telecloning

In this section, we describe the network and a simple initialization process. However, this initialization process allows certain vulnerabilities against the "Mingesz attacks" (see Section 7) especially during the first network-key generation run. In Section 7, we will describe a more time-demanding but totally secure initialization process. In Section 8, we describe the most economical and still satisfactory solution.

The KLJN network consists of electrically isolated Kirchhoff loops driven in a specific way, and a regular network with a special Coordinator-server. For the sake of simplicity, but without the restriction of generality, we consider a one-dimensional (chain-like) network where the units in the network are connected only to their two nearest neighbors (see Fig. 6).

First, we describe the KLJN cipher Units of the network. The connection between Units in the network is made by SC cipher channels (see Fig. 7). One SC connection is to a unit's neighbor on the left via the left-hand-side ports L, and the other is to the neighbor on the right-hand-side via the ports R. The initialization of the units is described below.

(i) In the simplest case, the initialization process here is similar to the SC cipher initialization described above (see also Sections 7 and 8). This is only done once — assuming the units are defended by the current/voltage method [3] (see Sections 7 and 8). The ports R_1^M and L_1^{M+1} of Units (M) and ($M+1$) are randomly driven (connecting one of their resistors) at each clock period until a secure bit is

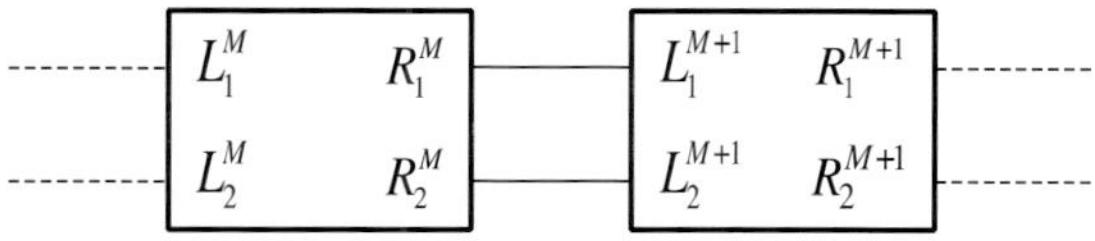

Figure 6. The M-th and ($M+1$)-th Units of the one-dimensional network and their external connections.

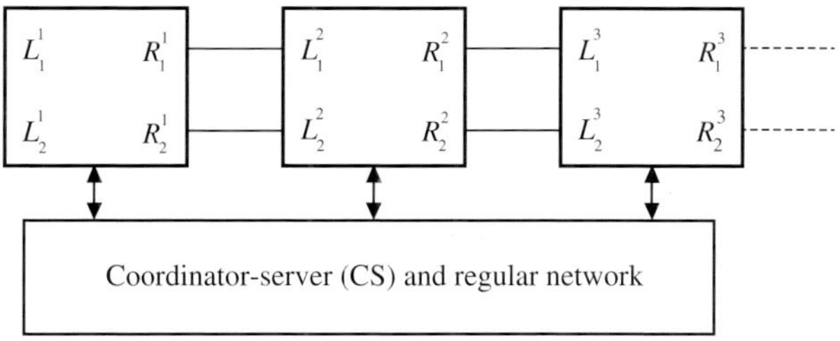

Figure 7. KLJN cipher based network for high-speed secret key generation and distribution via telecloning, with 100% fidelity, through chains of electrically isolated Kirchhoff loops. This system provides the highest speed when the defense is current/voltage based [3], see Sections 7 and 8. Note, a very good alternative can be realized with single wire R-L connections and regular KLJN ciphers (see Section 8).

Table 2. Truth table of sharing the port controls between two nearest neighbors in the chain-like KJLN network.

Situation of secure bits	Low (small resistor) at L_1^{M+1} High (large resistor) at R_1^M	Low (small resistor) at R_1^M High (large resistor) at L_1^{M+1}
Resulting port control	$R_1^M = R_2^M$ $L_1^{M+1} = -L_2^{M+1}$	$R_1^M = -R_2^M$ $L_1^{M+1} = L_2^{M+1}$

exchanged between them (intermediate *rms* noise voltage level). Then, for example, they can use the following pre-agreed table for sharing:

Now, let us see how the network works. The Units are connected via the secure lossless KLJN chain described above and they communicate through the regular network too (see Fig. 7). After the initialization described above, the Units generate random bits so that they satisfy their port control initialization rules and the actual internal logic connection rule (regenerated at each clock period). The Units report the logic relation F (+1 or −1) between their secure bit at their L-ports and their secure bit at their R-ports to the Coordinator-server (CS) of the regular network. Let us suppose that Unit (N) wants to clone the actual secure bit at the left-hand-side of Unit (M), where M < N. Then Unit (N) sends this request to the CS. The CS calculates

$$G_{M,N} = (-1)^{N-M} \prod_{j=M}^{N-1} F_j, \tag{1}$$

where $G_{M,N}$ is the logic relation between the secure bit at the L^M ports of Unit (M) and the secure bit at the $L^{(N)}$ ports of Unit (N); and F_i is the reported logic relation between the secure bit at the L^j ports and secure bit at the R^j ports of Unit (j). Thus Unit (N) will make the following calculation:

$$X_L^M = G_{M,N} X_L^N, \tag{2}$$

where X_L^M is the secure bit at the L^M ports of Unit (M) and X_L^N is the secure bit of the L^N ports of Unit (N). Because Unit (N) knows the secure bit values at his own ports and because he receives the $G_{M,N}$ information from the

CS, he can evaluate Eq. (3) and that is equivalent with the (tele)cloning of the X_L^M secure bit with 100% fidelity.

Let us see why this is a telecloning process. The X_L^M clone is the product of two information bits, the $G_{M,N}$ arriving via the regular network from CS, and the local secure bit X_L^N. Therefore, the information does not exist on the chain of intermediate Units. It looks like it is "teleported" via the chain built of electrically isolated Kirchoff loops. However, because the X_L^M bit stays intact at its original location during this operation, the proper term is telecloning, not teleportation.

We would like to mention that the expressions of "teleportation" and "telecloning" of information arise from the research on quantum communication (see for example [8, 9]) and that they had been considered as the unique properties of quantum systems, just like secure physical layers for communications. However, recently it was shown by Oliver Cohn [10] that the classical teleportation of classical information is also possible. In his model, he used boxes and coins and reached the teleportation of information by a total of 100% fidelity. In the present paper, we have also reported the telecloning of information with a total of 100% fidelity — a level which is impossible to reach in a quantum system due to fundamental mathematical constraints [9] of the telecloning operations and the extreme fragility of quantum information.

6. High-speed secret key distribution over the whole network

Because the CS collects all the logic relations between the secure bits at the L and R ports of each Unit, CS can provide the relevant $G_{k,m}$ bit for Unit (m) to teleclone the secure bit at the L ports of Unit (k). This information can be given to all Units simultaneously about all the other Units. In this way, each Unit can teleclone all the secure bits at the other Units, for each clock period — provided the regular network is fast enough. That means, if the network consist of N Units, an $(N-2)$ bit long secret key is generated and securely distributed over the whole network *at each clock period.* (We supposed that the L ports of Unit (1) and the R ports of the last Unit in the chain are not connected; this is the reason for the -2 correction of N). If the regular network is slow, the X bits can be recorded by the Units and they can make the telecloning later when they receive the relevant G bits.

For illustration, let us do a simple estimation. Let us suppose that 1% of New York's population (200 thousand people) installs a KLJN cipher based network card in their PC and they are connected by a pair of wires in a chain-like fashion (see Fig. 7). They are also connected to the internet and a central CS server with their regular Eternet card. Let us suppose that the distance between each user and their nearest neighbors is less then 1 km. It is then very convenient to run the KLJN ciphers with a bandwidth 1000 bit/second or greater. Thus the theoretical speed of secret key generation and distribution to each computer in the network would be 200 Megabits/second. *Such a network and key generation/distribution speed is not only far beyond the reach of any quantum communicator system, but also that of RSA and other software key distribution methods*. Of course, this speed is much higher than any PC can handle; thus the clock frequency of the KLJN ciphers could be reduced by orders of magnitude and the key generation and distribution could still be fast enough. The very important implication of this fact is that the requirements of the wire-pair connection of the KLJN cipher chain could be very low.

Alternatively, a much smaller network population could generate and distribute secret keys at sufficient speed. For example, 100 Units with a 1000/bit/second clock frequency would generate and share a secret key at the speed of 100 kbit/second.

Note that the above-described speed holds for the current/voltage based defense only. More economical solutions will result in a speed reduction by a factor of 3 (see Sections 7 and 8).

7. Enhancements for security with economic design to avoid the "Mingesz" attacks

One of us (Robert Mingesz) [11] pointed out two important vulnerabilities of the network described above. In this section, we describe the problem and show one possible solution.

(*i*) If during the initialization described in Section 5 at point (*i*) a brief man-in-the-middle-attack takes place and *if the system is defended against man-in-the-middle attacks by the economic solution* described in Section 3, the eavesdropper (with a small probability) may get the value of the secure bit because (on average) a 4-bit eavesdropping needs to occur for the systems to discover the man-in-the-middle.

(*ii*) A much more serious problem is that when the eavesdropper attacks the regular network at Unit (k), and because that network is not totally secure (at least at the beginning), she may get the $G_{j,k}$ functions for each j location. But that set of $G_{j,k}$ functions is either the key or its inverse, depending on the local bit at Unit (k).

These problems must be fixed for total security. and this needs a more elaborate initialization resulting in a reduction of the network speed in a factor of 2.

(*iii*) Either the SC communicator port initialization/distribution must be done after each N-bit key generation, or the man-in-the-middle defense should be the more expensive current/voltage based defense described in [3]. The fastest way of achieving part of the SC initialization/distribution is described here. During the secure port (re)distribution, the Units should be run such that all ports change randomly and independently at the beginning of each clock period until a new port distribution is reached in accordance with the truth table shown in Table 3 (confer Section 5, Table 2). This method results in a new port distribution after 4/3 clock periods, on the average.

(*iv*) As part of the initialization process, the very first N-bit secret key distribution has to be done in a slower way, without telecloning. Moreover, the Coordinator-server should also possess one Unit of the network. Let us suppose that the right ports of Unit (1) and the left ports of Unit (2) act as the first network-key generator. Unit (1)'s R ports and Unit (2)'s L ports generate, share and store an N-bit long key during N clock periods. At the same time, all other connected L and R ports also generate N-bit long keys for the communication between their nearest neighbors. Then by using the regular network and the key shared between Units (1) and (2), Unit (2) sends the N-bit long network-key to Unit (3). Then Unit (3) forwards the network-key to Unit (4) by using the regular network and the key generated between them, and so on... In this way, after N clock periods and the forwarding of the keys in $N-1$ subsequent steps via the fast regular

Table 3. Truth table of the high-speed secure sharing of the port controls between the two nearest neighbors, Unit (M) and Unit (M + 1), in the chain-like KJLN network. Only secure bit exchanges generate a new port distribution. The mean time needed is 4/3 clock periods because, from the 16 possibilities, 12 results in a new port distribution.

Situation of secure bits	Low at L_1^{M+1} High at R_1^M if $L_1^{M+1} - R_1^M$ are unsecure: Low at L_2^{M+1} High at R_2^M	Low at R_1^M High at L_1^{M+1} if $L_1^{M+1} - R_1^M$ are unsecure: Low at R_2^M High at L_2^{M+1}	All other cases
Resulting port control	$R_1^M = R_2^M$ $L_1^{M+1} = -L_2^{M+1}$	$R_1^M = -R_2^M$ $L_1^{M+1} = L_2^{M+1}$	Do it again (No port distribution)

network, all Units, including the Coordinator-server, will learn the first N-bit long network-key generated by Units (1) and (2). From that moment on, this key and its successors should be used for the communication between the Coordinator-server and the Units. The result is very close to a one-time-pad type encryption. Taking the New York example mentioned above, and assuming that the step of sending the network-key in the regular network to the nearest neighbor takes the same time as the clock period of the KLJN ciphers, this totally secure initialization would take 400 seconds. After this $\approx$7 minutes initialization process, the network begins the fast and continuous production of totally secure keys. Due to the redistribution of ports after each N-bit key distribution, the speed of secure key generation and whole distribution would be about a third of the speed mentioned above; about 70 Mbit/second.

8. The simplest and most economical network solution

As it is pointed out in point (*iii*) of (the preceding) Section 7, the inexpensive defense solution with the required SC port redistribution will need an extra 4/3 clock period time at each network key generation. This requirement will slow down the key generation/distribution by more than a factor of 2. Practically, the network key generation will need 3 clock steps. Therefore, it is worthwhile to consider the use a much simpler network based on simple KLJN loops instead of the SC type complementary loops. Indeed, in Fig. 7, we use only single right and left ports; the connected R_k and L_{k+1} ports represent a regular KLJN cipher, as described in [1]. They can switch randomly at each clock period and, on the average, every second clock period produces a secure bit. The network's initialization and running are basically the same (with minor differences) as that which was described for the SC cipher based network. This very simple system is at least as fast as the SC cipher based system with the economical defense (Section 3). Practically, every 3rd clock step will produce and distribute a network key of N bits. Therefore, the SC cipher based solution seems to have only one advantage: with the current/voltage based defense solution, it has the highest speed — about 3 times faster than that of the other solutions.

9. Remarks questions, perspectives

It is important to note that the network described in Fig. 7 is very different from the basically *point-to-point* key distribution methods used by quantum communication and software solutions. Even though the possibility of telecloning quantum states to multiple receivers has been pointed out by van Loock and Braunstein [8], the fidelity is poor ($<71\%$); and to reach an acceptable fidelity ($>50\%$) the number of Units has to be limited to 30. Moreover, each Unit has to be connected to all the other Units by a separate communicator and separate lines, which means $N(N-1)$ communicator devices and $(N^2-N)/2$ lines, indicating that the method is essentially a point-to-point communication type. For the New York example mentioned above, such a quantum telecloning solution requires about 200 *thousand communicator devices at each Unit* and that requires almost 40 billion communicator devices and nearly 40 billion optical cables. Moreover, these 40 billion connections are both short-distance and *long-distance* ones because we have to directly connect the farthest Units, too.

On the other hand, the KLJN-based network (Fig. 7) requires only two communicator devices for each intermediate Unit, and one for each end Unit. The intermediate Units have to be only connected to their respective two nearest neighbors. and the end Units to only one neighbor. That requires only $2 * (N - 1)$ communicator devices and $(N - 1)$ wires. In the New York example mentioned above, the two communicator devices at each Unit and two wires at each Unit result in only about 200 thousand communicator devices and about 200 thousand cable connections; moreover, these *connections are all of the short distance type*, connecting only the nearest neighbors. If the regular network and the CS are fast enough compared to the KLJN clock, the whole network receives an N-bit long key at every (second or third) KLJN clock period. On the other hand, the whole network will receive the same key. Therefore, the system is only totally secure against external attacks: *hackers from outside* the network. Within the network, the security, which can be added to the network-key security, is only a regular network security that protects the information sent to/from to the CS Unit. Thus, we have the same level of security *against hackers within the network*, as regular networks do.

This special situation generates a lot of different open questions. For example:

(*a*) What is the proper approach to encryption when we have this continuous high-speed key generation and simultaneous whole-network-key distribution?

(*b*) Can the generated secure bits be used to increase the security of the internal network?

(*c*) Higher dimensional network topologies and redundancy to protect against broken lines or Units down. This can, for example, be realized with server units with more than two (L and R) ports. The ports $P_i(k)(i = 1 \dots Q)$ of the k-th server can be connected to up to Q different Units. The Coordinator-server would collect the logic relations and evaluate Eq. (1), accordingly.

(*d*) How about network redundancy and coding in higher dimensional topologies (?)

(*e*) This network can alternatively be used to announce information to one or more Unit(s) or to the whole network simultaneously, in a totally secure way. Is there a need for this kind of solution?

10. Possible application of the idea in quantum communication

There are quantum key distribution schemes which produce entangled secure bits at the two ends of the line. Concerning the network solution described in Section 8, the same sequence of ideas, the same network topologies and initialization processes can be applied for such networks where the KLJN-based Units are replaced by quantum entangled state based Units. Such a network would produce the *multiple telecloning of classical bits via a quantum communicator network without telecloning the quantum states*. This can (theoretically) be a 100% fidelity process and could be a large chain-like network, for example. From these considerations, it is obvious that the telecloning of quantum states is an unnecessary luxury because it has many disadvantages, such as the need of separately connecting each units, and the very low fidelity. What is needed for practical applications is the secure transfer of the information bit, and that can also be done via the chain-like (or other) type of network of quantum channels, and a regular network with a Coordinator-server. Because there are situations where connection via wires is impossible (but an optical channel can be realized — like communication through space using a laser beam) this type of *quantum-based secure classical communication* may find feasible applications.

11. Conclusion

We have used electrically isolated KLJN ciphers to build enhanced ciphers and networks. We introduced the Secure Complementary (SC) KLJN chiper for top speed secure key generation and exchange. We built network Units from two SC ciphers with random internal logic relations. We built a one-dimensional chain network using the network Units and a regular inernet network with a Coordinator-server (CS). With such a network a high-speed, whole-network, one-step, secret key distribution can be achieved instead of the slow, point-to-point methods used by quantum communication or RSA-like key exchange methods. This is a teleportation-type multiple telecloning of the classical information bit because the information transfer can take place without the actual presence of the information bit at the intermediate points of the network. We have shown other, more economical and still fast-enough solutions, too.

Acknowledgments

LBK appreciates the discussions had with a number of security experts and scientists/engineers, and their valuable feedbacks and questions. Without the ability to provide a complete list, here are some of the commenters (regarding the KLJN crypto system) that LBK would like to thank: Derek Abbott, Steven M. Bellovin, Janos Bergou, Bob Biard, Terry Bollinger, Goong (Gordon) Chen, Adrian Cho, Walter Daugherity, David Deutsch, Julio Gea-Banacloche, Ryan Gerdes, Attila Geresdi, Phil Hemmer, Peter Heszler, Tamas Horvath, Andreas Klappenecker, Greg Kochanski, Frank Moss, Ashok Muthukrishnan, Imre Naggyorgy, Rainer Plaga, Charlie Strauss, Olivier Saidi, Bruce Schneier, Matthew Skala, Charlie Strauss, Christian Vandenbroeck, David Wagner, and Jonathan Wolfe. There have also been several other inspiring commenters on Bruce Schneier's security blog site LBK would like to thank; however, the users could not be named here as they were only identifiable by nicknames. LBK has also been inspired by Andreas Klappenecker's seminar series on quantum computing and network communications at TAMU.

References

[1] L.B. Kish, "Totally secure classical communication utilizing Johnson (-like) noise and Kirchoff's law", *Physics Letters A*, **352** (2006), 178–182.

[2] Adrian Cho, "Simple noise may stymie spies without quantum weirdness", *Science*, **309** (2005), 2148.

[3] L.B. Kish, "Protection against the man-in-the-middle-attack for the Kirchhoff-loop-Johnson(-like)-noise cipher and expansion by voltage-based security", *Fluctuation and Noise Letters*, **6** (2006), L57–L63.

[4] L.B. Kish, "Response to Scheuer-Yariv: "A Classical Key-Distribution System based on Johnson (like) noise — How Secure?", [online] physics/0601022. Available at: http://arxiv.org/physics/0602013.

[5] L.B. Kish, "Response to Bollinger: 'On the Impossibility of Keeping Out Eavesdroppers Using Only Classical Physics' and to Some Other Comments at Bruce Schneier's Blog Site" [online]. Available at: http://www.ece.tamu.edu/~noise/research_files/Response_Bollinger.pdf

[6] J. Bergou; statement in [2].
[7] Ch.H. Bennett, G. Brassard, C. Crepeau, R. Jozsa, A. Peres, W.K. Wootters, "Teleporting an Unknown Quantum State via Dual Classical and Einstein-Podolsky-Rosen Channels", *Physical Review Letters*, **70** (1993), 1895–99.
[8] P. van Loock and S.L. Braunstein, "Multipartite entanglement for continuous variables: A quantum teleportation network", *Physical Review Letters*, **84** (2000), 3482–3485.
[9] P. van Loock and S.L. Braunstein, "Telecloning of continuous quantum variables", *Physical Review Letters*, **87** (2001), 247901/1–4.
[10] O. Cohen, "Classical teleportation of classical states", *Fluctuation and Noise Letters*, **6** (2006), C1–C8.
[11] R. Mingesz, personal communication (March 8, 2006).

3.9. The Szegedin Whisper Project: Experimental Verification of KLJN

3.9.1. The name "Szegedin Whisper" was inspired by the Manhattan Project, which produced the nuclear bomb. In the February of 2006, the National Science Foundation (NSF) invited me to submit a grant proposal for the experimental demonstration of KLJN; but the program manager sent the proposal back without reviewing it (see the notes in Section 3.5). The turned-down proposal estimated a period of 3 years for the planning, building and execution of the demo project. Instead, 10 months later, the demonstration was carried out *without NSF support*, in Hungary, by colleagues/friends in the Noise Lab at the Department of Experimental Physics at the University of Szeged, which I founded in 1988. (In 2010, the Noise Lab was transferred to the Department of Technical Informatics, with Zoltan Gingl as Head of Department. Zoltan would go on to transform the Lab into a professional analog/digital stochastic signal and system circuitry lab.) It had happened repeatedly in my life; that *not winning* a grant made the actual project more successful because I could focus on the deeper questions instead of managing the finance and personnel of the project. The picture below was taken after the submission of the manuscript, in which we described the successful experiments and presented the data. At that time, submissions to the preprint database arXiv were problem-free — the manuscripts usually became public

(*Continued*)

(*Continued*)

on the same or next day. (This situation radically changed after the Wikipedia affair with quantum-fans, the very same issue that triggered the contract for this book; see in Section 4.4).

Figure 3.9.1. Happy moments. The heroes of the "*Szegedin Whisper*" Project: Robert Mingesz and Zoltan Gingl, in the Noise Laboratory, directed by Zoltan Gingl, at the Department of Experimental Physics, University of Szeged, December 2006.

Paper: Johnson(-like)-Noise-Kirchhoff-Loop Based Secure Classical Communicator Characteristics, for Ranges of Two to Two Thousand Kilometers, via Model-Line[1,2]

Abstract. A pair of Kirchhoff-Loop-Johnson(-like)-Noise communicators, which are able to work over variable ranges, were designed and built. Tests have

[1]R. Mingesz, Z. Gingl, L.B. Kish, *Phys. Lett. A*, **372** (2008), pp. 978–984.

[2]The results were featured in the technology section of the *New Scientist* magazine by Jason Palmer, "Noise keeps spooks out of the loop" [Online] (Published May 23, 2007). Available at: http://www.newscientisttech.com/article.ns?Id=mg19426055.300 (Accessed: May 2007).

been carried out on a model-line; performance characteristics were obtained for ranges beyond the ranges of any known direct quantum communication channel and they indicate unrivalled signal fidelity and security performance of the exchanged raw key bits. This simple device has single-wire secure key generation and sharing rates of 0.1, 1, 10, and 100 bits/second for corresponding copper wire of diameters/ranges of 21 mm/2000 km, 7 mm/200 km, 2.3 mm/20 km, and 0.7 mm/2 km, respectively, and performs with a 0.02% raw-bit error rate (99.98% fidelity). The raw-bit security of this practical system significantly outperforms raw-bit quantum security. Current injection breaking tests show that zero-bits of information may be obtained through eavesdropping without triggering the alarm; therefore no multiple measurements are needed to build error statistics to detect eavesdropping, unlike in quantum communication. Wire resistance based breaking tests of the *Bergou–Scheuer–Yariv* type give an upper limit of eavesdropped raw bit ratio, of 0.19%; this limit is inversely proportional to the sixth power of the cable diameter. *Hao*'s breaking method yields zero (below measurement resolution) eavesdropped information.

1. Introduction

Recently, a totally secure classical communication scheme, a statistical-physical competitor of quantum communicators, was introduced (see Fig. 1) [1, 2]: the Kirchhoff-loop-Johnson(-like)-Noise (KLJN) communicator, with two identical pairs of resistors and corresponding Johnson(-like) noise voltage generators. The KLJN communicators are utilizing the physical properties of an idealized Kirchhoff-loop and the statistical physical properties of thermal noise (Johnson noise). The resistors (low bit = small resistor, high bit = large resistor) and their thermal-noise-like voltage generators (thermal noise voltage enhanced by a pre-agreed factor) are randomly chosen and connected at each clock period at the two sides of the wire channel. A secure bit exchange takes place when the bit states at the two line ends are different; which is indicated by an intermediate level of the *rms* noise voltage on the line, or that of the *rms* current noise in the wire.

The most attractive properties of the KLJN cipher are related to its security [1, 3] and to the extraordinary robustness of classical information when compared to the fragility of quantum information. To provide security against arbitrary attacks, the instantaneous currents and voltages are measured at both ends by Alice and Bob, and are published and compared.

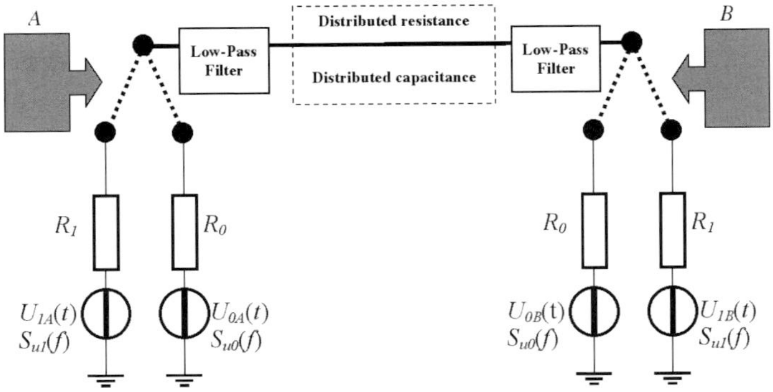

Figure 1. The outline of the KLJN communicator. The instantaneous current and voltage data are measured and compared at the two ends; in the case of deviance, the eavesdropping alarm goes off and the currently exchanged bit is not used. The low-pass line filters are necessary to protect against out-of-alarm-frequency-band breaking attempts and false alarms due to parasite transients. False alarms would occur due to any wave effect (transient or propagation effects), illegal frequency components or an external disturbance of the current-voltage-balance in the wire.

Figure 2. Zoltan Gingl's lab personnel during the *Szegedin Whisper Project*. Left to right: Robert Mingesz (PhD student), the Author (Laszlo Kish; external collaborator), Peter Makra (PhD student; dealing with other problems of Stochastics), Zoltan Gingl (Director).

In the idealized scheme of the KLJN cipher, the passively observing eavesdropper can extract zero bits (zero-bit security) of information, while the actively eavesdropping observer can extract at most one bit before getting discovered (one-bit security) [1]. The system has a natural zero-bit security against the *man-in-the-middle attack* which is a unique property among secure communicators [3]. The KLJN system has recently become network-ready [4]. This new property [4] opens a large scale of practical applications because the KLJN cipher can be installed as a computer card [4], like an Ethernet card. Other practical advantages (compared to quantum communicators) are the high speed, resistance to dust, vibrations and temperature gradients, and the low cost price to produce and integrate [1]. It has recently been shown [5] that using the KJLN communicator does not require new wire fittings; with proper filtering methods, currently used wire lines, such as power lines, phone lines, and internet wire lines may be used.

Concerning the security of the KLJN system, the basic rule to any physical secure communication method/device holds here, too: *If you compromise it, you lose it.* Though there have been several attempts to break into the KJLN line, so far, no proposed method has been able to challenge the total security of the idealized KLJN system. These proposed break-in attempts have been based on certain assumptions which directly violate the basic model of the idealized KLJN method. These assumptions are as follows:

i. Allowing non-negligible wire resistance (Bergou [2], Scheuer and Yariv [6]). For the evaluation of this claim at practical conditions, see the response in [7].

ii. Allowing a high-enough bandwidth for wave (transient propagation) effects (Scheuer and Yariv [6]). This possibility has been explicitly excluded since the original article [1]. For further refusal of this claim, see the response in [7]. Moreover (*and this is the most direct practical argument*), in this case, the standard current and voltage protection of the KLJN system would *immediately alarm and shut down the communication* because the existence of transient and wave effects *per-definition* yield different instantaneous current and/or voltage values at the two ends.

iii. Assuming inaccuracy of (noise) temperatures (Hao [8]). For the theoretical refusal of this claim, see [9].

iv. Utilizing practical inaccuracy of resistors (Kish [9]). For the mathematical evaluation of this claim at practical conditions, see [9].

Let us here consider an analogy in the field of quantum communications. Though all practical quantum communicators suffer from the inability of producing *totally single-photon* outputs, and this fact unavoidably compromises the security of practical quantum communicators, we do not say that idealized/mathematical quantum communicator schemes are insecure. In discussing the unconditional security of idealized quantum communication schemes, we suppose that single-photon sources do exist, which is a similar (though more difficult) claim than to suppose the existence of zero wire resistance (c.f. superconductors), and then we conclude that the quantum system is therefore unconditionally secure, at least conceptually. Therefore the practical claim that idealized single-photon sources do not exist cannot compromise the claim about the security of an idealized, conceptual quantum communicator. On the other hand, as soon as the security of a *practical communicator system* is discussed, the aspects of non-ideal single-photon sources, optical fiber absorption, etc, for quantum communicators (and similarly the aspects *i–iv* listed above for the KLJN systems) cannot be neglected any more, and the practical design must keep these effects under control.

In the present paper, we report the first experimental realization and tests of the KLJN secure communicator system. We test and analyze the most important breaking methods of the ones listed above. The tests indicate an unrivalled *beyond-quantum security level* by the realized system.

2. The realized secure classical physical communicator

The generic schematic of practical KLJN secure classical communicators is shown in Fig. 3. The logic-low (L) and logic-high (H) resistors, R_0 and R_1 respectively, are randomly selected at the beginning of each clock period and are driven by corresponding Johnson-like noise voltages. The actual

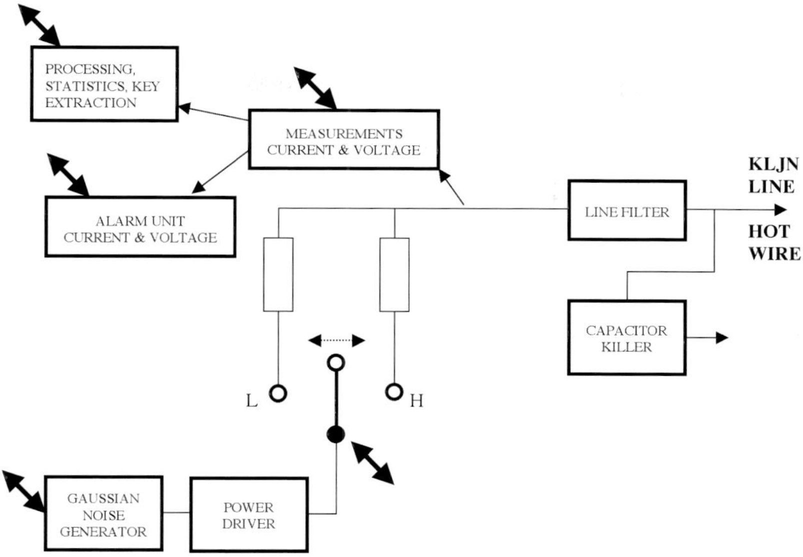

Figure 3. Generic schematics of practical KLJN secure classical communicators. The low (L) and high (H) resistors are randomly selected at the beginning of each clock period and are driven by corresponding Johnson-like noise voltages. The actual realizations contain more filters and amplitude control units (not shown here). The thick arrows mark computer control and data exchange. The controlling computer (not shown) has a regular network connection with the computer of the other KLJN communicator (not shown). Regarding the communication: a bare-wire line within the given ranges in the frequency band relevant to the given range with the driving resistor values described in the text, can be represented by a resistor and a capacitor; at below 1% of the ranges, it can be represented by a simple resistor; when built with a "capacitor killer" and twisted pair (two coaxial cables) arrangement, it can be represented by a single resistor in the relevant frequency bands.

realizations contain more filters and amplitude control units than shown in the figure-model.

The realized pair of KLJN communicators [1–4] were designed and built for variable ranges. The realized circuit used Digital Signal Processors (DSP) and analog technology (see the outline in Figs. 3 and 4). The details not described in this paper are part of intellectual property which we cannot disclose at this stage. The computer controlled parts of the communicator pair have been realized by ADSP-2181 type DSPs (Analog Devices). The digital and analog units were placed on two separate computer cards for easier variability during tuning. The communication line's current and voltage data were measured by AD-7865 type AD converters (Analog

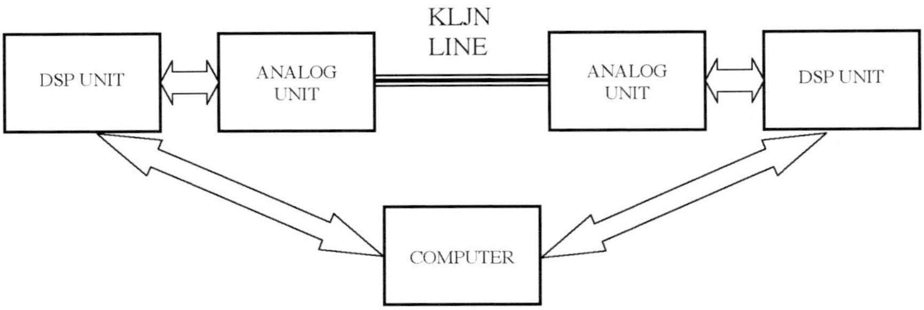

Figure 4. The schematics and arrangement of the realized and tested KLJN communicator pair. The KLJN line is a model line with capacitance compensation up to a 2000 km range. Alternatively, freestanding wires can be used up to 1% of the maximal "no-wave" range at the given bandwidth.

Devices) with 14-bit resolutions from which 12 bits were used. The Johnson-like noise was digitally generated in the Gaussian Noise Generator unit where digital and analog filters truncated the bandwidth in order to satisfy the KLJN preconditions of removing any spurious frequency components. The major bandwidth setting was provided by an 8-th order Butterworth filter with sampling frequency of 50 kHz. The remaining small digital quantization noise components were removed by analog filters.

The experiments were carried out on a model-line (a single resistor, representing delay-free lossy cable with capacitor-killer) with assumed cable velocity of light ($2*10^8$ m/s), in the capacitor killer arrangement, with a range up to 2000 km (which is far beyond the range of direct quantum channels or any other direct communication method via optical fibers). Under these conditions, and with the applicable KLJN bandwidths and copper wire diameters given below, the line model is a single resistor with a resistance of about 200 Ohms. The corresponding copper wire diameters are reasonable practical values for the different ranges; they measure 21 mm (2000 km), 7 mm (200 km), 2.3 mm (20 km) and 0.7 mm (2 km) respectively. Beyond 2000 km, the present realization would require non-practical wire diameters. Inductance effects are negligible with the selected resistance values R_0 and R_1 at the given ranges and the corresponding bandwidths. A bare-wire line within the given communication ranges, in the frequency band relevant to the given range, and with the driving resistor values described in the text, can be represented by a resistor and a capacitor;

and at below 1% of the ranges, as a simple resistor. With a "capacitor killer" and twisted pair (two coaxial cables) arrangement, it can be represented by a single resistor in the relevant frequency bands. The "capacitor killer" (Fig. 3) is a well-know electronic solution for capacitance compensation to remove capacitive cut-off effects in coaxial cables by feeding the coat of a coaxial cable with a follower driven by the hot wire input of the coaxial cable.

The voltage noises are band-limited Johnson-like noise with a power density spectrum of

$$S_u(f) = KR \tag{1}$$

within the noise bandwidth, and is zero elsewhere. The value of K depends on the noise bandwidth determined by the actual range and it is selected so that the noise voltage of the greater resistor is 1 Volt for all noise bandwidths. This resulted in $S_u(f)$ values of the greater resistor: 0.2, 0.02, 0.002, and 0.0002 V^2/Hz for the ranges of 2000, 200, 20 and 2 km, respectively. Wave effects in the stationary mode and information leaks due to them are avoided by the proper selection of the noise bandwidth. The noise bandwidth is selected so that the highest possible Fourier component in the line is at a frequency 10 times lower than the lowest frequency standing-wave mode in the line. That condition results in noise bandwidths of 5, 50, 500 and 5000 Hz for the ranges of 2000, 200, 20 and 2 km, respectively. This condition and the statistical sample size within the clock period determine the speed (bandwidth) of communication. The sampling rate satisfies the Shannon limit; thus the sampling frequency is two times greater than the noise bandwidth. The sample size within the clock period is equivalent to about 50 statistically independent data points (though the digital filtering uses an about one order of magnitude greater actual sampling rate) and that results in a secure bit rate that is 50 times lower than the noise bandwidth because only 50% of the communicated bits are secure. Thus the device has bit rates of 0.1, 1, 10, and 100 bits/second for the ranges of 2000, 200, 20 and 2 km, respectively.

The resistors pairs at Alice and Bob have been selected so that, with the practical wire diameters given above, the wire resistance is about 2% of the loop resistance during each secure bit exchange. The resistors pairs are $R_0 = 2$ kOhm and $R_1 = 11$ kOhm. The 11 kOhm resistor is composed by a serial 10 kOhm resistor and a 1 kOhm resistor with a 1 nF capacitor shunting their joint point to the ground to remove possible digital quantization noise. The 2 kOhm resistors are two serial 1 kOhm resistors also with a 1 nF capacitor shunting their joint point to the ground to remove possible digital quantization noise. These shunt capacitors were the first stages of the line filter. Because the security protection based on current and voltage comparison was effective up to a 50 kHz bandwidth, the shunt capacitors described above and additional 1 nF capacitors at the two ends of the model line were satisfactory line filters in the model line. Note: actual practical arrangements with real wires automatically provide much larger capacitances. Furthermore, these capacitors would have removed possible switching spikes originating from capacitive coupling in the analog switches due to possibly unbalanced parasitic capacitors; therefore there were no detectable switching transients in the line. The 1 kOhm resistor at the generator driven end was also used as a probe to measure the current in the line.

Transient wave effects at the end of each clock period are avoided in the Gaussian Noise Generator unit by driving the envelope of the time functions of noise voltage and current to zero before the switching of resisters using a linear ramp amplitude modulation (via 8% of the clock duration); and the reverse process is done at the beginning of the next clock cycle after the switch. The switching of resistors takes place within a zero-current-zero-voltage window at a duration that is 9% of the clock period and 0.8% before the start of the next clock cycle. Moreover a short pause (8% of the clock time) with no data collection, except for security checks — after the initial linear ramp at the beginning of stationary noise, is applied in order to avoid other possible types of transient/filter effects of stochastic nature. All these are done before the filtering process to avoid any spurious frequency components due to the linear ramp.

3. Experimental study of the security and fidelity of the realized communicators

During the communication and security tests, the exchange of secure bits took place between Alice and Bob with a 0.02% error rate, which is 99.98%

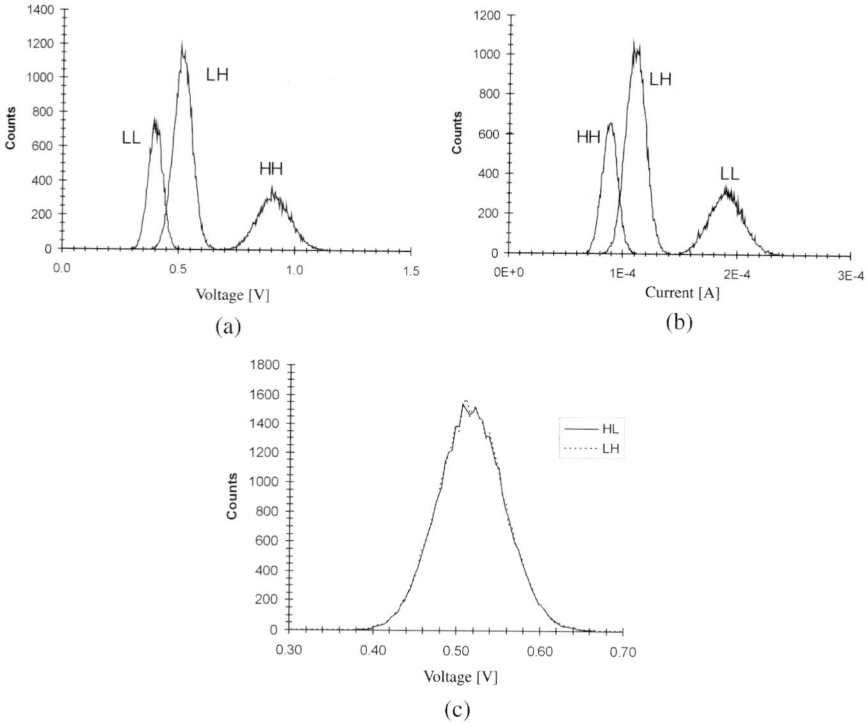

Figure 5. Empirical amplitude density functions (histograms) during the whole span of security checks utilizing 74497 clock cycles: (a) and (b) are voltage and current counts seen by Alice and Bob; (c) shows the voltage counts seen by Eve at one end of the line during the two different secure bit arrangements (*LH* and *HL*). These functions correspond to the situation when the bit arrangement is fixed (*LH* or *HL*) and then the two distribution functions are the voltage counts measured at the two ends of the line to execute a BSchY (Bergou–Scheuer–Yariv) type of attack. It is obvious from the strong overlap of the two curves that Eve has virtually zero information even with a fixed bit arrangement for 74497 clock cycles.

fidelity — a significant achievement beyond that of quantum channels. Figure 5 shows an illustrative example of the actual distribution functions and their statistics made over the total span of security checks (about 74497 clock cycles). Figure 6 shows an example of the actual counts (working statistics) during a single clock cycle.

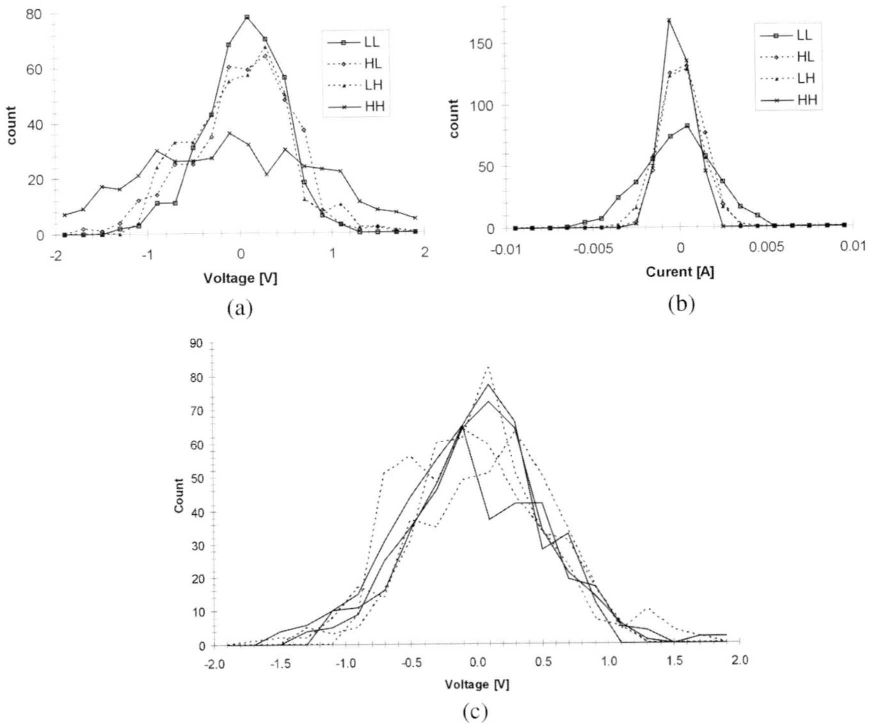

Figure 6. Empirical amplitude density functions (histograms) of the voltage (left window) and current (right window) during a single clock cycle: (a) and (b) are the voltage and current counts seen by Alice and Bob; (c) shows the voltage counts seen by Eve at end of the line at the two different secure bit arrangements, *LH* and *HL* — a total of three of each one. These functions correspond to the situation when the bit arrangement is fixed (*LH* or *HL*) and then the two distribution functions are the voltage counts measured at the two ends of the line to execute a BSchY (Bergou–Scheuer–Yariv) type of attack. The poor statistics seen in Figs. 6(a) and (b) are enough for Alice and Bob to identify secure bit alignment with a 0.02% error rate (99.98% fidelity). However when Eve tries to identify the bits from the two histograms recorded at the two ends of the line (see Fig. 6 (c)) she must work with these distributions which are very stochastic, almost identical and totally overlapping with a less than 1% shift of their centers [7], which results in a 0.19% eavesdropped bit per transmitted secure bit.

3.9.2. During these experiments, we used some combined voltage–current error removal methods; however, at that time, we did not know the most advanced tricks, see Section 4.3, which were later developed in 2013 by Yessica Saez, in her PhD program at Texas A&M. By then, error probabilities smaller by 10 orders of magnitude could easily be achieved. With some extra effort, drops of 20 orders of magnitude were also possible, though perhaps meaningless.

Table 1. Summary of security analysis of realized KLJN communicator pair.

Type of breaking	Measured number or ratio of eavesdroppable bits without triggering the current-voltage alarm (tested through 74497 bits)	Remarks
BSchY (*i*) [2, 6] attack in the present KLJN system	0.19%	0.00000019% when using a 10 times thicker wire (theoretical extrapolation). Can be arbitrarily enhanced by privacy amplification [12, 13] — the price for which is reduced speed.
Hao (*iii*) [8] attack in the present KLJN system	Zero bit	Below the statistical inaccuracy. Considering the 12-bit effective resolution of noise generation accuracy, it is theoretically: <0.000000006%
Kish (*iv*) [9] attack utilizing resistor inaccuracies in the present KLJN system	Zero bit	Below statistical inaccuracy. Theoretically, when pessimistically supposing a 1% resistance inaccuracy, it is: <0.01%
Current pulse injection (Kish) [1] in the present KLJN system	Zero bit	One bit can be extracted while the alarm is triggered; thus the bit cannot be used.

According to the security tests and the related evaluations, the raw-bit security of this practical system has been successfully designed to outperform raw-bit quantum security while remaining reasonably inexpensive. For the summarized security test evaluation, see Table 1.

Breaking tests (*i*) of the Bergou–Scheuer–Yariv (BSchY) type, utilizing wire resistance [2, 6], were carried out by measuring and comparing the *rms* voltages at the two ends of the cable line during secure bit transfers, supposing an *LH* or *HL* bit arrangement. We have been using Shannon's channel coding theorem to evaluate the information leak and fidelity [7]:

$$\frac{C_{eav}}{C_{trans}} = 1 + p\log_2 p + (1-p)\log_2(1-p) \tag{2}$$

where C_{eav} is the upper limit of eavesdropped bits (supposing optimal coding/decoding), C_{trans} is the total number of transferred secure bits, p is the probability of correct assumption, and $(1 - p)$ is the probability of erroneous assumption.

3.9.3. Since 2006, our knowledge and understanding of security measures have increased tremendously; thus some of the measures used in 2006 may look a bit simplistic. Fortunately, the bits of the KLJN key exchange are independent and identically distributed (IID) random variables; thus there is a one-to-one correspondence between Eq. (2) and the statistical (variational) distance, through Eve's successful guessing probability p of key bits (see Sections 4.4 and 5.1). The perfect security is indicated by $p = 0.5$, which is equivalent to Eve using a perfectly random coin instead of eavesdropping. In the experiments described here, the value was $p = 0.525$ in the (Bergou–) Scheuer–Yariv attack.

3.9.4. As mentioned earlier, it was shown in T. Horvath, L.B. Kish, and Y. Scheuer's "Effective Privacy Amplification for Secure Classical Communications" (*Europhysics Letters*, **94** (2011), pp. 28002-1–28002-6; available online at: http://arxiv.org/abs/1101.4264) that, by using a simple privacy amplification technique which does *not* induce correlations between key bits (and security risks by that), a large $p = 0.55$ value can be reduced to $p = 0.50005$ (a value sufficient for any application) for the price of a 4-times slowdown and a 4-times increase of error probability between Alice and Bob. An astronomically large information leak of $p = 0.85$ can be reduced to near-to perfect level, $p = 5.0001$, for the price of a 32-times slowdown. The negligible error probability of KLJN makes these improvements possible (see Section 4.3).

The experimental test of this practical information leak through 74497 clock cycles shows $p = 0.526$, which yields $C_{eav}/C_{trans} = 0.19\%$ and that is less than practical quantum bit-leaks when photons are randomly stolen, copied (noisy cloned) and replaced — a method by which Eve can easily extract 1% of the bits while staying hidden in practical situations. However to reach this 0.19% information leak Eve must know/use the optimal decoding of data in accordance with Shannon's channel coding theorem, and that is usually unknown. Moreover, the locations of correctly eavesdropped bits are random and that is also a significant deficiency compared to the random stealing of photons in quantum channels where the location of the correctly eavesdropped bit is known (with reasonable accuracy). A simple improvement possibility is served by the fact that the BSchY bit-leak scales with the reciprocal of the sixth power of the cable diameter [10]; that means that doubling cable diameter would reduce this 0.19% leak by a factor of 64, and a diameter increase of a factor of 10 (possible only for 20 km or shorter ranges due to practical limits of cable diameters) would yield a million-time reduction of the bit leak. Note: using additional software tools can provide an arbitrarily strong reduction of the bit-leak; this is the *privacy amplification* technique developed for quantum communication by Bennett *et al.* [12, 13]. However, the price to achieve that is reduced speed.

Hao's *(iii)* breaking method yields zero (non-measurable) information for the eavesdropper due to the 12-bit effective accuracy of current and voltage measurements and the limited statistics during the clock period. Considering the 12-bit effective resolution of noise generation accuracy, theoretical estimation based on Eq. (2) (see [7, 9]) yields an information leak of less than 0.000000006%.

The Kish *(iv)* [9] attack, utilizing resistor inaccuracies in the present KLJN system, yielded zero bits of extractable information because of the accurate choice of resistors at the two line ends. Supposing a 1% resistance inaccuracy, the bit-leak is less then 0.01%, according to estimations based on Eq. (2) (see [7, 9]).

Current pulse injection breaking attempts (Kish) [1] were performed by discharging a capacitor (1 nF, 5 V) on the line to inject a short current pulse (relevant for the line filter's high-frequency cutoff) and to use the current distribution to read out the resistance values. These breaking attempts are able to extract just a single bit of information while the current-alarm is triggered. Therefore, as expected, this type of break yields an extraction of zero useful bits. The smallest detectable current difference at the two ends of the line was 0.025% of the *rms* current in the line. Note, less sensitivity would have produced the same practical security level; however, due to the given resolutions of the AD converters, we evaluated the performance with this resolution.

Figure 7. The "Trinity" test. The Manhattan Project tested the first nuclear bomb at Trinity, New Mexico. The Szegedin Whisper Project also had a "Trinity" in the test; it was the trinity of three circuit boards comprising of the resistor pairs and the digitally controlled analog switches; the digital circuit board; and the analog circuit board.

Finally, we would like to note that, because breaking tests using the *man-in-the-middle attack* involved hardware and required significantly more effort, they will be performed later.

4. Comments on the speed of communication

> **3.9.5.** The comments below reflect on the 2006 situation. Since then, quantum communication has made progress — even though they are still limited by exponential speed cut-off due to absorption. KLJN has also evolved: the enhanced protocols allow much larger speeds at the same and/or heightened level of security (see Section 5.3). Currently (as of 2016), Kamran Entesari is exploring KLJN on a silicon chip, at TAMU.

So far (2007), according to trusted experimental data published, the speed of quantum communicators is generally very similar to that of the practical KLJN speed described above — though it is sometimes closer to its own theoretical single-line limit, which is about 10 times faster than the practical KLJN's. It is important to note that, in the literature, there are various claims about the speed of quantum communication; therefore one should only take seriously those based on experimental works — particularly those which are cautious about security, and that keep/use only the really secure bits. This process results in a significant bandwidth limitation.

Zeilinger and coworkers have recently reported state-of-the-art quantum communication with enhanced security, which requires entangled photon pairs, via the range of 144 km [15]. During a run of 75 seconds, they obtained 789 raw bits (coincidences) [15]. This is a speed of about 10 bits/second; however their information leak rate with these bits was 4.8% — 25 times higher than that of the KLJN communicator (0.19%) at the given wire resistance. After using their privacy amplification algorithm on these bits, they achieved a secure key of 178 bits [15] (final information leak data are not provided). This is about 2.4 bits/second, about 2 times faster than the projected speed of our KLJN communicator for the same distance. Note: the speed of KLJN communicator scales with the reciprocal of the range.

The speed of the present KLJN communicator is about 10 times slower than the theoretical limit for a single line (estimated in [1]). The speed can be increased and the theoretical limit approached by proper filtering, statistical tools, etc, to be described elsewhere. The only obvious way of significantly increasing the speed beyond the single line speed limit is to use silicon chip technology with a large number of parallel communicators and multi-wire cables [1] (such as screened phone cables with 500 wires, which are generally available). Such a solution can increase the speed by 2–3 orders of magnitude; however, due to cable costs, their practical use is very limited for very large distances, such as the case of 2000 km.

5. Conclusion

The realized KLJN communicator shows excellent performance through the model-line with communication ranges ten times beyond the range of known direct quantum communication channels with the longest range. The security claims of quantum communicators are usually theoretical, out of standard security measures provided by design and privacy applications, because of the expensive nature of relevant breaking tests. However, in our practical KLJN system we have been able to test security at a much deeper level by carrying out security tests with various types of breaking attempts.

The study indicates that the most important type of security leak, which may need the most resources (wire costs) to control, is the Bergou–Scheuer–Yariv type wire resistance effect [2, 6, 7]; the rest of the proposed practical security compromises, such as the transient/wave arguments of Scheuer-Yariv [6, 7], the noise strength argument of Hao [8] and the resistor inaccuracy argument of Kish [9], can be neglected.

However, it is important to note that the information leak in the KLJN communicator is less harmful than the information leak in quantum communicators; because the eavesdropper does not know which KLJN bit is actually eavesdropped successfully (while in quantum communication the eavesdropper has a very good estimate of the locations of the eavesdropped bits).

The results indicate unrivalled fidelity and security levels among existing physical secure communicators and, if it is necessary and resources are available, there are straightforward ways to further improve security, fidelity and range, such as a proper choice of resistors, thicker cables, enhanced statistical tools for bit decision, etc.

3.9.6. See Section 5.2 for the complete elimination of the (Bergou–) Scheuer–Yariv attack.

There is only one class of practical situations of secure communication where QKD looks unbeatable: the situations when wire communication is impossible or non-desired but communication with photons can be done, such as Zeilinger and coworkers' spatial quantum communication experiments mentioned above [15].

Acknowledgements

Comments by Michael Weissman and Peter Makra are appreciated. Zoltan Gingl is grateful for the Bolyai Fellowship of the Hungarian Academy of Sciences. The travel of LBK to the University of Szeged for the startup phase of the experiments was covered by the Swedish STINT foundation and the cost of his accommodation (10–15 December, 2006) was partially covered by the European Union's SANES grant. LBK's part of designing the KLJN system was partially covered by the TAMU Information Technology Task Force (TITF, grant 2002). Part of these results is also being reported in a plenary talk by the same authors at the Fluctuation and Noise Symposium (Florence, 2007) [14].

References

[1] L.B. Kish, "Totally secure classical communication utilizing Johnson (-like) noise and Kirchhoff's law", *Physics Letters A*, **352** (2006), 178–182.

[2] Adrian Cho, "Simple noise may stymie spies without quantum weirdness", *Science*, **309** (2005), 2148.

[3] L.B. Kish, "Protection against the man-in-the-middle-attack for the Kirchhoff-loop-Johnson(-like)-noise cipher and expansion by voltage-based security", *Fluctuation and Noise Letters*, **6** (2006). L57–L63.

[4] L.B. Kish and R. Mingesz, "Totally secure classical networks with multipoint telecloning (teleportation) of classical bits through loops with Johnson-like noise", *Fluctuation and Noise Letters*, **6** (2006), C9–C21.

[5] L.B. Kish, "Methods of Using Existing and Currently Used Wire Lines (power lines, phone lines, internet lines) for Totally Secure Classical Communication Utilizing Kirchoff's Law and Johnson-like Noise", draft, http://arxiv.org/abs/physics/0610014. Extended version published as L.B. Kish, F. Peper, "Information Networks Secured by the Laws of Physics", *IEICE Transactions on the Fundamentals of Communications, Electronics, Information & Systems*, **E95-B** (2012), 1501–1507.

[6] J. Scheuer, A. Yariv, "A classical key-distribution system based on Johnson (like) noise — How secure?", *Physics Letters A*, **359** (2006), 737.

[7] L.B. Kish, "Response to Scheuer-Yariv: 'A Classical Key-Distribution System based on Johnson (like) noise — How Secure?'", *Physics Letters A*, **359** (2006), 741–744.

[8] F. Hao, "Kish's key exchange scheme is insecure", *IEE Proceedings on Information Security*, **153** (2006), 141–142.

[9] L.B. Kish, "Response to Feng Hao's paper 'Kish's key exchange scheme is insecure'", *Fluctuation and Noise Letters*, **6** (2006), C37–C41.

[10] L.B. Kish, R. Mingesz and Z. Gingl, "How Secure, How to Design, and How to Run Practical Kirchhoff-loop-Johnson-like-Noise Ciphers — I Theory", in preparation.

[11] R. Mingesz, Z. Gingl and L.B. Kish, "How Secure, How to Design, and How to Run Practical Kirchhoff-loop-Johnson-like-Noise Ciphers — II Practice", in preparation.

[12] C.H. Bennett, G. Brassard, C. Crepeau and U.M. Maurer, "Generalized Privacy Amplification", *IEEE Transactions on Information Theory*, **41** (199), 1915–1923.

[13] A. Berzanskis, "Applications of Quantum Cryptography in Government Classified, Business, and Financial Communications", Supercomputing'05, Seattle November 12–18, 2005.

[14] L.B. Kish, R. Mingesz, Z. Gingl, "Thermal noise informatics: Totally secure communication via a wire; Zero-power communication; and Thermal noise driven computing", plenary talk, Proc. 4th SPIE Symposium on Fluctuation and Noise, (Florence, 2007), to be published.

[15] R. Ursin, F. Tiefenbacher, T. Schmitt-Manderbach, H. Weier, T. Scheidl, M. Lindenthal, B. Blauensteiner, T. Jennewein, J. Perdigues, P. Trojek, B. Ömer, M. Fürst, M. Meyenburg, J. Rarity, Z. Sodnik, C. Barbieri, H. Weinfurter and A. Zeilinger, 'Entanglement-based quantum communication over 144 km', *Nature Physics*, (3 June 2007) doi: 10.1038/nphys629.

3.10. Securing Computers, Instruments, Games, etc.

Paper: Unconditionally secure computers, algorithms and hardware, such as memories, processors, keyboards, flash and hard drives[1,2]

Abstract. In the case of the need for extraordinary security, Kirchhoff-loop-Johnson-(like)-noise (KLJN) ciphers can easily be integrated on existing types of digital chips in order to provide secure data communication between hardware processors, memory chips, hard disks and other units within a computer or other data processor system. The secure key exchange can take place at the very first run and the system can renew the key later on at random times in an authenticated fashion to prohibit man-in-the-middle attacks. The key can be stored in flash memories within the communicating chip units at hidden random addresses among other random bits that are continuously generated by the secure line but are never actually used. Thus, even if the system is disassembled, and the eavesdropper

[1]L.B. Kish, O. Saidi, *Fluct. Noise Lett.*, **8** (2008), L95–L98. Available onlline at: http://www.arxiv.org/abs/0803.4479.

[2]For a recent, relevant advanced feature, see Section 5.6: how to use KLJN to make unique, *ultra-strong, physically unclonable function* (PUF) hardware keys.

can have direct access to the communication lines between the units, or even if she tries to use a man-in-the-middle attack, no information can be extracted. The only way to break the code is to learn the chip structure, to understand the machine code program, and to read out the information during runs by accessing the proper internal ports of the working chips. However such an attack needs extraordinary resources and even that can be prohibited by a password lockout. The unconditional security of commercial algorithms against piracy can be provided in a similar way.

Keywords: Secure memories, secure microprocessors, secure hard drives, secure hardware.

3.10.1. Even we were using the incorrect word, “cipher”, at that time. It took years to realize this mistake.

When a computer is lost, even if it is password protected, after disassembling it, the information stored in it will be accessible; the same applies to unsecured flash memories and portable hard drive backups. To defend against these problems, data can be encrypted by software tools. However, these software encryption keys have *only conditional* security and offer *zero information-theoretic security* (see Section 1.6).

The solution described in this short note offers a much higher level of data security by providing unconditionally secure data communication between the hardware elements of interest. Quantum encryption is an example where, with the help of the laws of physics, secure data communication is achieved [1, 2]. However, those instruments are bulky and the price of a pair of communicators is around $100k, making them insufficient for this task.

Recently, a statistical physics based circuitry, the *Kirchhoff-loop-Johnson-(like)-noise* (*KLJN*) cipher was proposed as an electronic competitor of quantum communicators [3, 4] (see Fig. 2). To detect the eavesdropper, the KLJN system does *not* need to build statistics (the way quantum communicators do), and the communication of even a single bit is secure [2, 3]. This is the only communicator which is naturally secure against man-in-the-middle attacks [5]. The fully protected theoretical KLJN scheme has unconditional security against passive and active eavesdropping. The secret key is generated and shared by randomly choosing and connecting one resistor from both ends; one by Alice and

Figure 1. Olivier Saidi (Courtesy of O. Saidi). He is currently the Vice Chair for Clinical and Healthcare Effectiveness at Mount Sinai Health System, New York, USA. His rich CV shows a wide range of experiences including Commander in the Second Information Technology Division of the Strategic Nuclear Air Traffic Control, in the French Strategic Nuclear Air Force. Olivier has a creative imagination. He emerged around 2008, and expressed a great interest in KLJN. We had a number of interactions via email and over the phone. He brought the video game industry into the picture by telling me the story of "Bunny", the PhD student who cracked the Xbox using simple means. That brought the focus on the need to protect inter-chip communications, computers and instruments in an unconditionally secure way, which meant that key refreshments must use KLJN because quantum encryption cannot be integrated onto a microchip. The discussion resulted in this joint paper.

another by Bob, and then measuring the total loop resistance with their thermal noise or a properly enhanced noise [3]. Eve may actively try to extract information by injecting a current at the middle (see Fig. 2). However, constantly monitoring, broadcasting and comparing the current and voltage values at the two ends uncovers the eavesdropping, and Eve can extract no bits without setting off the alarm [3]. If Eve breaks the line and executes a man-in-the-middle attack, that turns out to be an even weaker type of attack due to the statistical independence of the generators at the two ends [5]. In secure KLJN line communicators, the public channel is protected by broadcasting the information in the public channel.

A system of communicator pairs can be expanded into an unconditionally secure data communication network [6]. Recently, a laboratory KLJN model was realized (see Fig. 3) and tested via a model line and parameters up to 2000 km range were obtained. The prototype was experimentally tested with all the known/proposed breaking methods and the (raw-bit) security of the KLJN test model was found to be superior to theoretical (raw-bit) estimations of quantum security. The KLJN prototype pair was placed on

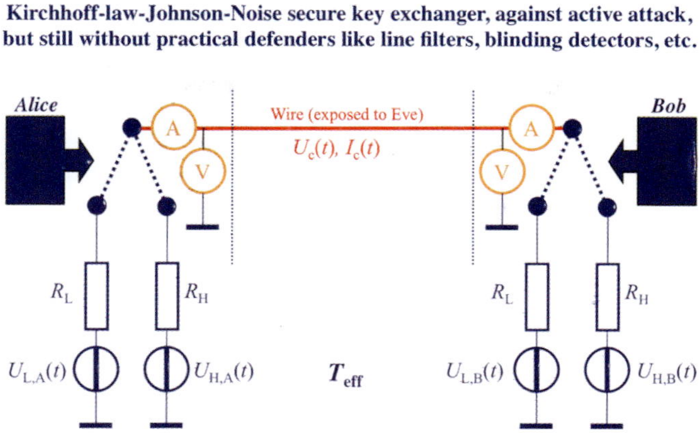

Figure 2. The fully protected theoretical KLJN scheme with unconditional security against passive and active eavesdropping. The secret key is generated and shared by randomly choosing and connecting one of the Alice and Bob's resistors and then measuring the total loop resistance with their thermal noise or a properly enhanced noise [3]. Eve may actively try to extract information by injecting a current at the middle. However, constantly monitoring, broadcasting or exchanging and comparing the current and voltage values at the two ends via authenticated communication, uncovers the eavesdropping and Eve can extract no bits without setting off the alarm [3]. If Eve breaks the line and executes a man-in-the-middle attack, that turns out to be an even weaker type of attack due to the statistical independence of the generators at the two ends [5]. In the secure hardware, after the first key is made, the public channel should carry data encrypted with the shared key together with a hidden changing signature for authentication.

Figure 3. The realized KLJN network unit (2006) with a pair of KLJN communicators [7]. The computer chip version can be much simpler because of the very short cable distances in the computer; providing idealistic conditions to approach the theoretical limits of performance.

two computer cards (see Fig. 3); however, simpler versions could have been made by integrating them on a chip.

In the case of the need for extraordinary hardware, algorithm or data security [9], KLJN ciphers can easily be integrated on existing digital chips in order to provide secure data communication between hardware processors, memory chips, hard disks and other units within a computer or other data processor systems [10]. Therefore, we propose that each secured computer hardware part has a sufficient number of integrated KLJN communicator units. In the simplest case, each unit separately communicates and exchanges a key with all the others. In this case, the required number of KLJN communicators in each hardware element is $N - 1$ when N elements (chips) are connected securely (see Fig. 4).

Alternatively, if the number of units is too high, the secure network solution can be applied [6] which requires at least 2 KLJN communicators in each unit, but requires a more sophisticated protocol to run [6].

The secure key exchange can take place at the very first run and the system can renew the key later at random times (in an authenticated fashion to prohibit man-in-the-middle attacks). In the secure hardware, after the first

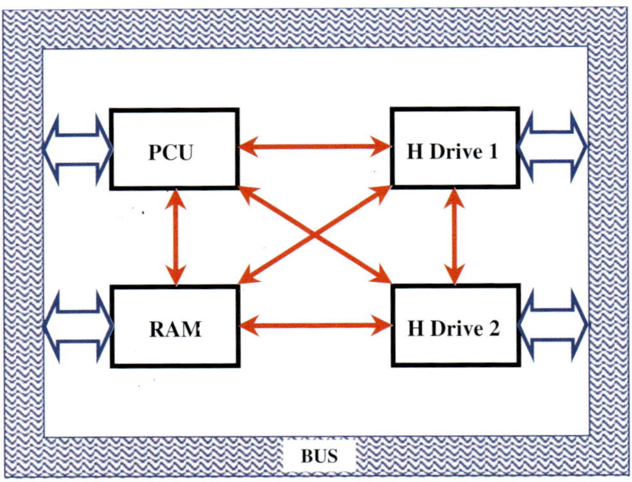

Figure 4. Example of securing a subsystem of a PCU, a RAM and two hard drives. Solid arrows: KLJN connections; Block arrows: classical data bus connections. Each unit has 3 KLJN communicators integrated on their chip. The required number of KLJN units/chip is $N - 1$ when N chip gets connected securely.

key is made, the public channel should carry data encrypted with the shared key together with a hidden changing signature for authentication. Thus after the first few milliseconds of the very first run, the communication between chips becomes unconditionally secure. The key can be stored in flash memories within the communicating chip units at hidden random addresses among other random bits that are continuously generated by the secure line but are never actually used. Thus, even if the system is disassembled, and the eavesdropper can have direct access to the communication lines between the units, or even if she is trying to use a man-in-the-middle attack, no information can be extracted. The only way to break the code is to learn the chip structure, to understand the machine's code program, and to read out the information while the system is running by accessing the proper internal ports of the working chips. However such an attack needs extraordinary resources.

3.10.2. Here, this paper misses the notion of unconditional (information-theoretic) security. It does not matter how much resources Eve needs; if, with extraordinary resources at her disposal, Eve could break into the communication and extract information that *would render the security of the communicators to be only conditionally secure*, (i.e. not unconditional security). *Why is the system still unconditionally secure then?* Because drilling into the chip and finding out the location of randomly hidden information and reading it out have nothing to do with cracking the secure key exchange process. This is a matter pertaining to *tamper-proof design*, not to key exchange security. In security theory, *tamper-proof communicators are always assumed*, otherwise no unconditional security can be defined. Chipmakers design *self-destroying chips where either the chip or its information is destroyed* as soon as Eve tries to drill into the chip or attempts any other invasive action. See Section 5.4 for more advanced hardware concepts (PUF) utilizing KLJN.

Acknowledgments

In 2006, during a discussion about the potentials of KLJN ciphers, Olivier Saidi came up with a closely related idea to use KLJN to secure algorithm in digital games; however, unfortunately, at that time the issue was not worked out further.

References

[1] Wikipedia, "Quantum Cryptology", *Wikipedia*. [Online]. Available at: http://en.wikipedia.org/wiki/Quantum_cryptography [Accessed 24 April 2008].

[2] L.B. Kish, R. Mingesz, Z. Gingl, "Unconditionally secure communication via wire", *SPIE Newsroom* 2007 [Online]. Available at: http://spie.org/x16669.xml

[3] L.B. Kish, "Totally secure classical communication utilizing Johnson (-like) noise and Kirchhoff's law", *Physics Letters A*, **352** (2006), 178–182; also available at http://arxiv.org/physics/0509136.

[4] Adrian Cho, "Simple noise may stymie spies without quantum weirdness", *Science*, **309** (2005), 2148.

[5] L.B. Kish, "Protection against the man-in-the-middle-attack for the Kirchhoff-loop-Johnson(-like)-noise cipher and expansion by voltage-based security", *Fluctuation and Noise Letters*, **6** (2006), L57–L63; also available at http://arxiv.org/physics/0512177.

[6] L.B. Kish and R. Mingesz, "Totally secure classical networks with multipoint telecloning (teleportation) of classical bits through loops with Johnson-like noise", *Fluctuation and Noise Letters*, **6** (2006), C9–C21.

[7] R. Mingesz, Z. Gingl, L.B. Kish, "Johnson(-like)–Noise–Kirchhoff-loop based secure classical communicator characteristics, for ranges of two to two thousand kilometers, via model-line", *Physics Letters A*, **372** (2008), 978–984.

[8] Jason Palmer, "Noise keeps spooks out of the loop", May 23, 2007 [Online]. Available at: http://www.newscientisttech.com/article.ns?Id=mg19426055.300

[9] Andrew "Bunnie" Huang, "Keeping Secrets in Hardware: the Microsoft XBoxTM Case Study", *AI memo*, 2002–008. [Online]. (Last updated: May 26, 2002). Available at: www.ai.mit.edu

[10] L.B. Kish, "Noise Driven Informatics", invited talk at PRANACOPM Workshop, Rice University, Houston, April 3, 2008.

Chapter 4

Developments and Fall-Out of KLJN (2009–2014)

4.1. The initial media transients decayed and the basic features were clarified by 2008, and KLJN, as a secure key exchanger, has become accepted by those who could follow the papers. Since 2009, KLJN has been in a new period: further developments took place; KLJN inspired other security tools (including a software-based one), and even a new class of advanced computing hardware schemes has been created. *Noise-based logic* is a hardware scheme where the logic information is carried by noise — the superposition and/or products of stochastic signals. It has the same information depth and parallelism as theoretical quantum computing schemes; an exponentially richer and exponentially faster scheme than classical logic. Big differences compared to quantum computers are that, due to the robustness of classical noise quantities, no decoherence problems disturb the system and it can work at room temperature. In the same period, KLJN had its highest-profile attack because *i*) the manuscript mathematically claimed (but could not prove) that KLJN offered zero security; and because *ii*) the senior author was computer scientist *Charles Bennett* from IBM, one of the famous fathers of quantum informatics. While we proved that their manuscript was flawed at all levels and Eve would need an infinitely long time to make them right, the challenge was very valuable because it forced me to dig deep into the related issues of physical security, in Sweden, during my 2013 summer research.

Figure 4.1. David Ferry (Courtesy of D. Ferry). He is not only a famous quantum-heretic but also a prominent solid state physicist and engineer, and Regents' Professor at Arizona State University, USA. We have fought on the same side of physical informatics debates on several occasions — the most recent time being in 2013 when we co-organized (with He Wen) a new conference: *Hot Topics in Physical Informatics* (HoTPI) (see its open-access proceedings on the internet, in Volume 33 of the *International Journal of Modern Physics: Conference Series*), with three public debates to challenge some of the myths regarding quantum informatics and Landauer's Principle. Dave has written numerous books and received many honors from both Physics and Electrical Engineering establishments. In the 1980s, in a famous debate in *Physical Review Letters*, he and his postdoctoral researcher, Wolfgang Porod (who is now Professor at the University of Notre Dame, USA) fought against Rolf Landauer and Charles Bennett, claiming that the reversible computer idea violates the laws of physics. Later, he expressed doubts that *exponential-superposition-based quantum computers* can ever fly, and also raised objections against the current interpretation and implications of Bell theorem of quantum physics too. Who is right? I don't know. However, a negative change in the behavior of the scientific community (since then) is that, while it was the establishment fighting him in the 1980s (see the above-mentioned debate), today, his deeply critical works meet with silence from the entire community (not just the establishment) — a sign of the dying of science's key purpose: *truth-seeking*.

4.2. And, as KLJN gained credibility, the fights about it intensified on forums such as Wikipedia that potential sponsors of quantum informatics were reading. In 2008, Derek Abbott created the "Kish Cypher" Wiki site. The site was never good, especially because neither its creator and supporters, nor its quantum-fan

(*Continued*)

(*Continued*)

objectors understood KLJN. In 2012, I requested for the deletion of the Kish Cypher page but my request was turned down. In response, I signed the contract for the present book to tell the truth because Wikipedia's anonymous system made it possible for vandals and hackers to freely destroy information as they liked. Some known names could *seemingly* be identified in the 23 pages long wiki Talk page. However the names on the Talk page *could also have been faked* — perhaps as impersonations of credible names — since Wikipedia is fundamentally anonymous with no way to check real identities. The basic idea of Wikipedia, that anonymity and/or the proof of credibility of authors are unimportant as the majority will eventually reach a consensus, is fundamentally flawed when the issue is scientific, emotional, or strongly controversial.

(*i*) First: the *majority of whom*? We are taking about science, not a popularity contest. Remember the book organized by the Nazis in 1931, *A Hundred Authors Against Einstein*? Albert Einstein's response to it resonates with me; "If I were wrong, then one would have been enough!"

(*ii*) Second: is *it really a majority*? A single person can create an unlimited number of identities by using different IP addresses (e.g. by connecting from different wireless hotspots).

4.3. The Kish Cypher Wikipedia page was deleted in 2013 with my strong support when my debate with Charles Bennett went public (see the screen grab of the last page of the 23 page long Wiki Talk page below). Its termination was a relief to me, because my name and results had been used and misused in football-fan type debates where most of the participants did not understand the game they were fighting about. The fights were in violation of Wikipedia's basic rules because the page should have contained a balanced survey of results from peer-reviewed papers instead of personal views and arguments.

4.4. *Scholarpedia*, the professional scientific alternative of Wikipedia, took over the case. This time, things were better controlled: I went to San Diego and wrote a draft for Gabor Schmera for a simple Scholarpedia survey about some of the key developments until 2012. However, Gabor was worried about becoming the Curator due to his SPAWAR/Navy affiliation. Then, Derek Abbott accepted the function of Curator (with Gabor as coauthor), which was the natural continuation of his former Wiki page creator role. The Editors asked me if I wanted to be the Curator; however, I refused the role. I prefer to focus on the

(*Continued*)

(Continued)

science of the matter instead of running the show. KLJN is still too strongly connected to my name.

4.5. It is like radio waves: *A wave cannot propagate and spread until it gets disconnected from the antenna (the source). This fact is valid for scientific discoveries and inventions, too!*

4.6. Observe below that the Kish cypher Wikipedia Talk page mostly contained debates between people (on both sides) who did not understand the KLJN system. It was already 23 pages long when the entry was finally deleted!

Proposal to delete the Kish cypher page

Dear All,

This is Laszlo B. Kish, the originator of the Kirchhoff-Law-Johnson-Noise (KLJN) secure key exchange system with the popular but incorrect name "Kish cypher". I have been informed that there is a proposal to delete the "Kish cypher" page.

I want to express that I fully and strongly support this proposal. Already, in March 2012, I asked Wikipedia Editor Jake Ocaasi to delete that page because my name has been used in fights where the supporters of both the quantum-informatics and the KLJN systems have been often wrong and fighters at both sides often misrepresented our results.

I am somewhat sympathizing with their failure because I myself also went through such a learning phase after publishing the first KLJN paper, where I used the word "totally" instead of the proper "perfectly", etc. However, for me, it has always perfectly clear that what we are discussing here is measurement information in a system of thermal equilibrium, which most debate participants seem to miss. They also miss elementary terms such as information theoretic security (=unconditional security); perfect security; imperfects security, etc, etc.

Anyway, to cut it short, public debates about scientific matters should be done by professionals at conferences and in peer-reviewed journals, and not by enthusiastic supporters on Wikipedia. While I am not wiki expert, my understanding is that Wikipedia should represent the peer-reviewed scientific literature in a balanced fashion, not the personal opinions of enthusiastic supporters of the Pro and Con sides. The history of the "Kish cypher" page has proven that these rules cannot be enforced therefore I strongly support the termination of this page, which is related to my name and is misrepresenting the research results of our team.

Note, while I fully support deletion, I must express that I disagree with the personal opinion of Skyppido that the KLJN system is "almost totally ignored by the cryptographic community" and is " fringe research by cryptographic/security standards, and totally dead in terms of citations/potential research". I encourage Skyppido to take a look at the history of the first 10 years of quantum key distribution, and see how many papers they published, how large team they were, how many citations they received that time, and at which type of journals (often only at conferences) they succeeded to publish because they were turned down elsewhere. If you do that comparison, KLJN is actually much better off already during its first 8 years. Future will show how these will further evolve.

For those, who are interested in deeper debates than wiki I recommend to read quantum-cryptography-founder Charles Bennet's recent manuscript about KLJN, where he and his coworker (coauthor Jess Riedel) are attempting to "debunk" the KLJN claim, see the manuscript at http://arxiv.org/abs/1303.7435 . While we disagree with almost all the claims in that manuscript, we are very happy about it because this is the way a scientific debate should be. We encourage interested people to read also our detailed response, which will soon be available. All the debate materials will be available at the KLJN page http://www.ece.tamu.edu/~noise/research_files/research_secure.htm at Texas A&M University.

Finally, for those who are interested in a *scientific* debate about the question if quantum key distribution is really secure or not, I recommend the forthcoming meeting *Hot Topics of Physical Informatics* http://www.ece.tamu.edu/%7Enoise/HotPI_2013/HotPI_2013.html and its conference proceedings, where the major challengers of quantum security will be present and debate with the representatives of quantum security in a *public debate session* that will also be published. Also, the KLJN system will be featured and the arguments of Bennett and others addressed. An open-access proceedings will be published.

To conclude this long message in short: I support Skippydo's proposal to delete the Kish cypher page. Thanks for the attention.

Laszlo Kish http://www.ece.tamu.edu/People/bios/bkish.php Laszlo B Kish (talk) 17:35, 14 April 2013 (UTC)

Retrieved from "http://en.wikipedia.org/w/index.php?title=Talk:Kish_cypher&oldid=550337670"

http://en.wikipedia.org/wiki/Talk:Kish_cypher Page 23 of 23

4.1. To Beat the Quantum Computer Dreams. Noise based Logic and Computing

4.1.1. Selecting the papers to show in this section from the many publications we had about noise-based logic since I introduced the first schemes and their hyperspace in 2008 [see Kish, L. B., "Noise-based Logic: Binary, multi-valued, or fuzzy, with optional superposition of logic states", *Physics Letters A*, **373** (2009), pp. 911–918. Also available at: http://arxiv.org/abs/0808.3162] was not an easy choice. In the end, I decided to go with this short review of the principles leading to exponentially increased information depth and computational speed, such as, in the theory of quantum computers.

4.1.2. Note for experts: Noise based logic is a universal logic when it emulates Boolean logic. However, its exponentially enhanced (hyperspace-based) version is not universal; only special-purpose — just like quantum computers and analog computers. For more, the Reader can run a Google search using the terms "noise based" or "noise-based logic".

Paper: Noise based logic: why noise? A comparative study of the necessity of randomness out of orthogonality[1]

Abstract. Noise based logic, like the brain, uses different random noises to represent different logic states. While the operation of brain logic is still an unsolved problem, noise-based logic shows potential advantages of reduced power dissipation, and possesses the ability of large parallel operations with low hardware and time complexity. However, a fundamental question remains: is randomness really needed out of orthogonality? Orthogonal signal systems (similar to orthogonal noises) can also represent multidimensional logic spaces and superpositions. So, does randomness add any advantage to orthogonality or is it disadvantageous due to the required statistical evaluation of signals? In this talk, after some general physical considerations, we show and analyze some specific examples to compare the computational complexities of logic systems based on orthogonal noise, and sinusoidal signal systems, respectively. The conclusion shows that, in certain special-purpose applications that are particularly relevant for mimicking quantum informatics, noise-based logic is exponentially better than its sinusoidal counterpart: its computational complexity (time and hardware) can be exponentially smaller, and still perform the same task.

Keywords: Noise based logic; random telegraph waves; sinusoidal signals; orthogonality; randomness

1. Introduction

Recently, new, non-conventional ways of deterministic (non-probabilistic) multi-valued *noise based logic* (*NBL*) systems [1], stealth communications [2], and unconditionally secure communications [3] inspired by the brain's use of noise and its statistical properties for information processing, have been introduced as lower energy consumption and higher complexity parallel operations in post-Moore-law-chips.

Noise based logic (NLB) uses electronic noise as an information carrier, as shown in Fig. 2, where the logic 0 and 1 signals are represented by independent stochastic noise sources. In Fig. 2, an orthogonal system of random noise processes forms the reference signal system (orthogonal base) of logic values.

[1]Expanded version of H. Wen, L.B. Kish, "Noise based logic: why noise? A comparative study of the necessity of randomness out of orthogonality, *Fluct. Noise Lett.*, **11** (2012), 1250021. Also available at: http://arxiv.org/abs/1204.2545. Also presented at ICCAD 2012.

Figure 1. He Wen and his wife Jing (front) at Corpus Christi beach in Texas (March 2012). He Wen is an Associate Professor at Hunan University, Changsha, China; while Jing is an accountant. They spent a year in College Station, Texas, because He had joined me as a visiting scholar that year, to study noise-based logic. It was a fruitful visit; we explored several new schemes and wrote this paper, too. He also created an animation to demonstrate the exponential information-depth and exponentially accelerated speed of computation by noise-based hyperspace. We used Shankar Bhattacharyya's (Fig. 4) Indian music for "spiritual" support. For more information on the Fluctuation and Noise Exploration Laboratory at Texas A&M University, USA, see http://www.ece.tamu.edu/~noise/research_files/noise_based_logic.htm.

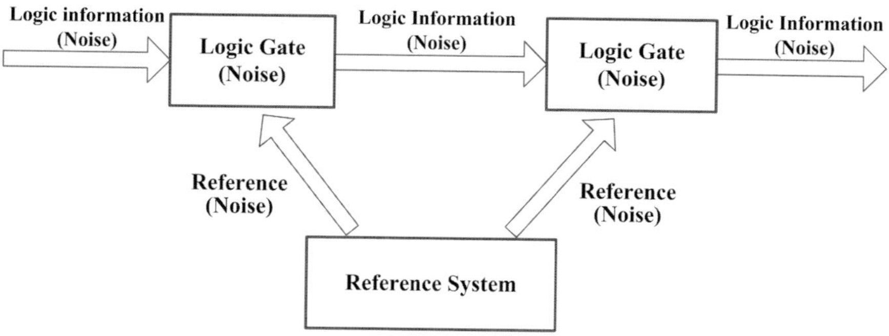

Figure 2. Basic structure of a noise-based logic system.

Within the *NBL* framework, *instantaneous noise-based logic* (*INBL*) [4,5] has been introduced where the logic values are encoded into nonzero, bipolar, independent *random telegraph waves* (*RTW*). The *RTWs* used in the *INBL* are random square waves, which take the value of +1 or −1 with a probability of 0.5 at the beginning of each clock period, and stay with this value during the rest of the clock duration. In the non-squeezed *INBL*, for the *r-th* noise bit, there are two reference *RTW*s, $H_r(t)$ and $L_r(t)$, representing its respective logic values.

Moreover, logic schemes based on other orthogonal processes, e.g. sinusoidal signals [6], have also been proposed to identify a signal without ambiguity. Thus, although the *NBL* has several potential advantages, such as reduced error propagation and power dissipation, it is still interesting to answer the question: why noise? Is randomness really needed out of orthogonality? In this paper, we compare the *RTWs* and sinusoidal signals to illustrate the necessity and importance of randomness when representing logical values with orthogonal signals. For the sake of simplicity, this paper will concentrate on the instantaneous logic scheme and special-purpose applications.

In the rest of the paper, we will compare *NBL* with an alternative orthogonal deterministic realization using sinusoidal signals. It was already pointed out for universal correlator-based *NBL* that sinusoidal representations are inferior [1] and a subsequent circuit-analysis-based study confirmed the negative expectations [7]. However no analysis or comparison exists between instantaneous *NBL* and its sinusoidal alternatives.

2. General arguments: the case of entropy

Preliminary considerations of the energy dissipation issue in biological informatics have been given in [8–10]. Noise-based logic was inspired by the stochastic neural signals in the brain [1], and a particular brain logic scheme utilizing stochastic signals was also proposed [9,10]. The human brain consumes some 10–20 Watts of power while it operates with noise (stochastic neural spike sequences) as an information carrier. The natural question occurs: is there a physical principle that favors stochastic processes for energy-friendly computation?

In this section, we consider the related aspects for noise-based logic by incorporating Brillouin's negentropy principle [11–12]. When energy dissipation is an issue, the problem of entropy can provide a generic answer to why certain types of noises may serve better than deterministic signals. Let us imagine an isolated system in thermal equilibrium (see Fig. 3).

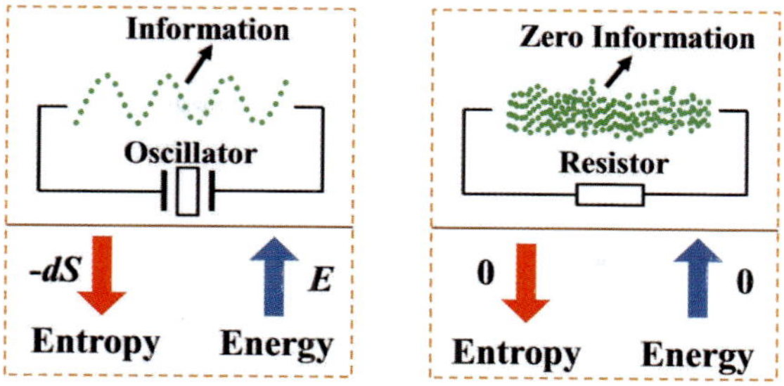

Figure 3. Entropy increase and energy dissipation in an isolated thermodynamical system (Left: generation of a deterministic signal by an oscillator; Right: generation of thermal noise by a resistor)

According to Brillouin's negentropy principle [11–12] if we set up an oscillator to generate a deterministic signal in a subvolume of an isolated thermodynamical system, that will unavoidably lead to an entropic production of heat that is emitted to the rest of the system. According to Brillouin's negentropy principle, the deterministic signal has negative entropy (negentropy) $-dS$ due to its reduced relative uncertainty. According to the second law of thermodynamics, the total entropy of the whole system cannot decrease, which means that the entropy in the rest of the system (outside of the signal channel) must increase by dS. That means generating heat. Moreover, oscillators have loss, which means a continuous energy injection is needed to keep the oscillation going, which results in a heating power. Therefore, if we set up a system of $2N$ sinusoidal oscillators with different frequencies to provide orthogonal signals for the N-bits reference system, having the signal system will require an upfront energy dissipation and a subsequent steady power dissipation.

On the other hand, spontaneous fluctuations of thermal origin (thermal noise) are present in the system. As a relevant example for noise-based logic, we can utilize these thermal fluctuations as reference signals. For example, we can use the thermal noise of $2N$ resistors as the orthogonal signals for the reference signal system of N-bits noise-based logic. Having this signal system requires zero energy dissipation, granted by the existence of thermal fluctuations with zero energy requirement, as shown in Fig. 3.

Even though, the above considerations are only about the reference signal system, and computations would require energy dissipation even with noise, this simple example indicates that noise can have a key role when energy-friendly computation is necessary.

3. Hardware complexity

A general problem with today's binary logic is that, the DC voltage levels representing the different bits and bit values can be considered one-dimensional vectors [1]. These vectors are not orthogonal to each other; thus, the classic binary logic cannot perform high parallelism with low hardware complexity.

Noise Based Logic (NLB), however, can represent a multidimensional vector system utilizing the orthogonality of independent noises [1].

Figure 4. Shankar Bhattacharyya. He is an endowed professor at Texas A&M University, an Elected Member of the Brazilian Academy of Science, and a famous artist of Indian music (he's a sarod player). Shankar is an old friend and we share the same politically incorrect views about many things, such as this expression he made: "Research grant moneys are not the output, but are the input, of research". Shankar is a distinguished "guru" of not only music but also of Control Theory. He permitted the use of his music in He Wen's noise-based logic animation (the project in which I acted as the music editor). The spirituality of Shankar's *sarod* improvisation and that of Gourisankar's drum waves are in harmony with the noise-based hyperspace and noise based logic's ability to recognize or complete patterns and emulate intuition.

Generally, $2N$ independent (orthogonal) noises form N noise-bits, where in a single noise bit X, one noise represents the L and another the H value of the bit, respectively. Products of the base vectors lead out from the base and form a *hyperspace* with 2^N dimensions [1,13]. A logic signal that propagates in a single wire can be constructed by the binary superposition (on or off) of hyperspace vectors, which results in $2^{(2\wedge N)}$ different logic values [2]. Consequently, N noise-bits correspond to 2^N classical bits in a single wire [13].

For applications mimicking quantum computing, the N-bit long hyperspace vectors (product states), $W = X_1 X_2 \ldots X_N$ and their superpositions have key importance [12], where these product states represent the basis vectors of the 2^N dimensional Hilbert space of quantum informatics. Then, there are 2^N different products in an N-bit logic system. Thus, in a classical computer or utilizing the sinusoidal signals, the corresponding 2^N dimensional *universe*, which is the uniform superposition of all the N-bit long product states has a complexity of $O(N^*2^N)$ 13].

For applications, it is essential that the specific universe of flat distribution of N-bit long products can be synthesized with low-complexity operations. With the Achilles' heel method, it can be generated [13] by the following operation that has only $O(2N)$ complexity

$$Y(2^N) = \prod_{r=1}^{N} (L_r + H_r) \tag{1}$$

The easy synthesis of the universe and the resulting high parallelism when manipulating it with simple, low-complexity operations, is very attractive. For example, when the number of noise-bits $N = 200$, a proper algebraic operation acting on this superposition in a single wire can potentially achieve a (special-purpose) parallel operation that would require a $200 * 2^{200} = 200^{208}$ bits operation in classical computing [14].

4. Bandwidth (time complexity) of product strings

As described above, to set up a hyperspace with 2^N dimensions, $2N$ independent (orthogonal) noises are required to form N noise-bits. Similarly, sinusoidal signals can also be used to represent the noise-bit values [1,7]. In this section, we compare the time complexity of *RTW* hyperspace based logic with the corresponding sinusoidal one. It will be shown that the required bandwidth of the sine-based product string (hyperspace vector) will scale with 2^N to avoid the *degeneration* of the system.

Time complexity in these special-purpose applications has two sides:

(a) The time complexity needed to set up the hyperspace vectors.
(b) The time complexity needed to analyze/decode the result which, in practically important cases, is a single product string; because while large-scale parallel calculations are being executed on the superposition of hyperspace vectors, the final result on which the actual measurement is done is typically a single vector.

4.1. *Time complexity needed to set up the hyperspace vectors*

When *RTWs* are used to represent the bit values results, the complexity increases in no time when setting up a hyperspace vector because the *RTW* representing the hyperspace vector has the same clock frequency as its components.

Concerning the sinusoidal signal system — First, let us suppose that we fill up the "harmonics space" linearly with the bit values. For example, the subsequent odd harmonics represent the subsequent L-bit values and the subsequent even harmonics represent the subsequent H values (*linear frequency representation*; see Table 1). Thus to represent N "sine-based" bits, $2N$ different harmonics are needed. In this case, we can define the logic values of the *r-th* bit as

$$L_r(t) = e^{j2\pi(2r-1)f_0 t} \tag{2}$$

$$H_r(t) = e^{j2\pi 2r f_0 t} \tag{3}$$

In this case, the highest frequency in the system will be the sum of all the utilized harmonics from 1 to $2N$, that is, $N(2N+1)f_0$, which scales with N^2. However, this system will be *degenerated* because of the frequency-overlap that occurs when forming the hyperspace vector. For example, the product $L_1 H_2$ and $H_1 L_2$ will produce the same harmonics, which will cause significant information loss and make the system virtually useless (i.e. degenerate).

Table 1. The logic values and corresponding frequencies for both linear and exponential frequency representations.

		Frequency	
Bit	Logic Value	Linear Representation	Exponential Representation
1st	L_1	f_0	f_0
	H_1	$2f_0$	$2f_0$
2nd	L_2	$3f_0$	$4f_0$
	H_2	$4f_0$	$8f_0$
...	...	...	...
Nth	L_N	$(2N-1)f_0$	$2^{2N-2}f_0$
	H_N	$2Nf_0$	$2^{2N-1}f_0$

Second, let us suppose that we fill up the "harmonics space" exponentially (exponential frequency representation) where the frequencies of the harmonics form a 2-based geometric sequence. Then the subsequent bit values are represented by harmonics that scale with the power of 2 (see Table 1 also):

$$L_r(t) = e^{j2\pi 2^{2r-2} f_0 t} \tag{4}$$

$$H_r(t) = e^{j2\pi 2^{2r-1} f_0 t} \tag{5}$$

As a result, the highest frequency in the system will be the sum of all the utilized harmonics, which is $(2^{2N} - 1)f_0$, and that scales with 2^{2N}. In conclusion to set up a single hyperspace vector of a sine-based system requires exponential time complexity $O(2^{2N})$.

4.2. *Time complexity needed to analyze/decode the result (hyperspace vector)*

Reading out the bit values of this hyperspace vector requires a Fourier analysis and a time window of $1/f_0$. However, this time window contains frequency components as high as $(2^{2N} - 1)f_0$, which means that the time complexity of the read-out operation is also $O(2^{2N})$.

Concerning the reading out or the noise-bit values in a hyperspace vector, Stacho [15] developed an algorithm that provides an exponential speedup compared to a "brute force" determination. Stacho's algorithm provides that, when utilizing less than $t_\varepsilon = N(\log_2 N)^{1+\varepsilon}$ clock periods, the probability P that the measurement algorithm fails scales as $P \propto 2^{-N}$ in the limit of a large N, where for $N \to \infty$, $\varepsilon \to 0$.

In time-shifted *NBL*, there is a polynomial (linear in $2N$) time complexity in setting up the hyperspace vector and also the same $O(N)$ complexity to read out the values; specifically, $2N \log_4(N/P)$ time steps are needed, which in that system means $\log_4(N/P)$ clock periods [14].

5. Conclusions

Our goal was to answer the following question: Does noise-based logic really need "noise" (stochasticity) or is an orthogonal system of sinusoidal signals with a similar computational complexity enough?

The answer is that "noise" must be an essential component of these special-purpose, orthogonal-signal-based logic systems for various reasons including low power dissipation (low entropy production) and low computational complexity.

Acknowledgements

Discussions with Sergey Bezrukov, Andreas Klappenecker and Laszlo Stacho are appreciated.

References

[1] L. B. Kish, "Noise-based logic: Binary, multi-valued, or fuzzy, with optional superposition of logic states," *Physics Letters A*, **373** (2009), pp. 911–918.

[2] L. B. Kish, "Stealth communication: Zero-power classical communication, zero-quantum quantum communication and environmental-noise communication", *Applied Physics Letters*, **87** (2005), pp. 234109.

[3] L. B. Kish, "Totally secure classical communication utilizing Johnson (-like) noise and Kirchoff's law", *Physics Letters A*, **352** (2006), pp. 178–182.

[4] L. B. Kish, S. Khatri, F. Peper, "Instantaneous noise-based logic", *Fluctuation and Noise Letters*, **9** (2010), pp. 323–330.

[5] F. Peper, L. B. Kish, "Instantaneous, non-squeezed, noise-based logic", *Fluctuation and Noise Letters*, **10** (2011), pp. 231–237.

[6] M. Ueda, M. Ueda, H. Takagi, M. J. Sato, T. Yanagida, I. Yamashita and K.Setsune, "Biologically-inspired stochastic vector matching for noise-robust information processing", *Physica A: Statistical Mechanics and its Applications*, **387** (2008), pp. 4475–4481.

[7] K.C. Bollapalli, S.P. Khatri, L.B. Kish, "Implementing digital logic with sinusoidal supplies". *in Design, Automation & Test in Europe Conference & Exhibition (DATE)*, 2010, Dresden, Germany, 2010, pp. 315–318.

[8] S.M. Bezrukov, L. B. Kish, "How much power does neural signal propagation need?", *Smart Materials and Structures*, **11** (2002), pp. 800–803, Sept., 2002.

[9] S. M. Bezrukov, L. B. Kish, "Deterministic multivalued logic scheme for information processing and routing in the brain", *Physics Letters A*, **373** (2009), pp. 2338–2342.

[10] Z. Gingl, S. Khatri, L.B. Kish, "Towards brain-inspired computing", *Fluctuation and Noise Letters*, **9** (2010), pp. 403–412.

[11] L. Brillouin, *Scientific Uncertainty and Information*, Academic Press, New York (1964).

[12] L. Brillouin, *Science and Information Theory*, Academic Press, New York (1962).

[13] L.B. Kish, S. Khatri, S. Sethuraman, "Noise-based logic hyperspace with the superposition of 2(N) states in a single wire", *Physics Letters A*, **373** (2009), pp. 1928–1934.

[14] H. Wen, L.B. Kish, A. Klappenecker, F. Peper, "New noise-based logic representations to avoid some problems with time complexity", *Fluctuation and Noise Letters*, **11** (2012), pp. 1250003. Also available at: http://arxiv.org/abs/1111.3859.

[15] L. Stacho, "Fast measurements of hyperspace vectors in noise-based logic", *Fluctuation and Noise Letters*, **11** (2012), pp.1250001.

4.2. Liu-attacks and Responses. Circulators and a Deeper Analysis of KLJN

4.2.1. I was visiting Tamas Horvath at Fraunhofer (Bonn area), Germany, to discuss noise based logic which had been freshly created, when I received an invitation to referee Pao-Lo Liu's paper from Charlie Doering, for *Physics Letters A*. So Tamas and I focused on that instead. This resulted in the paper reproduced in this section. It is easy to recognize the different styles; Tamas wrote the last section (4) of this paper, which covers his important contribution about high-risk bits. For me, writing this was a demanding work and I could only complete it several weeks after Tamas. I had to understand what circulators were and then design my own circulator circuitries and schemes. But it was worthwhile because sooner or later we will face attackers using circulators or similar active circuitry and this work was the first essential step to approach that problem. Pao-Lo brought a very important aspect into the picture.

4.2.2. The funny part of this paper is the identification of Pao-Lo's mistake with the wave impedance during his simulation, which *seemingly* cracked KLJN, and the unphysical large cable diameters. It is obvious from reading even the abstract below that I was a bit upset; and thus enjoyed grilling poor Pao-Lo. The reason was that I had wanted to focus on noise based logic at that time, but this paper had to be written first before I could return to my intended study. Of course, that fact does not justify my sin; it only explains it.

Paper: Notes on Recent Approaches Concerning the Kirchhoff-Law-Johnson-Noise-based Secure Key Exchange[1]

Abstract. We critically analyze the results and claims in [1].

We show that the strong security leak that appeared in the simulations is only an artifact, and is not caused by "multiple reflections". Since no wave modes exist at the cable length of 5% of the shortest wavelength of the signal, no wave is present to reflect it.

In the high wave impedance limit, the conditions used in the simulations are heavily unphysical (requiring cable diameters up to 28000 times greater than the

[1]L.B. Kish, T. Horvath, *Physics Letters A*, **373** (2009), 2858–2868. Also available at: http://arxiv.org/abs/0903.2071.

measured size of the known universe) and the results are modeling artifacts due to the unphysical values.

At the low cable impedance limit, the observed artifacts are due to violating the recommended (and tested) conditions by neglecting the cable capacitance restrictions and using an about 100 times longer than recommended cable, without a cable capacitance compensation arrangement.

We implement and analyze the general circuitry of Liu's circulator [1] and confirm that they are conceptually secure against passive attacks. We introduce an asymmetric and more robust version without feedback loop. Then we crack all these systems using an active attack: a circulator-based man-in-the middle attack.

Finally, we analyze the proposed method to increase security by dropping only high-risk bits. We point out the differences between different types of high-risk bits and show the shortage of this strategy for some simple key exchange protocols.

1. Introduction

In a recent paper [1] a circulator-based development and computer simulations about the secure key exchange system utilizing Kirchhoff's Law and Johnson-like noise (KLJN cypher) was published [2–6]. This system is, at the moment, the only classical physical system offering unconditional security in the idealized limit. If the claims in [1] were valid, they could have a high impact on the future of KLJN type systems. In the present paper, we give a critical analysis of the results and claims in [1], especially about the circulator arrangement, the theory, the computer simulations of the original system proposed in [2] and their interpretations, and about the proposal of dropping high-risk bits to enhance security.

We first briefly survey the necessary background to help the reader follow the arguments. Figure 1 shows the model of the idealized KLJN cipher designed for secure key exchange in [2]. The resistor R_L and R_H represent the low, L (0), and high, H (1), bits, respectively. At each clock period, Alice and Bob randomly choose one of the resistors and connect it to the wire line. The situation LH or HL represents secure bit exchange [2], while LL and HH are insecure. The Gaussian voltage noise generators (white noise with publicly agreed bandwidth) represent a corresponding thermal noise at the publicly agreed effective temperature T_{eff} (typically $T_{eff} > 1$ billion Kelvins [3]). The power density spectra $S_{u,L}(f)$ and $S_{u,H}(f)$ of the voltages $U_{L,A}(t)$ and $U_{L,B}(t)$ supplied by the voltage

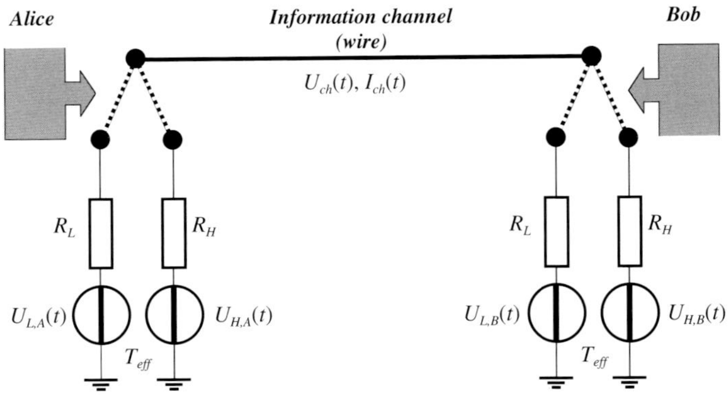

Figure 1. Model of the idealized KJLN cipher (secure key exchange system) [2].

generators in R_L and R_H are given by:

$$S_{u,L}(f) = 4kT_{eff}R_L \quad \text{and} \quad S_{u,H}(f) = 4kT_{eff}R_H, \tag{1}$$

respectively.

In the case of secure bit exchange (*LH* or *HL* situations), the power density spectrum of channel voltage $U_{ch}(t)$ and channel current $I_{ch}(t)$ are given as (see [2] for further details):

$$S_{u,ch}(f) = 4kT_{eff}\frac{R_L R_H}{R_L + R_H} \tag{2}$$

and

$$S_{i,ch}(t) = \frac{4kT_{eff}}{R_L + R_H} \tag{3}$$

Observe that during the *LH* or *HL* situation, due to linear superposition, Eq. (2) is the sum of the spectrum of two situations: When only the generator in R_L is running:

$$S_{L,u,ch}(f) = 4kT_{eff}R_L\left(\frac{R_H}{R_L + R_H}\right)^2 \tag{4}$$

and when the generator in R_H is running:

$$S_{H,u,ch}(f) = 4kT_{eff}R_H\left(\frac{R_L}{R_L + R_H}\right)^2 \tag{5}$$

The ultimate security of the system against passive attacks is provided by the fact that the power $P_{H \to L}$ with which resistor R_H is heating R_L is equal to the power $P_{L \to H}$ with which resistor R_L is heating R_H [2]. This follows the second law of thermodynamics and the energy conservation law; the proof can also be derived from Eq. (3) for a frequency bandwidth of Δf by:

$$P_{L \to H} = \frac{S_{L,u,ch}(f)\Delta f}{R_H} = 4kT_{eff}\frac{R_L R_H}{(R_L + R_H)^2} \tag{6a}$$

$$P_{H \to L} = \frac{S_{H,u,ch}(f)\Delta f}{R_L} = 4kT_{eff}\frac{R_L R_H}{(R_L + R_H)^2} \tag{6b}$$

The equality $P_{H \to L} = P_{L \to H}$ (see Eqs. (6a–b)) is in accordance with the *second law of thermodynamics*; violating this equality would mean not only violating basic laws of physics, but also that the eavesdropper (Eve) could utilize the voltage-current cross-correlation $\langle U_{ch}(t) I_{ch}(t) \rangle$, which is normally zero, to extract the bit [2].

Any deviations from this circuitry, including parasitic elements, inaccuracies, non-Gaussianity of the noise, etc. will cause potential information leak [7–10]. For further analysis, design aspects and comparisons, see [3,8,10]. The situation is similar to quantum key exchange systems, which are the only unconditionally secure competitors of the KLJN system. However, they assume idealistic situations (e.g. single photon source, a noise-free channel) which are unreachable in a real physical system. The experimental results in [3] clearly demonstrate that the KLJN cipher outperforms the raw-bit security of conceptual quantum communicators. Because of the non-zero information leak due to non-ideality effects, both the KLJN and quantum communicators require privacy amplification [11] of the raw bits, which is a proper algorithmic transformation making a shorter key of high security from the longer raw key of low security.

To provide unconditional security against invasive attacks, the fully armed KLJN cipher system monitors the instantaneous current and voltage values at both ends (i.e., at both Alice and Bob's ends) [3,12], and these values are compared either by broadcasting them or via an authenticated public channel. It is important to note that these current and voltage data contain all the information Eve can have. This implies that Alice and Bob have full knowledge about the information Eve may have; a particularly important property of the KLJN system that can be utilized in secure key exchange. This situation implies the following important features of the KLJN system:

1. The KLJN system is *always unconditionally secure*, even when it is non-ideal, in the following sense. The current and voltage data inform Alice and Bob about the exact information leak. Hence, they can always decide either to shut down the communication or to take the risk. Notice that this is a stronger property than that of quantum communicators.
2. Even when the communication is jammed by invasive attacks or inherent non-idealities in the KLJN system, the system remains secure because no information can be eavesdropped by Eve without the full knowledge of Alice and Bob about this potential incidence, and without the full information Eve might have extracted (see Section 3 also).
3. The KLJN system is naturally and fully protected against the man-in-the-middle attack [12], *even during the very first run of the operation* when no hidden signatures can be applied yet. This feature is provided by the unique property of the KLJN system that only zero bits of information can be extracted during a man-in-the-middle attack because the alarm goes off before the exchange of a single key bit has taken place [12].

The outline of the prototype of the KLJN cipher [3] is shown in Fig. 2. The various non-idealities have been addressed by different tools with the aim that the information leak due to non-idealities should stay below 1% of the exchanged raw key bits. It is important to note that while 1% is about the lowest limit an idealistic quantum communicator may achieve, for the KLJN cypher, the best attack achieved only 0.19% of the exchanged key bits [3]. In the present paper, we are only concerned about two of these tools (see Section 2 also):

(i) The role of the line filter (and that of the band limitation of the noise generator) is to provide the no-wave limit in the cable (see Fig. 2). That is, the shortest wavelength component in the driving noises should be much longer than the double of the cable length in order to guarantee that no active wave modes and related effects (e.g., reflection, invasive attacks at high frequencies, etc.) take place in the cable. While in [2] an upper limit for the cable length of 10% of the shortest noise wavelength

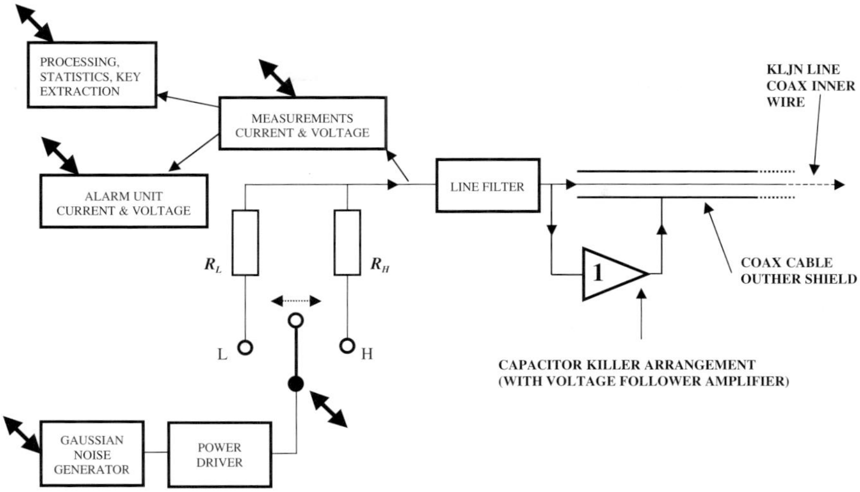

Figure 2. The tested KLJN cipher [3]. Double-ended arrows symbolize computer control.

was proposed, the system was tested in [3] with parameters relevant to a cable length of 5% of the shortest noise wavelength.

(ii) Another tool to fight non-idealities is the cable capacitance compensation (capacitor killer) arrangement (see Fig. 2). With practical cable parameters and their limits, there is a more serious threat of the security: the cable capacitance shortcuts part of the noise current, resulting in a greater current at the side of the lower resistance; yielding an information leak. This effect can be avoided by a cable-capacitor-killer [3]; using the inner wire of a coax cable as the KLJN line while the outer shield of the cable is driven by the same voltage as the inner wire. However, this is done via a follower voltage amplifier with zero output impedance. The outer shield will then provide all the capacitive currents toward the ground and the inner wire will experience zero parasitic capacitance. Without the capacitor killer arrangement and practical bare-wire line parameters, the recommended upper limit of cable length is much shorter depending on the driving resistor values R_L and R_H. For example, the reported measurements in [3] used a 5 kHz noise bandwidth and 2 km range for the capacitor killer arrangement. That was 5% of the wavelength corresponding to the 5 kHz cut-off frequency. However, without a capacitor killer, the allowed upper limit for the bare-wire arrangement would have been only 1% of the range limit mentioned above (see page 980 in [3]); that is, only 0.05% of the wavelength, which is 20 meters.

2. Security: theory, circulators, cables and modeling

In this section, we focus on the physical aspects and implications of the claims in [1].

2.1. *The "no-wave" frequency regime*

First, we provide the valid physical picture about cables in the no-wave limit. These considerations have been well known for a very long time and we use standard textbook material, as well as well-established classical theories and pictures [13–15]. We first address the theory describing cable reflections in [1] and [7].

Electromagnetic waves are described as wave solutions of Maxwell's Equations [13], where half of the wave energy is carried by the electrical field and the other half by the magnetic field. Stationary wave solutions in a finite sized cable are given by a Fourier series consisting of propagating forward and backward wave modes. The lowest frequency wave mode's wavelength is double the cable length. Since no lower frequency wave modes exist in a cable, lower frequency signal components do not propagate as a wave, but as a retarded (no-wave) potential. or a near field around an antenna. Thus, such a propagating potential lacks basic wave properties, such as interference and reflection.

Therefore, Eqs. (1–6) in [1] containing reflection and transmission coefficients, which are intended to describe the original system [2] and the new circulator-based setup [1], are *invalid* because no reflection can exist at the given conditions of a no-wave regime: where the cable length is. at most. 5% of the shortest signal wavelength — the essential regime of security [2,3,8]. A simple example from optics clearly illustrates this situation: using a 500 nanometers visible light for communication, the maximal allowed length of the optical fiber would be $0.05 * 500 = 25$ nanometers. This means that only about 100 atoms would separate the outputs of Alice and Bob: they are virtually collocated!

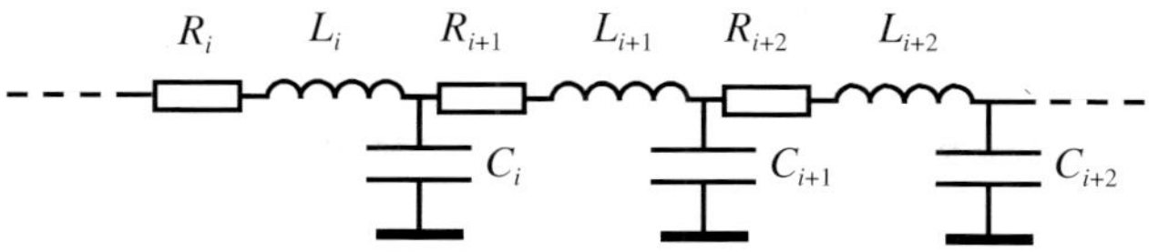

Figure 3. The cable is a distributed RLC network. Here, a wave represents a propagating disturbance where half of the energy is carried by the capacitors and the other half by the inductors. Other types of disturbances do not have wave properties, such as interference and reflection.

2.2. *Cables and their simulation in the no-wave limit*

Cables in the high-frequency limit can be described by a distributed *RLC* network (see Fig. 3). In this picture, a wave represents a propagating voltage/current disturbance with half of the energy in the capacitors due the voltage of the wave, and half in the inductors due to the current of the wave. For example, driving a cable at one end with a voltage generator in the no-wave (low-frequency) limit with no load at the other end means there is no current because the impedance of the capacitors is large; thus there is no current and related energy in the inductors. Hence, all the electrical energy is present only in the capacitors and are therefore not in a wave mode. Time delay effects in such cases are no-wave types of retarded potentials and thus, they neither get reflected nor do they interfere like waves do.

In the cable simulations presented in [1] the cables were characterized by their wave impedance Z_0. The wave impedance of the cable in the high-frequency limit is given as

$$Z_0 = \sqrt{\frac{L_0}{C_0}}, \tag{7}$$

where L_0 and C_0 are the characteristic cable inductance and capacitance relevant for 1 meter cable length, respectively. It is important to recognize that the wave impedance alone does not determine the cable behavior at such short distances and provides only the ratio L_0/C_0 , but not the actual values of L_0 and C_0. There is an *infinite* number of pairs of cable capacitance and inductance for a given wave impedance because only the ratio of L and C is determined. This has important implications. The resultant generator resistance driving the cable at the two ends will determine if the cable

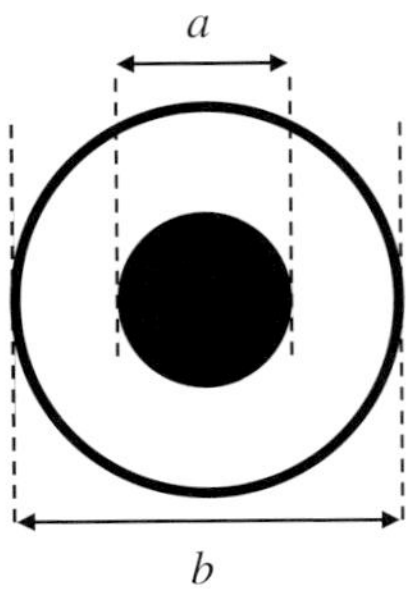

Figure 4. Cross-sectional view of the coaxial cable and the related definitions: a is the diameter of the inner wire and b is the internal diameter of the outer shield.

shows a capacitive or inductive response. For example, in the case of high-resistance drive and short cable (no-wave regime), cable capacitance effects will dominate the transfer characteristics and the wave impedance does not contain enough information to determine the behavior because it leaves the actual capacitance value undetermined. Thus, for simulations in such a case like in [1], the wave impedance alone is irrelevant (see Section 2.5 also).

In this paper, we focus on coaxial (coax) cables (see Fig. 4) because of their potential for a capacitor killer arrangement. However sometimes we will also show that two-wire lines result in a similar conclusion. The approximate characteristic capacitance and inductance of coax cables are given as [14,15] :

$$C_0 = \frac{2\pi\varepsilon_r\varepsilon_0}{\ln(b/a)} \quad \text{(Farad/meter)}, \tag{8}$$

$$L_0 = \frac{\mu_r\mu_0 \ln(b/a)}{2\pi} \quad \text{(Henry/meter)}. \tag{9}$$

With the practical values of the relative permeability $\mu_r = 1$, the wave impedance is [14,15]:

$$Z_0 = \frac{138}{\sqrt{\varepsilon_r}} \ln(b/a)(\Omega/\text{meter}) \tag{10}$$

and, in the case of the given wave impedance, where the inner wire has a diameter of a, and relative dielectric constant ε_r, the required inner diameter of shield (the lower limit of coax cable diameter) is:

$$b = a \exp \frac{Z_0\sqrt{\varepsilon_r}}{138} \quad \text{(meters)} \tag{11}$$

Supposing a coax cable and that $Z_0 = 100\Omega$ (see Fig. 4a in [1]), and that dielectric constant $\varepsilon_r = 1$, $a = 1$ mm, we obtain $d = 2.1$ mm. The characteristic inductance (Eq. (9)) is $L_0 = 1.48 * 10^{-07}$ Henries/meter and the total inductance of the simulated 2-km line in [1] is $L = 2.96 * 10^{-4}$ Henries. The impedance of that inductance at the allowed highest frequency in the noise bandwidth is $Z_L(5\,kHz) = 9.29\Omega$, which is far below the value $R_w = 200\Omega$ of practical resistance of the cable in [3]. Thus, the voltage drop in the wire is purely dominated by the resistance (the Bergou–Scheuer–Yariv effect [3]).

The lower limit ($\varepsilon_r = 1$) of characteristic capacitance is $C_0 = 8*10^{-11}$ Farads/meter, which is at the lower end of practical values. Thus, the lower limit of total cable capacitance for the 2-km length in [1] is $C = 1.6 * 10^{-7}$ Farads.

To show the non-existence of reflections from another aspect, we give a pessimistic estimation about the possibility of oscillations by concentrating the distributed *RLC* elements of the cable into discrete circuit elements (see Fig. 5). The quality factor Q of the concentrated *RLC* serial resonator is:

$$Q = \frac{L/R}{\sqrt{LC}} = 0.0228 \tag{12}$$

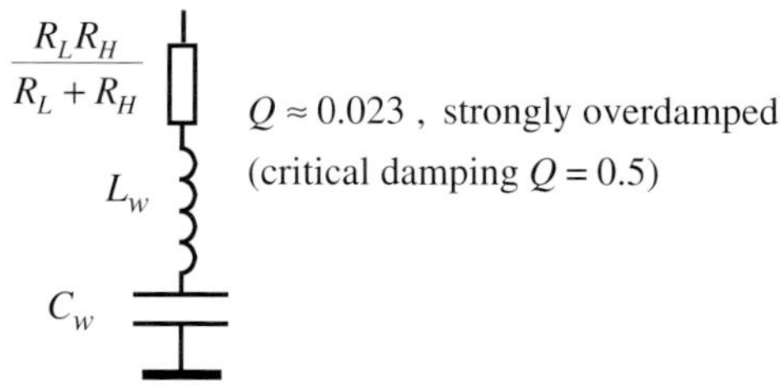

Figure 5. The cable capacitance, cable inductance, cable resistance and driving resistance can be concentrated into lumped circuit elements when the cable length is much shorter than the signal wavelength. When the driving resistance is much greater than the wave impedance of the cable, the cable acts as a capacitor and the inductance can be neglected.

where R is now the sum of $R_w = 200\Omega$ and the resultant diving resistance ($1.69\,k\Omega = 2\,k\Omega \| 11\,k\Omega$) is used by Alice and Bob.

Such an extremely low Q factor means that the cable with these parameters is a totally overdamped system with neither reflections nor oscillations because the critical limit of the overdamped (non-oscillating) range is $Q = 0.5$. Below this limit, the system shows no oscillation but a monotonic relaxation. This is another, phenomenologic interpretation of why reflections cannot occur in the frequency range of the no-wave regime. This holds not only for the concentrated circuitry, but also for the cable with the distributed elements. The reason is that, for the distributed elements, the $(2\pi\sqrt{L_i C_i})^{-1}$ eigenfrequencies are higher; and thus they are much further beyond the noise bandwidth than those of the concentrated elements. When the driving resistance is much greater than the wave impedance of the cable, like here, the cable acts as a capacitor and the inductance can be neglected. Therefore, if any oscillation had been observed during the simulations that would have been an artifact (e.g. errors caused by stray frequency components due to aliasing issues, etc.).

For practical free-standing two-wire lines [14,15] we have the corresponding quantities (see Fig. 6 for definitions):

$$C_0 = \frac{\pi \varepsilon_r \varepsilon_0}{\ln[(d - r_0)/r_0]} \text{ Farad/meter} \tag{13}$$

$$Z_0 = 276 \log_{10}(d/r), \quad \text{thus } d = r\, 10^{Z_0/276}, \tag{14}$$

with similar implications.

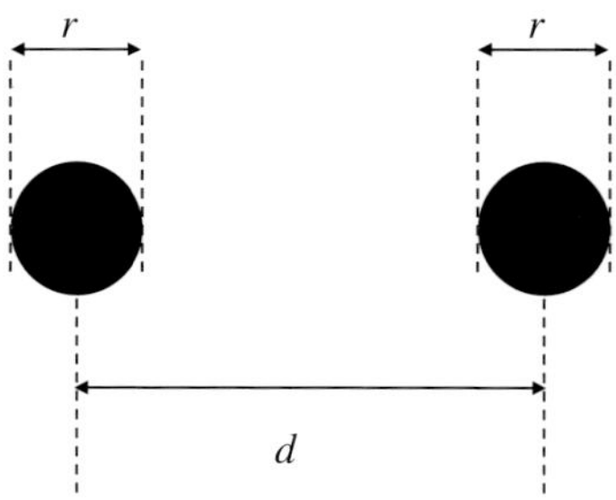

Figure 6. Cross-sectional view of the two-wire cable and the related definitions; where r is the diameter of the inner wire and d is the distance between the centers of the wires ($d > r$).

2.3. *High-impedance simulation results in [1]: heavily unphysical parameters*

Here we explore the implications of the cable parameters used for the simulations in [1]. In Section 2.2, we have already pointed out that wave impedance is a completely irrelevant parameter for the given combination of cable length and frequency band. We now show that the simulations used in [1] in the high wave impedance regime ($2\,k\Omega < Z_0 < 11\,k\Omega$) inherently assume heavily unphysical cable parameters. Below we suppose $\varepsilon_r = 1$ and $\mu_r = 1$ which are around the practical cable values. Using materials with $\varepsilon_r = 30000$, the value around the maximum in ferroelectric ceramics, are non-usable for cables; the predicted sizes below will shrink by a factor of $\sqrt{30000} \approx 170$ (see Eq. (11)) and they still stay unphysical or unpractical. We arrive at similar conclusions for high μ_r materials, too. Furthermore, using $\varepsilon_r = 30000$ would further enhance the capacitance problems discussed below (in Section 2.4) and that 30000-times-greater cable capacitance would kill the secure communication, even at 1% of the present range and at any wave impedance.

With a coax cable of $Z_0 = 11k\Omega$ (see [1]), inner wire diameter of 1 millimeter and $\varepsilon_r = 1$, according to Eq. (11) in Section 2.2, the required coax shield diameter is $4.15 * 10^{31}$ meters. This is more than 28000 times greater than the size of the known universe (156 billion light years $= 1.48 * 10^{27}$ meters [16])!

At $Z_0 = 4.7k\Omega$ (see the simulation data in Fig. 4b in [1]), the required cable diameter is $6.18 * 10^{11}$ meters which is about the same size as the mean distance of the Earth and Jupiter ($7.79 * 10^{11}$ meters) [17].

At $Z_0 = 2k\Omega$ [1], which is the boundary case of the two observed types of behaviors of current and voltages in [1], the required cable diameter is 1.97 km, which is basically the same size as the 2-km range of communication; that is, the length of the cable in [1].

Though our focus is on coax cables, we note that the situation of two-wire lines produces even more unphysical values. With $r = 1$ millimeter and $Z_0 = 11k\Omega$, Eq. (14) yields $d = 7.16 * 10^{36}$ meters, which is more than *one billion times* greater than the known size of the universe (see above). With $Z_0 = 4.7\,k\Omega$ (Fig. 4(b) in [1]) $d = 1.1 * 10^{14}$ meters, which is about 20 times greater than the size of the solar system (the mean distance of the Sun and Pluto is $5.9 * 10^{12}$ meters [18]). With $Z_0 = 2\,k\Omega$ [1], the distance d between the cables is 17.6 km — 8.8 times longer than the range of communication (2 km); i.e. the modeled cable length [1].

Therefore, we regard all the simulation results for the high impedance range, $2\,k\Omega < Z_0 < 11\,k\Omega$, including Fig. 4(b) in [1], as massively non-physical. They are unphysical because, among others, the cable equations and simulation programs are all based on the key assumptions that the *length of the cable is much greater than its diameter.*

It is an open question how the applied simulation program LTSpice accepted and processed the massively unphysical cable parameters. The clarification of this question is out of the scope of our paper. Though the unphysical nature of the large wave impedance parts of the simulations makes further considerations unnecessary, we also note that if the data in the LTSpice program did not get corrupted by the non-physical parameters then the observed difference of the voltages can be simply due to the cable inductance which would be huge at such diameters. When the cable inductance is large enough so that its impedance $Z_L(f) = 2\pi L$ becomes comparable to the resultant resistance of driving by Alice and Bob (see below), the inductance will separate Alice and Bob and a similar security breach takes place as with the wire resistance issue [4,7,8].

2.4. *Low-impedance simulation results in [1]: cable capacitance artifacts*

As mentioned above, according to earlier estimations (see page 981 of [3]) using the driving resistances discussed above, without a capacitor killer arrangement only 1% of the nominal communication length (i.e. 0.05% of the wavelength) could have been utilized as range in [3]. Cable parameters in the simulations in [1] with the low wave impedances (see Fig. 4(a) in [1]) violate this standard practical recommendation and assumption about handling the cable capacitance problems. The same high driving resistance values as in the experiments in [3] (i.e. $2k\Omega$, $11k\Omega$) are used in the simulations in [1] for the same range (2 km), but without a capacitor killer. Notice that this is a 100 times longer cable than allowed. Thus the low wave-impedance simulation results are cable capacitance artifacts (see point (ii) in Section 1) and hence, in contrast to the conclusion in [1], they do *not* imply the insecurity of the KLJN cipher. Of course, not surprisingly, the KLJN cipher may become insecure in cases of inappropriate use. Note, however, that remark 1 about unconditional security given in Section 1 still holds because Alice and Bob have full knowledge of Eve's information (see Section 4 also).

With the simulated arrangements in [1], where no capacitor killer was applied and the length was 2 km, the resistors of Alice and Bob were virtually shortcut by the cable capacitance in the high-frequency range, where the dominant portion of the noise power is located. The simple model of this case is shown in Fig. 7, together with the relevant equations for the high-capacitance limit. The impedance of the capacitance is much less than R_L, therefore the capacitor represents a short circuit toward the ground and the currents at Alice and Bob's ends are simply the thermal noise currents of the noise voltage generators.

The current will be greater at the end with the lower resistance, just like it was observed in [1]. In the worst case scenario, the ratio of the currents of the two sides is R_H/R_L.

Supposing that Alice has the $2\,k\Omega$ and Bob the $11\,k\Omega$ resistance, the mean-square current at Alice's side can be 5.5 times greater than Bob's in the worst case. This implies that Bob's effective current amplitude is only 42% of the current at Alice's end. The current amplitudes shown in Fig. 4(a) in [1] support this observation, even though we have no information about the value of the actual cable capacitance LTSpice used for the simulation.

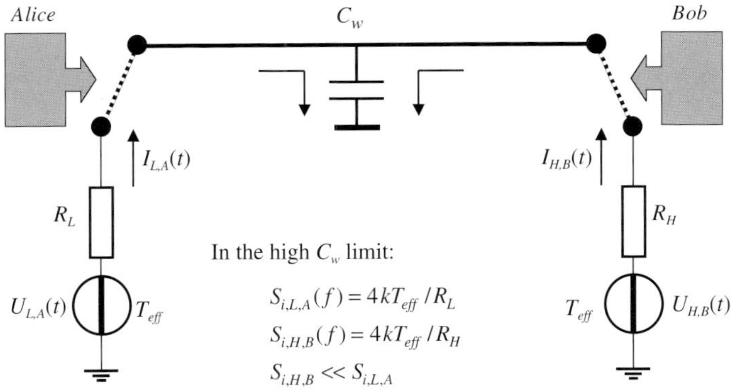

Figure 7. Illustration of the cable capacitance artifact leading to an incorrect conclusion about the security of the KLJN cipher during the low wave impedance simulations in [1].

Let us make a more accurate estimation with the practical cable data used in Section 2.2 by using the lower practical limit of total cable capacitance for the 2 km length ($C = 1.6 * 10^{-7}$ Farads). For the resultant driving resistance $R = 1.69\,k\Omega$, we get $f_c = 1/2\pi RC \approx 0.59\,kHz$. Since the total noise bandwidth is $5\,kHz$, the band $0.59\,kHz < f < 5\,kHz$ is cut off by the cable capacitance. This means that only about 12% of the total noise power is used for communication; the remaining 88% is shorted by the cable capacitance. This implies a serious security risk. Using higher dielectric constants would make this error progressively worse and, at the maximal value of $\varepsilon_r = 30000$ mentioned above, no security would exist even at 1% of the 2 km range.

On the other hand, using the usual $\varepsilon_r \approx 1$, at the 1% of the 2 km length (20 meters) would result in a cutoff frequency of 59.63 kHz and guarantee that the capacitive shunt currents stay sufficiently low to achieve the aimed raw-bit information-leak in the order of 1% [3,8,10]. Similarly, using a capacitor killer arrangement designed to compensate for 99% of the cable capacitance at 2 km distance would result in the same satisfactory situation.

Finally, note that Eve could use much more efficient tools than in [1], e.g. by calculating the mean product (crosscorrelation) of the time-derivative of the line voltage and the currents at the two ends of the line. The side where this product is greater has the L bit.

3. Analyzing and generalizing the circulator-based key exchange proposed in [1]

In [1] a new secure key exchange system utilizing circulators to emulate the KLJN system was introduced. Due to the lack of a detailed circuit analysis in [1], the security aspects and weaknesses of the circulator-based system were not clear. In this section we show a generalized circulator circuitry, deduce its relevant equations and introduce such secure communicators. Surprisingly, the knowledge of the Johnson noise equations is unnecessary for the design. We show a major weakness of all circulator-based key exchange systems: *their absolute vulnerability against a circulator-based man-in-the-middle attack*. Finally, we point out that the original, circulator-free, Johnson noise based KLJN system is secure even against circulator based man-in-the-middle attacks.

3.1. *The circulator circuitry and how it works*

Circulators are usually passive circuit elements for high-frequency applications. In [1], an active, circuit-based system is cited (reference [19] here) which emulates most of the essential properties of a circulator. We have analyzed the function of the circuitry in [19] and constructed a *general* circulator circuitry (see Fig. 8). Despite its simplicity with respect to the system in [19], it has the same function. Another advantage of the circuit in Fig. 8 is that it does not change the sign of the voltage between ports, which is an artifact in [19].

The function of the circulator in Fig. 8 is easy to follow. A current induced through a port by an external voltage will yield a voltage at the next port; however, an internal voltage at the output of a VA will not result in any voltage at the next port as the following will explain. The input/output resistance of the ports is Z_0, and they should be driven by the same external resistance value. The external resistances are grounded in this figure representing zero voltage drive. When one of them (Port 1 in Fig. 9) is not grounded, but a voltage U drives it via the external Z_0, a voltage of

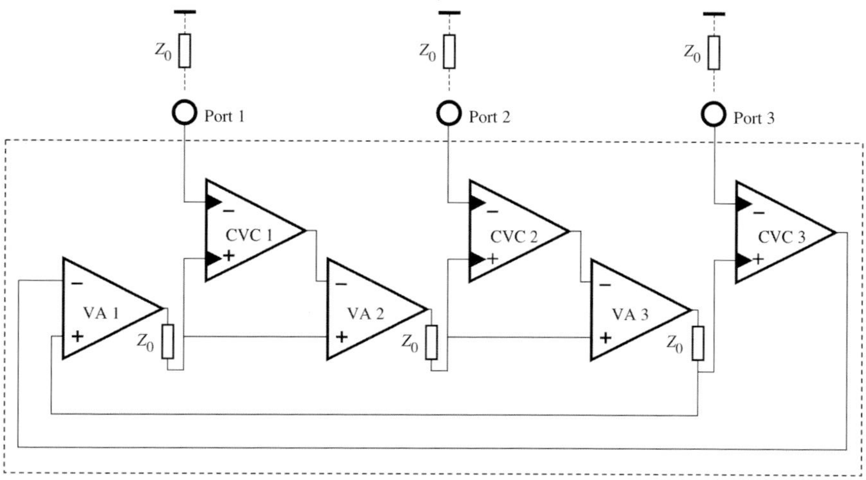

Figure 8. Generalized circuit scheme to emulate passive circulators (within the dashed line area). An improved version of the circuitry cited in [1] (Ref [19] here). VA stands for differential voltage amplifier with an amplification of 2, and CVC stands for current-voltage converter with an amplification of Z_0, respectively.

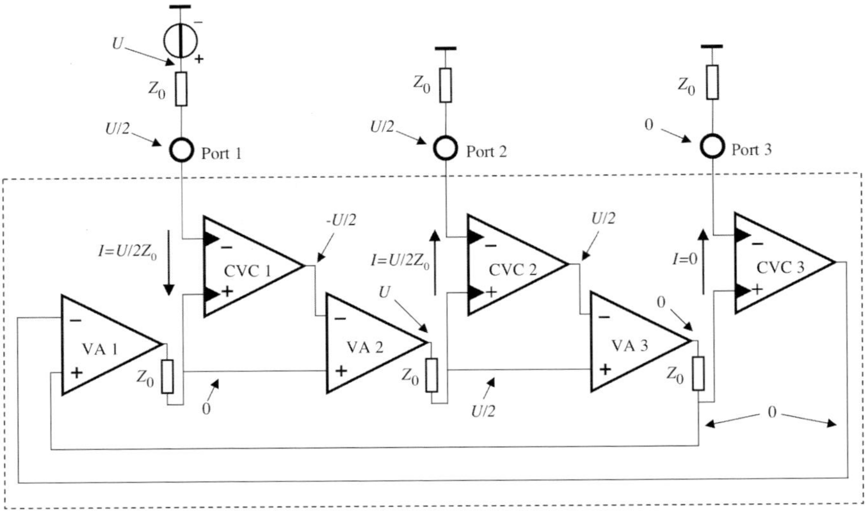

Figure 9. Explanation of the function of the general circulator circuit. Port 1 is driven by a voltage generator of voltage U and internal resistance Z_0. Then a voltage of $U/2$ will drop on Port 1 and, driven by the current I flowing into Port 1, the circulator yields the same $U/2$ voltage on the next port (Port 2). The subsequent port (Port 3) stays at zero voltage because the current and voltage signals driving Port 2 will compensate to zero.

$U/2$ drops on that port. The same voltage ($U/2$) will be outputted on the next port, too. On the other hand, the corresponding current flowing out at the next port via the related CVC yields an internal voltage component at the output of that CVC which kills further voltage propagation at the input of the VA driving the subsequent (third port). Thus the third port stays at zero voltage. Because the system is linear, the ports can be run independently and the circulators will satisfy the linear superposition theorem. However, it is essential to recognize that these idealistic working conditions of the circulator can never be reached. Depending on the practical inaccuracies of the resistance values and amplification, the same order of relative errors will occur in signals on the relevant output ports. Furthermore, stray signals with the same order of relative amplitude will occur on ports which should show zero voltage at ideal conditions. Thus, in the secure key exchange schemes analyzed below, practical inaccuracies in the circulator will lead to additional information leak, out of the known ones, such as the non-zero cable resistance.

3.2. *Generalized circulator schemes (unrelated to Johnson noise) for secure key exchange*

In Fig. 10, the generalization of Liu's circulator-based communicator (the Liu-cipher) is shown. The Liu-cipher works by emulating the Johnson noise situation; that is, the net power flow between Alice and Bob is zero. It is unnecessary to make Johnson noise calculations because the system can be securely designed without the knowledge of the corresponding Johnson noise situation. We will point out that the best practical situation is when only one side is using a circulator, and the other just uses a resistance Z_0 and a noise voltage generator.

The generalized circulator-cipher in Fig. 10 consists of a noise generator at each side ($U_A(t)$ and $U_B(t)$, respectively), a circulator, and a voltage amplifier. The voltage amplifier outputs, with the circulator output voltage originating from the other end of the line, are added to the noise

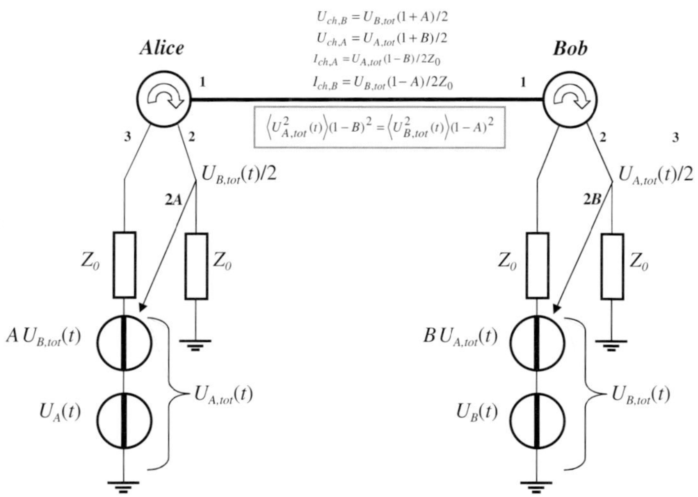

Figure 10. The generalized circulator-based secure key exchange to replace the Liu-cipher proposed in [1]. Voltage amplifier outputs with the voltage from the other end are added to the noise generator voltages. With proper amplification factors A, B and noise generator voltages $U_A(t)$ and $U_B(t)$, the system can run with zero net power flow — the condition to maintain security. The Johnson noise equations are not needed for the design. In the best circulator-based system either A or B is zero, which means that one of the circulators and amplifiers can be left out (see Fig. 11).

generator voltages. The resultant channel voltage and current contributions originating from Alice are given by

$$U_{ch,A} = U_{A,tot}(1 + B)/2 \quad \text{and} \quad I_{ch,A} = U_{A,tot}(1 - B)/2Z_0, \qquad (15)$$

respectively. The resultant channel voltage and current contributions from Bob are:

$$U_{ch,B} = U_{B,tot}(1 + A)/2 \quad \text{and} \quad I_{ch,B} = U_{B,tot}(1 - A)/2Z_0. \qquad (16)$$

With proper amplification factors A, B, and noise generator voltages $U_A(t)$, $U_B(t)$, the security is ensured by running the system with zero net power flow. The security condition, that is, the requirements for a zero-average power flow between Alice and Bob, are as follows (the linear superposition theorem and the statistical independence of the noise generators of Alice and Bob are supposed):

$$\begin{aligned} &\langle U_{ch,A} I_{ch,A}\rangle_t = \langle U_{ch,B} I_{ch,B}\rangle_t \\ &\quad \Rightarrow \langle U_{A,tot}^2(t)\rangle (1 - B)^2 = \langle U_{B,tot}^2(t)\rangle (1 - A)^2 \end{aligned} \qquad (17)$$

One might argue that the line-node resistors of Alice and Bob's circulators dissipate equal energy; however this is irrelevant because that neglects the fact that the real issue is the power balance between the correlated voltage generators at the two ends. The line-node resistors always have the same current, because this is a single loop. Therefore, their power dissipations are always trivially equal, even when their total power balances are different.

To determine the needed noise voltage generator contributions by Alice and Bob, we must realize that the system has a *positive feedback loop* at nonzero A and B because the voltage generated at one side will be outputted at the other side and fed into the line there. However, that voltage will also be re-processed by the other circulator and added to the original voltage at the first side; and so on. Positive feedback amplifies the incoming noise voltages $U_A(t)$ and $U_B(t)$ and the total voltage contributions of the two sites are given as:

$$U_{A,tot}(t) = \frac{U_A(t)}{1 - AB} \quad \text{and} \quad U_{B,tot}(t) = \frac{U_B(t)}{1 - AB}. \qquad (18)$$

Equation (18) is deduced from

$$U_{A,tot}(t) = U_A(t) + AU_{B,tot}(t) \quad \text{and} \quad U_{B,tot}(t) = U_B(t) + BU_{A,tot}(t). \tag{19}$$

by considering one of the cases of $U_A(t) = 0$ or $U_B(t) = 0$. It can be seen from Eq. (19) that the positive feedback's loop-amplification (AB) will enhance the input noise generator voltages of Alice and Bob by the same factor.

After these preparations, the design of a secure circulator-based system is straightforward by using Eqs. (14–18) — the knowledge and application of the *Johnson noise equations* (current and voltage [2]) *are unnecessary* unless the task is to exactly emulate a concrete situation with Johnson noise.

Though the conceptual/idealized Liu-cipher (Fig. 12) is secure, except against the man-in-the-middle attack (see Section 3.3) the positive feedback loop in the circulator system with $A > 0$ and $B > 0$ coefficients raises serious concerns about the practical security. It follows from well-known facts in basic electronics that any non-idealities, such as nonlinearity, frequency-dependence, wire resistance, and wire capacitance, will be enhanced by a positive feedback loop. This situation will help Eve to

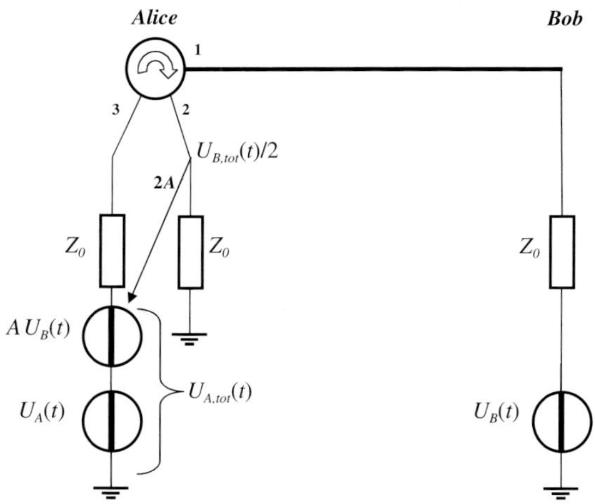

Figure 11. The simplest circulator-based cipher which is also the most robust because there is no positive feedback loop in this system therefore the nonideality, bandwidth and transient problems are kept at the lowest level here.

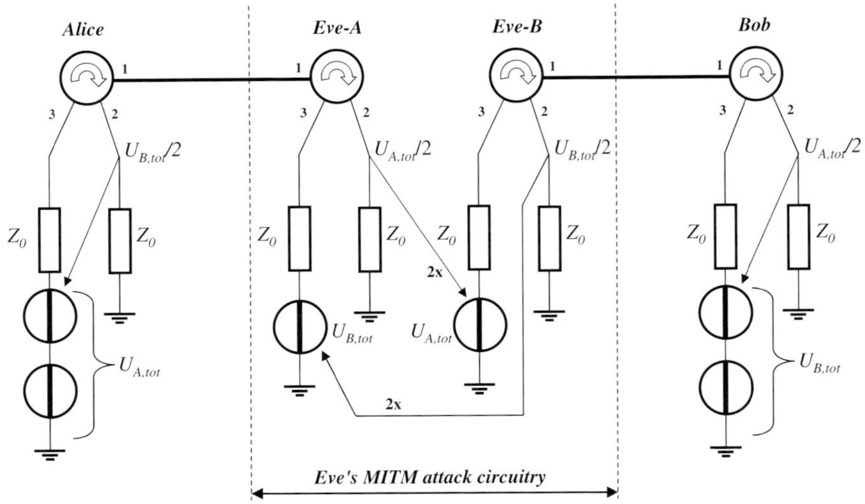

Figure 12. Cracking all circulator-based communicator schemes (including the Liu-cipher, the generalized cipher and the cipher on Figure 10) with a circulator-pair based man-in-the-middle attack.

enhance the information leak normally present in any practical KLJN cipher. Moreover, it opens new information leak channels due to the circulator circuit non-idealities. Therefore, these effects must be thoroughly analyzed and addressed in an actual design, and the necessary protection tools installed.

However, this situation can be improved by avoiding a positive feedback loop with the circulator being located at only one end. The modified circulator-based cipher is shown in Fig. 11. In the figure, only Alice has a circulator; Bob's side only has a noise generator and Z_0. To enhance security, the decision of whether to use a circulator should be made randomly at each clock period. Clearly, only the asymmetric situations of Alice and Bob count as secure bit.

3.3. *Cracking all the circulator-based key exchange schemes*

We now show that the Liu-cipher and its generalizations discussed above are vulnerable against even a simple attack utilizing circulators. It is important to note that the attack described below does not address small information leaks like the ones generally present in any practical (classical and quantum) communicator (see Sections 1 and 2) and which can be removed by privacy amplification. The issue here is fatal. We show that Eve can extract exactly the same amount of information that Bob and Alice share. That is, no secrecy remains implying that no privacy amplifier or other method can be used to guarantee security. The major conceptual disadvantage of any circulator-based ciphers (including the Liu-cipher) is that they are absolutely vulnerable to man-in-the-middle attacks executed by similar circulators (see Fig. 12). This is because Eve can break the line and install a properly wired circulator pair, which is basically a circuit acting as a *repeater* in both directions. Node-1 of circulator Eve-A is connected to Alice via the line and Eve-A yield half of Alice's total voltage ($U_{A,tot}/2$) at its node-2. That node, after a voltage amplification of a factor of 2, will drive node-3 of Eve's other circulator, i.e. Eve-B. Thus, idealistically, Node-1 of Eve-B will exactly emulate the original voltage ($U_{A,tot}$) and current contribution of Alice toward Bob. The same holds in the reverse direction, too. Node-1 of circulator Eve-B is connected to Bob via the line and Eve-B yield half of Bob's total voltage ($U_{B,tot}/2$) at its node-2. That node, after a voltage amplification of a factor of 2, will drive node-3 of Eve's other circulator, Eve-A. Thus, idealistically, Node-1 of Eve-A will exactly emulate the original voltage ($U_{B,tot}$) and current contribution of Bob toward Alice. In this way, the feedback loop is closed and the original functions of the circulators of Alice and Bob are restored by Eve's repeater. However, Eve can measure $U_{A,tot}$ and $U_{B,tot}$ and therefore identifies the shared bit with the same accuracy as Alice and Bob; while Eve is extracting the information, the current and voltage based alarm system does not recognize the attack situation.

In conclusion, the circulator-based key exchange protocols have zero protection against a circulator-based man-in-the-middle attack, as stated.

3.4. *Robustness of the security of the original Johnson noise based scheme*

It can be shown that, using two *different* pairs of properly modified (asymmetric) circulator circuits, or two differently wired asymmetric circulator pairs, corresponding to situations of the different resistance values used by Alice and Bob, a modified circulator-based man-in-the-middle attack can also be carried out against the original KLJN cipher. Asymmetric circulators, circulators with different node resistances, are needed to make a repeater because of the asymmetric resistor values of Alice and Bob during secure bit exchanges. However, because Eve does not know which resistor Alice and Bob will choose, Eve does not know which arrangement of circulators should be used. Her chance of succeeding at drawing a single bit is 25% because there are 4 distinct possibilities. That is, there is a 75% probability of a mismatch, during which the current/voltage based alarm will go off (if Eve attempts a single bit extraction in a mismatched situation). Thus the probability that Eve can stay hidden for N bits is $(0.25)^N$. Note, only half of those bits will be useful because the *LL* and *HH* cases can't be used. Thus Eve's chances of successfully extracting 10 key bits without igniting the alarm is less than one-trillionth. In conclusion, the original Johnson-like noise based KLJN system is secure even against a circulator-based man-in-the-middle attack.

4. Key Exchange by Dropping High-Risk Bits

We now comment on a second remark in [1] stating that: Alice and Bob "*can drop high risk bits from the stream to decrease the risk [2]. Such an approach will reduce the effective key generation rate and is subject to further crypto-analysis.*"

We first note that the high-risk bits destined to be dropped in [2] are restricted only to the *LL* and *HH* bits because the state of the system is always publicly known for these types of bits. Accordingly, the locations of *all* such bits are also publicly known and hence, these bits can be dropped without loss. Clearly, this transformation will halve the key generation rate, as the *LL* and *HH* bits form about 50% of all the bits. We also note that dropping the *LL* and *HH* bits is done automatically at the very beginning and is a trivial but essential step of the secure key exchange method described in [2]. Neither Alice and Bob nor Eve will try to use these bits anymore. Thus, in what follows, we focus only on sequences not containing any *LL* and *HH* bits. Such sequences will be referred to as *basic* sequences.

Besides *LL* and *HH* bits, we have to consider the other type of high-risk bits causing information leaks due to non-idealities, stray components, inaccuracies, etc. This type of high-risk bits is of a completely different nature (see [1] and [2] for a detailed discussion). To comment on the remark cited above from [1] for these types of high-risk bits, we first recall that Alice and Bob have full knowledge of the system's state, as well as of the whole information (current/voltage data) Eve can have. To the best of our knowledge, this is a unique characteristic feature of the *KLJN* cipher among the existing physical secure key exchange systems that can be utilized in improving the security. Moreover, Alice and Bob also know the best strategy for the most efficient bit estimation. Since they can assume without loss of generality that Eve uses this best strategy, Alice and Bob also know her guess, or equivalently, they know whether or not Eve's guess is correct. On the other hand, Eve knows the probability P (>0.5) by which she can successfully guess the *LH* and *HL* bits in the underlying system. However, she never knows if her guess of a particular *LH* or *HL* bit is correct or not. We recall that the success probability P can be derived from the system parameters; we consider the worst-case scenario: that these parameters are publicly known. Below we assume the much worst case: that Eve knows not only P, but also the number M of correctly guessed *LH* and *HL* bits. This assumption enables a simplified analysis of the number of potential keys. For the rest of this section, 'high-risk bits' will mean the *LH* and *HL* bits correctly guessed by Eve. Furthermore, we will always assume that the underlying key exchange protocol is also known by Eve. Before going into the details, we note that if Alice and Bob use the entire basic sequence as the final key then, for all basic sequences, the number of potential keys for Eve is exactly $\binom{N}{M}$. This number will be regarded as the *baseline* measure for the risk.

(a) Consider first the key exchange protocol that Alice and Bob drop *all* high-risk bits from the basic sequence. Since Eve knows N and M, as well as the protocol, she can obtain the set of potential keys for this case by dropping M bits from the basic sequence in all possible ways. Thus, $\binom{N}{M}$ remains an upper bound on the number of potential keys and hence, this protocol does not decrease the baseline risk. It can, however, increase it because the sharpness of the above bound depends on the basic key eavesdropped by Eve. As an extreme example, consider the case that Eve's basic sequence consists of only ones. Clearly, she can easily reason that the final key is the sequence consisting of $N - M$ zeros. Thus, dropping *all* high-risk bits might only *increase* the baseline risk.

(b) We now show that there exists no *deterministic* strategy based on dropping any subsets of the high-risk bits from the basic sequence such that it results in an increase in security (i.e, in a decrease in the baseline risk). In fact, any such deterministic strategies may result only in a decrease in security (see the case above for an extreme situation). More precisely, suppose that Alice and Bob determine the final key by a function T, assigning a sequence k to the pair (k_{AB}, k_E), where

a. k_{AB} is the basic sequence communicated by Alice and Bob,
b. k_E is the sequence eavesdropped by Eve, and
c. k is obtained from k_{AB} by dropping some high-risk bits.

For this protocol, Eve can guess k_{AB} from her sequence k_E in $\binom{N}{M}$ in different ways and then, as she knows T by the worst-case assumption, apply T to (k'_{AB}, k_E) for every guess k'_{AB}. The number of potential keys is therefore also at most $\binom{N}{M}$. In other words, even this more general strategy based on dropping any of the subsets of the high-risk bits does not decrease the risk. Notice, however, that the above bound is sharp only when T is injective in its first argument; all other choices of T do increase the baseline risk. We also note that the argument above holds for any arbitrary function T, i.e., T need not be restricted to dropping high-risk bits.

(c) Finally, we consider the case that Alice and Bob do not fix some deterministic strategy in advance and generate the final key from the basic sequence by (i) first selecting the high-risk bits to be dropped with some unknown strategy and (ii) then exchanging the positions of the bits kept. As an example of this protocol, consider the case that Alice first selects the high-risk bits to be dropped by some stochastic method (e.g., a simple coin toss) and then sends these bit positions to Bob. (For authentication, she can again utilize the KLJN cipher, as it is robust against the "man-in-the-middle" attack [12].) We again regard the worst-case scenario; that Eve is aware of the number M' ($M' < M$) of high-risk bits dropped from the basic sequence, as well as of the $N–M'$ bit positions kept from the basic sequence. Using these assumptions, in this case, she can determine the set of potential keys by

(1) first dropping the same M'bits from her eavesdropped sequence,
(2) choosing $N - M + M'$ positions in the sequence obtained, and
(3) inverting each bit in all selected positions for every choice.

Since different choices in step (2) result in different keys in step (3), the number of potential keys is equal to the number $\binom{N-M'}{N-M+M'} = \binom{N-M'}{M-M'}$ of possible choices. As $\binom{N}{M} > \binom{N-M'}{M-M'}$ always holds for every $M' > 0$, dropping only high-risk bits increases the risk for this protocol as well and is therefore a non-optimal strategy.

A more generalized and practical crypto analysis of the privacy amplification of the non-idealistic KLJN cipher, including the situation when Eve works with a combination of the various strategies and utilizes system parameter fluctuations, is currently in progress by the present authors.

Conclusions

After analyzing the relevant cable and circuit systems, we have refuted most of the claims in [1] concerning the security of the circulator-based key exchange system and the theory of the KLJN cipher. In particular, in contrast to the conclusion of [1], we have shown that:

(i) the circulator-based key exchange system described in [1] is insecure and
(ii) the arguments in [1] do not break the security of the KLJN system [2].

We have also studied a remark in [1] about a security-enhancing method based on dropping high-risk bits and shown that that this method does not increase the security of protocols. Our unpublished preliminary results, however, indicate that this method could still be useful if not only high-risk bits are dropped from the sequence communicated by Alice and Bob.

Acknowledgements

The author would like to thank and appreciate Pao-Lo Liu, Cam Nguyen, Zoltan Gingl, Robert Mingesz, Olivier Saidi, and Bob Nevels for the discussions the author enjoyed with the above-named.

References

[1] Pao-Lo Liu, "A new look at the classical key exchange system based on amplified Johnson noise", *Physics Letters A*, **373** (2009), 901–904.

[2] L.B. Kish, "Totally secure classical communication utilizing Johnson (-like) noise and Kirchhoff's law", *Physics Letters A*, **352** (2006), 178–182.

[3] R. Mingesz, Z. Gingl, L.B. Kish, "Johnson(-like)–Noise–Kirchhoff-loop based secure classical communicator characteristics, for ranges of two to two thousand kilometers, via model-line", *Physics Letters A*, **372** (2008), 978–984.

[4] Adrian Cho, "Simple noise may stymie spies without quantum weirdness", *Science*, **309** (2005), 2148.

[5] D. Jason Palmer, "Noise keeps spooks out of the loop", *New Scientist*, **2605** (2007), 32. Also available at: www.newscientisttech.com/channel/tech/mg19426055.300. Accessed on: May 2007).

[6] L.B. Kish and R. Mingesz, "Totally secure classical networks with multipoint telecloning (teleportation) of classical bits through loops with Johnson-like noise", *Fluctuation and Noise Letters*, **6** (2006), C9–C21.

[7] Scheuer, A. Yariv, "A classical key-distribution system based on Johnson (like) noise—How secure?", *Physics Letters A*, **359** (2006), 737.

[8] L.B. Kish, "Response to Scheuer-Yariv: "A Classical Key-Distribution System based on Johnson (like) noise — How Secure?"", *Physics Letters A*, **359** (2006), 741–744.

[9] F. Hao, "Kish's key exchange scheme is insecure", *IEE Proceedings on Information Security*, **153** (2006), 141–142.

[10] L.B. Kish, "Response to Feng Hao's paper "Kish's key exchange scheme is insecure"", *Fluctuation and Noise Letters*, **6** (2006), C37–C41.

[11] C.H. Bennett, G. Brassard, C. Crepeau and U.M. Maurer, "Generalized Privacy Amplification", *IEEE Transactions on Information Theory*, **41** (199), 1915–1923.

[12] L.B. Kish, "Protection against the man-in-the-middle-attack for the Kirchhoff-loop-Johnson(-like)-noise cipher and expansion by voltage-based security", *Fluctuation and Noise Letters*, **6** (2006), L57–L63.

[13] J.D. Jackson, "Classical Electrodynamics", Second Edition, John Wiley & Sons, Inc. New York (1975).

[14] Jerry C. Whitaker, "The Electronics Handbook", CRC Press (1996).

[15] Edmund Laport, "Radio Antenna Engineering", McGraw-Hill (1952).

[16] Robert Roy Britt, "Universe Measured: We're 156 Billion Light-years Wide!", Space.com (2004, May 24). Accessed January 2009. Not available anymore.

[17] http://www.nasa.gov/worldbook/jupiter_worldbook.html. Accessed January 2009. Not available anymore.

[18] http://www.swivel.com/data_columns/show/2202452. Accessed January 2009. Not available anymore.

[19] C. Wenzel, "Low frequency circulator/isolator uses no ferrite or magnet", *RF Design*, July 1991. http://www.wenzel.com/pd?les1/pdfs/RFDesign3.pdf. Accessed January 2009. Not available anymore.

4.3. The Outstanding Fidelity of Alice and Bob and How to Make It Better

Figure 4.3.1. Yessica Saez. She was a PhD student at Texas A&M University, USA, where she defended her PhD Thesis in April 2015. In 2013, she asked to join my group as her former supervisor (in software-based security) had left the department, causing her to lose 3 years as a result. She had had a governmental scholarship from Panama for a PhD thesis then. The condition of acceptance I laid down was that we had to restart her PhD research from scratch (because KLJN is physics-based). I had a few unfinished ideas that looked suitable and Yessica turned out to be a hard worker who, during these two years, successfully published four journal papers and two conference papers. She defended the first Texas A&M PhD Thesis in KLJN research!

4.3.1. The paper below summarizes the pessimistic estimations of error probabilities in the current and voltage based measurements, and introduces a combined error removal procedure that does not give out information to Eve and does not introduce correlations between key bits. An exceptionally low error probability of 10^{-12} can effortlessly be reached, and optimized systems can offer error probabilities of many orders of magnitude less.

4.3.2. Already in the Szegedin Whisper Project during the first experimental demonstration of KLJN (see Section 3.9), we had to use combined error removals of a similar nature; however, at that time we did not know that an even simpler and much more efficient trick existed — that which is described and analyzed in the paper below.

Paper: Current and voltage based bit errors and their combined mitigation for the Kirchhoff-law–Johnson-noise secure key exchange[1]

Abstract: We classify and analyze bit errors in the current measurement mode of the Kirchhoff-law–Johnson-noise (KLJN) key distribution. The error probability decays exponentially with increasing bit exchange periods and fixed bandwidth, which is similar to the error probability decay in the voltage measurement mode. We also analyze the combination of voltage and current modes for error removal. In this combination method, the error probability is still an exponential function that decays with the duration of the bit exchange period, but it has superior fidelity to the former schemes.

Keywords: Information theoretic security, Second Law of Thermodynamics, Statistical physical key distribution, Unconditional security, Secure key distribution via wire.

1. Introduction

Information theoretic security, often referred to as "unconditional security" [1], means that the security measures are determined by information theory or, in physical systems, by measurement theory. These security measures

[1]Y. Saez, L.B. Kish, R. Mingesz, Z. Gingl, C.G. Granqvist, *J. Computational Electronics*, **13** (2014), 271–277.

can be perfect or imperfect and are determined by the eavesdropper's ("Eve's") supposed optimum conditions for extracting the maximum amount of information. In other words, Eve's information is calculated by assuming that she has unlimited computational power and that her measurement accuracy and measurement speed are limited only by the laws of physics and the protocol's conditions.

Quantum key distribution (QKD) [2] was the first scheme based on the laws of physics that claimed to possess unconditional security. However, this claim is not uncontested and there is an ongoing debate [3–7] about the security inherent in existing QKD schemes. This discussion was initiated by quantum security experts Horace Yuen [3–4, 7] and Osamu Hirota [5], who agreed in their claim that the achievable level of security in QKD schemes is questionable. Renner [6] later entered this debate to defend the foundations of quantum cryptography and to validate existing security proofs.

From a practical point of view, one observes that several communicators, including commercial and laboratory-type QKD devices, have been successfully cracked, as shown in numerous publications [8–22]. These demonstrated flaws of the QKD devices — and also some practical issues, such as limited communication range and high price — have inspired new initiatives that involve non-QKD schemes utilizing alternative types of mechanisms to achieve security [23,24].

Recent studies have shown that a system employing two pairs of resistors, with Gaussian voltage noise generators to imitate and enhance their Johnson noise, can be used for secure key distribution [25–30]. This system is known as the Kirchhoff-law–Johnson-noise (KLJN) secure key distribution and provides information theoretic security [26,27]. It is based on Kirchhoff's loop law of quasi-static electrodynamics and the fluctuation–dissipation theorem of statistical physics [25–31]. The KLJN scheme has potential applications including physical uncloneable function hardware keys [32]; unconditional security within computers, hardware and other instruments [32, 33]; and unconditionally secure smart grids [34–36].

Figure 1 shows the fundamental KLJN system [25–30] without defense elements against active (invasive) attacks and vulnerabilities represented by non-ideal building elements. Under practical conditions, this system utilizes enhanced Johnson noise with high noise temperature, obtained from Gaussian noises generated electronically so that quasi-static and

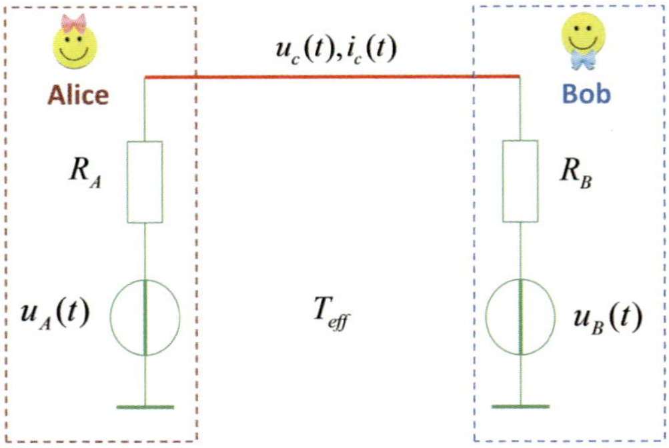

Figure 1. Outline of the core KLJN secure key exchange scheme without defense circuitry (current/voltage monitoring/comparison) against invasive attacks or attacks utilizing non-ideal components and conditions. T_{eff} is the effective noise temperature, R_A, $u_A(t)$, R_B, and $u_B(t)$ are the resistor values and noise voltages at Alice and Bob, respectively. $u_c(t)$ and $i_c(t)$ are channel noise voltage and current, respectively.

thermodynamic characteristics are emulated as accurately as possible, in order to approach perfect security [31]. The core KLJN channel is represented by a wire line to which the two communicating parties, "Alice" and "Bob", connect their resistors R_A and R_B, respectively. These resistors are randomly selected from the set $\{R_0, R_1\}$, with $R_0 \neq R_1$. The resistor R_0 indicates the low (0) bit and the resistor R_1 indicates the high (1) bit, respectively [25]. At the beginning of each clock period or bit exchange period, Alice and Bob, who have identical pairs of resistors, randomly choose one of these resistors and connect it to the wire line. The Gaussian voltage noise generators represent either the Johnson noises of the resistors or external noise generators delivering band-limited white noise with publicly known bandwidth and effective noise temperature T_{eff} [25–26, 30]. According to the fluctuation–dissipation theorem, the enhanced Johnson noise voltages of Alice's and Bob's resistors — denoted $u_A(t)$ and $u_B(t)$ respectively, where $u_A \in \{u_{0,A}(t), u_{1,A}(t)\}$ and $u_B \in \{u_{0,B}(t), u_{1,B}(t)\}$ — generate a channel noise voltage $u_c(t)$ between the wire line and ground as well as a channel noise current $i_c(t)$ in the wire.

Within the bit exchange period, Alice and Bob measure the mean-square channel noise voltage and/or current amplitudes $\langle u_c^2(t)\rangle$ and/or $\langle i_c^2(t)\rangle$. By applying Johnson's noise formula and Kirchhoff's loop law, it follows that the theoretical values of the mean-square noise voltage and current for a given channel noise bandwidth B_{KLJN} and temperature T_{eff} are [25,26]

$$\langle u_c^2(t)\rangle = S_{u,c}(f)B_{KLJN} = 4kT_{eff}R_{\|}B_{KLJN}, \tag{1a}$$

$$\langle i_c^2(t)\rangle = S_{i,c}(f)B_{KLJN} = 4kT_{eff}\frac{1}{R_{loop}}B_{KLJN}. \tag{1b}$$

Here $\langle\,\rangle$ represents ideal infinite-time average, $S_{u,c}(f)$ is the power density spectrum of the channel voltage noise, $S_{i,c}(f)$ is the power density spectrum of channel current noise, k is Boltzmann's constant, $R_{\|} = R_A R_B/(R_A + R_B)$, and $R_{loop} = R_A + R_B$.

The resistance values $R_{\|}$ and/or R_{loop} can be publicly known by comparing the result of the measurement of the mean-square channel noise voltage and/or current amplitudes with the corresponding theoretical values obtained from Eq. (1). Alice and Bob know their own chosen resistors, and hence the total resistances $R_{\|}$ and/or R_{loop} allow them to deduce the resistance value and actual bit status at the other end of the wire.

The cases when Alice and Bob use the same resistance values — *i.e.*, the 00 and 11 situations — represent *non-secure* bit exchange. Eve will then be able to find the resistor values, their location and the status of the bits, because the total resistance will either be the lowest or the highest value of the three possible magnitudes of the total resistance. The situations when Alice and Bob use the resistance values 01 and 10 signify a *secure* bit exchange event because these resistances cannot be distinguished by measured mean-square values. Alice and Bob will know that the other party has the inverse of his/her bit, which implies that a secure key exchange has taken place.

The KLJN key distribution scheme has statistical errors due to the finite duration time τ of the bit-exchange period [30,31]. Specifically, an experimental demonstration of the KLJN scheme, conducted recently by Mingesz *et al.* [30], yielded that the fidelity of the KLJN key exchange was 99.98 %, corresponding to a bit error probability of 0.02 %.

The bit errors were analyzed recently by Saez and Kish [31] for the case of the mean square noise *voltage* being utilized for key exchange. The bit error probability showed exponential decay vs. τ. In the present paper we analyze the bit errors in the *current* measurement mode, as well as the *combination* of voltage and current modes for error mitigation.

2. Bit interpretation of the measured channel current

We suppose ideal components/conditions and proceed as in earlier work [31]. Alice and Bob obtain the total loop resistance by measuring the mean-square channel noise current amplitude $\langle i_c^2(t)\rangle_\tau$, where $\langle\ \rangle_\tau$ indicates a finite-time average over random fluctuations around the exact mean-square noise current. Figure 2 illustrates the three possible levels of the measured mean-square channel noise current. The 11, 01/10 and 00 bit situations result in mean-square channel noise currents $\langle i_{11}^2(t)\rangle_\tau$, $\langle i_{01/10}^2(t)\rangle_\tau$ and $\langle i_{00}^2(t)\rangle_\tau$, respectively.

Thresholds determine the boundaries between the different interpretations of the measured mean-square channel voltages [31]. In the present paper, we use threshold values Δ_3 and Δ_4 to interpret the measured mean-square channel current over the time window τ, as indicated in Fig. 2.

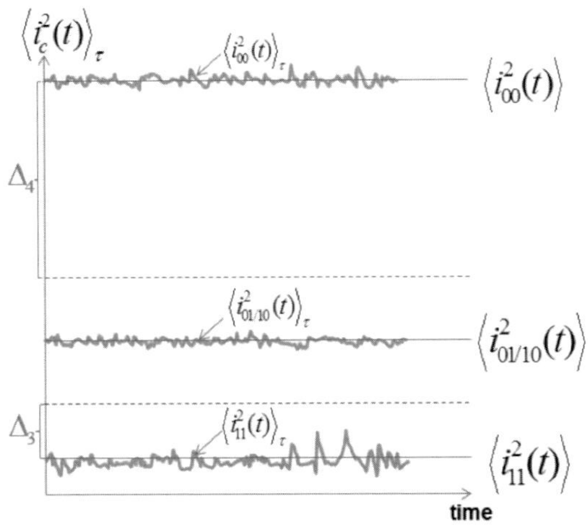

Figure 2. Illustration of statistical fluctuations of the finite-time mean-square current levels around their mean values for the 11, 01/10 and 00 bit situations. The scales are arbitrary. Solid lines denote exact (infinite) time average results. Δ_3 and Δ_4 are thresholds for bit interpretation.

The interpretation is 11 when $\langle i_c^2(t)\rangle_\tau < \langle i_{11}^2(t)\rangle + \Delta_3$, and 00 when $\langle i_c^2(t)\rangle_\tau > \langle i_{00}^2(t)\rangle - \Delta_4$, respectively. The secure case 01/10 is interpreted as such when $\langle i_{11}^2(t)\rangle + \Delta_3 \leq \langle i_c^2(t)\rangle_\tau \leq \langle i_{00}^2(t)\rangle - \Delta_4$. In earlier work [31], the corresponding voltage-based threshold values Δ_1 and Δ_2 were chosen, for normalization purposes, to be proportional to the related mean-square voltages, namely, $\Delta_1 = \beta\langle Du_{00}^2(t)\rangle$ with $0 < \beta < 1$ and $\Delta_2 = \delta\langle Du_{11}^2(t)\rangle$ with $0 < \delta < 1$, for the bit situations 00 and 11, respectively. We choose Δ_3 and Δ_4 in a similar way below.

3. Error probabilities due to statistical inaccuracies in noise current measurements

Bit errors occur when the protocol makes incorrect bit interpretations due to statistical inaccuracies in the measured mean square noise current, and an error analysis for voltage-based operation was presented before [31]. There are different types of error situations, as shown in Table 1.

Similarly to the voltage-based case [31], two types of errors need to be addressed for current-based measurements: the 11==>01/10 errors, *i.e.*, errors when the actual situation 11 is interpreted as 01/10, and the 00==>01/10 errors when the actual situation 00 is interpreted as 01/10. The probabilities for these types of errors are estimated below in a similar way as before [31].

Table 1. Types of errors in the KLJN bit exchange scheme for voltage-based operation [31].

		Actual Situation		
		00	11	01/10
Measurement Interpretation (Decision)	00	Correct	Error removed (automatic)	Error removed (automatic)
	11	Error removed (automatic)	Correct	Error removed (automatic)
	01/10	Error*	Error*	Correct

*The paper addresses these errors and their probability.

3.1. *Probability of 11==>01/10 type errors in current-based measurement*

We set $R_0 = R$ and $R_1 = \alpha R$, with $\alpha \gg 1$. The mean-square channel noise current for infinite-time average at the 11 bit situation is given by

$$\left\langle i_{11}^2(t) \right\rangle = S_{i,11}(f) B_{KLJN}, \tag{2}$$

where $S_{i,11}(f)$ is the power density spectrum of the channel current at the bit situation 11. From Eq. (1) we obtain

$$\left\langle i_{11}^2(t) \right\rangle = 4kT_{eff} \frac{1}{(1+\alpha)R} B_{KLJN}. \tag{3}$$

Figure 3 shows a block diagram for the measurement process in the 11-bit situation. The channel current first enters a squaring unit. For typical practical applications, the output signal is a voltage, because the squaring unit employs voltage-signal-based electronics. However, for the sake of simplicity and without loosing generality, we assume that the numerical values of the voltage correspond to the measured current.

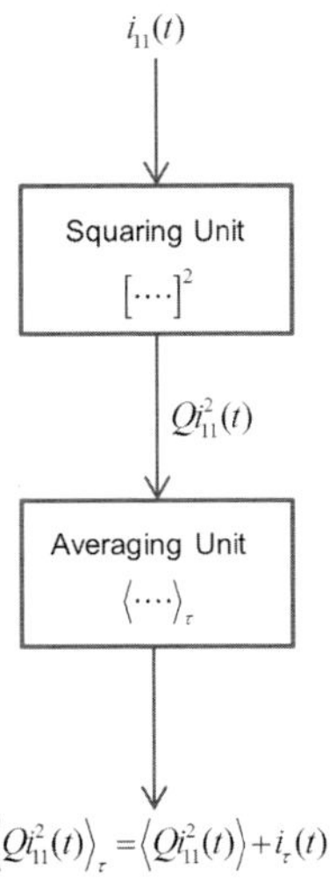

Figure 3. Measurement process at bit situation 11. $\langle\ \rangle$ denotes infinite time (exact) average, Q is transfer coefficient of a hypothetical squaring device, and $i_\tau(t)$ is the noise component of the finite-time average of the square of the current.

Thus we keep the current-based notation as if the electronics would be a current-based signal system. In other words, the voltages are calibrated so that the numerical values are the same as those of the current. The numerical value of this instantaneous amplitude is expressed as $Qi_{11}^2(t)$, where the constant $Q = \frac{1}{Amper}$ denotes the transfer coefficient of the hypothetical multiplier device providing a volt unit also for the square value [37]. This instantaneous amplitude then enters an averaging unit and, after averaging for the finite duration τ, the measurement result is mathematically expressed as $\langle Qi_{11}^2(t)\rangle_\tau = \langle Qi_{11}^2(t)\rangle + i_\tau(t)$, where $i_\tau(t)$ is the AC component remaining after the finite-time average of $Qi_{11}^2(t)$. This averaging process can be represented as low-pass filtering with a cut-off frequency f_B inversely proportional to τ, *i.e.*, $f_B \approx 1/\tau$.

The AC component $i_\tau(t)$ of the finite-time average is Gaussian, which follows from the Central Limit Theorem because τ is much larger than the correlation time for the AC component $i_{2,11}(t) = Qi_{11}^2(t) - \langle Qi_{11}^2(t)\rangle$ of $Qi_{11}^2(t)$ since $f_B \ll B_{KLJN}$. Thus the probability of 11==> 01/10 type of errors is the probability that $i_\tau(t)$ is beyond the threshold, *i.e.*, $i_\tau(t) > \Delta_3$. This probability can be evaluated from the error function, but such a procedure requires numerical integration. However, one can achieve an analytic solution by using Rice's formula [38,39] of threshold crossings, as discussed next.

Rice's formula can be employed to compute the mean frequency by which $i_\tau(t)$ crosses the threshold value Δ_3 [31]. If we define $S_{i,\tau}(f)$ as the power density spectrum of$i_\tau(t)$, the mean frequency of level crossing can be expressed as

$$\nu(\Delta_3) = \frac{2}{\hat{i}_\tau} \exp\left(\frac{-\Delta_3^2}{2\hat{i}_\tau^2}\right)\sqrt{\int_0^\infty f^2 S_{i,\tau}(f)df}, \tag{4}$$

where $\hat{i}_\tau$ denotes the RMS value of $i_\tau(t)$ and is given by $\hat{i}_\tau = \sqrt{\langle i_\tau^2(t)\rangle} = \sqrt{\int_0^\infty S_{i,\tau}(f)df}$. For normalization purposes, we define the threshold value Δ_3 as a fraction of the measured mean-square channel noise current, *i.e.*,

$$\Delta_3 = \lambda\big\langle Qi_{11}^2(t)\big\rangle = \lambda Q S_{i,11}(f) B_{KLJN}, \quad \text{for } 0 < \lambda < 1. \tag{5}$$

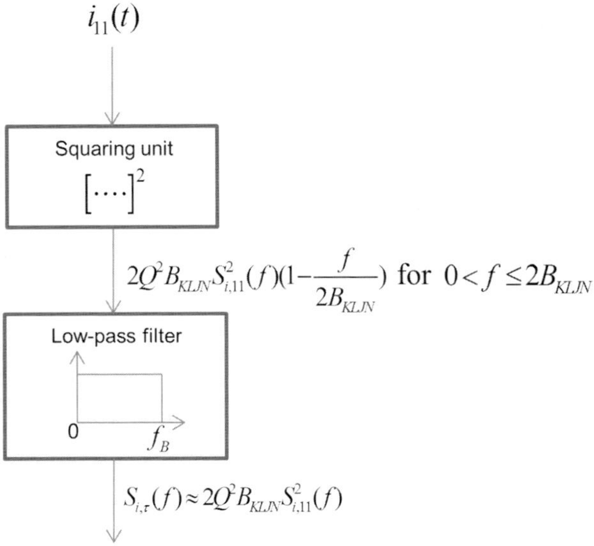

Figure 4. Spectra at bit situation 11. f_B is the cut-off frequency for low-pass filtering, $S_{i,11}(f)$ and $S_{i,\tau}(f)$ are power density spectra of the channel current at the bit situation 11 and of the noise component $i_\tau(t)$, respectively.

The power spectral density $S_{i,2,11}(f)$ for the AC component $i_{2,11}(t)$ of $i_{11}^2(t)$ is considered next. According to previous work [31,37], and also as given in Fig. 4, $S_{i,2,11}(f)$ can be written as

$$S_{i,2,11}(f) = 2Q^2 B_{KLJN} S_{i,11}^2(f)\left(1 - \frac{f}{2B_{KLJN}}\right), \quad \text{for } 0 \le f \le 2B_{KLJN}, \tag{6}$$

and $S_{i,2,11}(f) = 0$ otherwise. The low-pass filtering effect of the time averaging cuts off this spectrum for $f > f_B$ but keeps the $S_{i,2,11}(f)$ spectrum for $f < f_B$. Considering that $f_B \ll B_{KLJN}$, the value of $S_{i,2,11}(f)$ can be approximated by its maximum, *i.e.*, $S_{i,\tau}(f) \approx S_{i,2,11}(0)$. Setting $\gamma = B_{KLJN}/f_B$, one obtains

$$\hat{i}_\tau = \sqrt{\int_0^\infty S_{i,\tau}(f)df} \approx \sqrt{f_B S_{i,2,11}(0)} = \sqrt{2Q^2\gamma f_B^2 S_{i,11}^2(f)} \tag{7}$$

The frequency for unidirectional level crossings, $\nu(\Delta_3)$, which is half of the value given by Rice's formula, is

$$\nu_{\uparrow}(\Delta_3) = \frac{1}{\hat{i}_\tau} \exp\left(\frac{-\Delta_3^2}{2\hat{i}_\tau^2}\right) \sqrt{\int_0^\infty f^2 S_{i,\tau}(f) df}, \tag{8}$$

where

$$\Delta_3 = \lambda Q S_{i,11}(f) \gamma f_B. \tag{9}$$

Using Eqs. (7) and (9) one obtains

$$\nu_{\uparrow}(\Delta_3) = \frac{f_B}{\sqrt{3}} \exp\left(\frac{-\lambda^2 \gamma}{4}\right). \tag{10}$$

Thus the probability $\varepsilon_{i,11}$ of the 11==>01/10 type of errors is

$$\varepsilon_{i,11} \approx \nu_{\uparrow}(\Delta_3)\tau \approx \frac{\nu_{\uparrow}(\Delta_3)}{f_B} = \frac{1}{\sqrt{3}} \exp\left(\frac{-\lambda^2 \gamma}{4}\right). \tag{11}$$

It should be noted that this error probability is an exponential function of the parameters λ and γ, which is consistent with earlier results [31]. The dependence on γ shows that the error probability decays exponentially with increasing bit exchange period τ.

3.2. *Probability of 00==>01/10 type errors in current-based measurements*

In order to compute this probability, we introduce ρ to define the threshold Δ_4 as a fraction of the measured mean-square channel noise. Thus

$$\Delta_4 = \rho \langle Q i_{00}^2(t) \rangle = \rho Q S_{i,00}(f) B_{KLJN}, \quad \text{for } 0 < \rho < 1, \tag{12}$$

where $S_{i,00}(f)$ is the channel noise spectrum at the 00 bit situation.

Following the same procedure as above, the probability $\varepsilon_{i,00}$ of the 00==>01/10 type of errors is again found to be exponentially scaling according to

$$\varepsilon_{i,00} = \frac{\nu(\Delta_4)}{f_B} = \frac{1}{\sqrt{3}} \exp\left(\frac{-\rho^2 \gamma}{4}\right), \quad \text{for } 0 < \rho < 1. \tag{13}$$

3.3. *Illustration of results with practical parameters*

Setting $\gamma = 100$ and $\lambda = 0.5$, the probability $\varepsilon_{i,11}$ for 11==>01/10 type of errors is

$$\varepsilon_{i,11} = \frac{1}{\sqrt{3}} \exp\left(\frac{-\lambda^2 \gamma}{4}\right) \approx 0.001. \tag{14}$$

Increasing the parameter γ, and consequently τ, by a factor of two reduces the error probability to $\varepsilon_{i,11} \approx 10^{-6}$. The bit error probability $\varepsilon_{i,00}$ for the 00==>01/10 type of errors can be computed analogously to the bit error probability $\varepsilon_{i,11}$. In our case of $\alpha \gg 1$, the mean square noise level at 11 is much closer to the value at 01/10 than to the value at 00 (*cf.*, Fig. 2 as an illustration). Therefore, the bit error probability $\varepsilon_{i,00}$ will be significantly smaller than the bit error probability $\varepsilon_{i,11}$. This situation is the opposite for the case of the voltage-based method [31]. Accordingly the experimental test of the KLJN scheme [30] used either the voltage or the current data for decision, depending of which scheme gave the smaller bit error probability.

4. An effective error removal method

Below we show a new error removal strategy, utilizing both voltage and current measurements without applying any error correction algorithm, which is superior to the method used in earlier work [30].

Let us assume that Alice and Bob measure both $\langle u_c^2 \rangle_\tau$ and $\langle i_c^2 \rangle_\tau$. In an ideal error-free situation, the same bit interpretations ensue from both mean-square channel noise amplitudes. However, the bit interpretations can differ when there are errors, because the current and voltage amplitudes are statistically independent due to the Second Law of Thermodynamics (*cf.* Eq. 6) and the Gaussian nature of the noises (when the crosscorrelation between two Gaussian processes with zero-mean is zero, the two processes are statistically independent). To eliminate errors, we select the cumulative measurement output that has the smallest error associated with it; see Fig. 5 and Table 2. We make use of the fact that, in the bit situation when the current evaluation method has maximum error probability, the voltage-based method has minimum error probability, and *vice versa*. Figure 5 shows the three possible mean-square channel noise current and voltage levels.

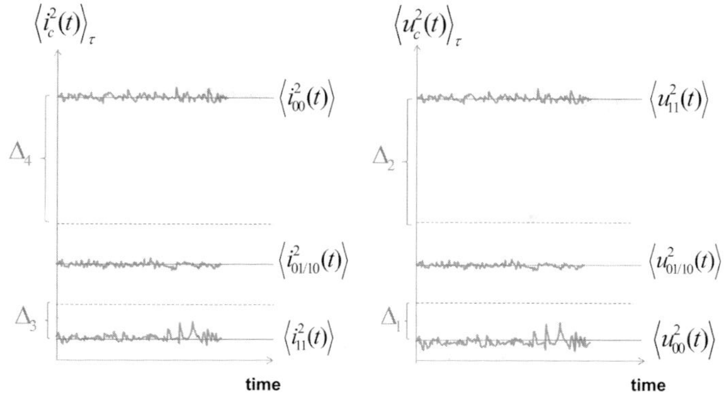

(a) (b)

Figure 5. Mean-square channel noise measurements of current (a) and voltage (b). (Δ_1, Δ_2) and (Δ_3, Δ_4) are the thresholds for interpreting the measured mean-square voltage and current values, respectively. ($\langle i_{11}^2(t)\rangle_\tau$, $\langle i_{01/10}^2(t)\rangle_\tau$, $\langle i_{00}^2(t)\rangle_\tau$) and (, $\langle u_{11}^2(t)\rangle_\tau$ $\langle u_{01/10}^2(t)\rangle_\tau$, $\langle u_{00}^2(t)\rangle_\tau$) are the mean-square channel noise current and voltage at the 11, 01/10 and 00 bit situations, respectively.

Table 2. KLJN error removal method with combined current and voltage analysis.

		Voltage measurement interpretation		
		00	11	01/10
Current measurement interpretation	00	00 (Insecure/ Discard)	Discard (check attack)	00 (Insecure/ Discard)
	11	Discard (check attack)	11 (Insecure/ Discard)	11 (Insecure/ Discard)
	01/10	00 (Insecure/ Discard)	11 (Insecure/ Discard)	01/10 (Secure)

The threshold values Δ_1, Δ_2, Δ_3 and Δ_4 again provide the boundaries for interpreting the measured mean-square voltage and current values.

The only output that is kept is when *both* the current and voltage bit interpretations are secure, *i.e.*, when both are 01/10. For instance, suppose that the bit interpretation obtained from the current measurement is 00 and that the bit interpretation for the voltage measurement is 01/10. In this case, we assume 00 as the correct bit interpretation and hence discard the bit.

5. Error probabilities in the combined current–voltage analysis method

The current and voltage noises are independent as a consequence of the Second Law of Thermodynamics and the Gaussianity of thermal noise, [27,29], and hence the probability of errors in the combined current–voltage analysis method is the product of the error probabilities of the current-based and voltage-based methods.

5.1. *Probability of 00==>01/10 type errors in combined current–voltage analysis*

The probability ε_{00} of the 00==>01/10 type of errors in the voltage-based method is $\varepsilon_{00} = \frac{1}{\sqrt{3}} \exp\left(\frac{-\beta^2\gamma}{4}\right)$ for $0 < \beta < 1$, as reported before [31], and the probability $\varepsilon_{i,00}$ of the 00==>01/10 type of errors in the current-based method is $\varepsilon_{i,00} = \frac{1}{\sqrt{3}} \exp\left(\frac{-\rho^2\gamma}{4}\right)$ for $0 < \rho < 1$, as shown above. Thus the probability $\varepsilon_{t,00}$ of the 00==>01/10 type of errors in the combined method is given by

$$\begin{aligned}\varepsilon_{t,00} &= \varepsilon_{00}\varepsilon_{i,00} \\ &= \frac{1}{3}\exp\left(\frac{-\gamma(\beta^2+\rho^2)}{4}\right), \quad \text{for } 0 < \beta < 1 \text{ and } 0 < \rho < 1. \end{aligned} \tag{15}$$

This error probability is again an exponential function of the parameters.

5.2. *Probability of 11==>01/10 type errors in combined current–voltage analysis*

By following the same procedure as above, we find that the probability $\varepsilon_{t,11}$ of the 11==> 01/10 type of errors in the combined voltage and current measurements is also exponential

$$\begin{aligned}\varepsilon_{t,11} &= \varepsilon_{11}\varepsilon_{i,11} \\ &= \frac{1}{3}\exp\left(\frac{-\gamma(\delta^2+\lambda^2)}{4}\right), \quad \text{for } 0 < \delta < 1 \text{ and } 0 < \lambda < 1. \end{aligned} \tag{16}$$

5.3. *Illustration of results with practical parameters*

To demonstrate the results for the bit error probability, we assign practical values to the parameters β, ρ and γ. For $\gamma = 100$ and $\beta = \rho = 0.5$, we find that $\varepsilon_{t,00}$ is

$$\varepsilon_{t,00} = \varepsilon_{00}\varepsilon_{i,00} = \frac{1}{3}\exp\left(\frac{-\gamma(\beta^2 + \rho^2)}{4}\right) = 1.24 \times 10^{-6}. \tag{17}$$

If the duration of the bit exchange period, *i.e.*, γ, is increased by a factor of two (meaning that the speed is decreased by the same factor), the total bit error probability $\varepsilon_{t,00}$ is decreased to $\varepsilon_{t,00} \approx 4.6 \times 10^{-12}$.

6. Conclusion and final remarks

We classified and evaluated the types of errors that occur in the current-based scheme of the KLJN key exchange. These error probabilities showed an exponential dependence on the duration of the bit exchange, which is analogous to the result for the corresponding voltage-based scheme as discussed in earlier work [31].

Furthermore, we presented an error mitigation strategy based on the combination of voltage-based and current-based schemes: only those exchanged bits that are indicated to be secure by both the current and voltage methods are kept. The resulting error probability of this combined strategy is the product of the error probabilities of the two methods, which follows from the statistical independence of the current and voltage measurements. As a consequence, the KLJN scheme can operate without error-correcting algorithms, thereby preserving the independence of the exchanged bits of the secure key. Thus the key bits remain independently and identically distributed random variables, which is an important advantage for secure communication [27].

Acknowledgments

Discussions with Elias Gonzalez are appreciated. Y. Saez is grateful to IFARHU/SENACYT for supporting her PhD studies at Texas A&M. R. Mingesz's contribution is supported by the European Union and the European Social Fund. Project #TÁMOP-4.2.2.A-11/1/KONV-2012-0073.

References

1. Liang, Y., Poor, H.V., Shamai, S., "Information theoretic security", *Foundations Trends Commun. Inform. Theory*, **5** (2008), 355–580. DOI: 10.1561/0100000036
2. Bennett, C.H., Brassard, G., Breidbart, S., Wiesner, S. "Quantum cryptography, or unforgeable subway tokens", *Advances in Cryptology: Proceedings of Crypto '82*, pp. 267–275, Plenum Press, Santa Barbara. (1982).
3. Yuen, H.P., "On the foundations of quantum key distribution — Reply to Renner and beyond" (2012). Available online at: arXiv:1210.2804.
4. Yuen, H.P., "Unconditional security in quantum key distributions" (2012). Available online at: arXiv: 1205.5065v2.
5. Hirota, O., "Incompleteness and limit of quantum key distribution theory" (2012). Available online at: arXiv:1208.2106v2.
6. Renner, R., "Reply to recent skepticism about the foundations of quantum cryptography" (2012). Available online at: arXiv:1209.2423v1.
7. Yuen, H.P., "Security significance of the trace distance criterion in quantum key distribution" (2012). Available online at: arXiv:1109.2675v3.
8. Merali, Z., "Hackers blind quantum cryptographers", *Nature News* (29 August 2009). DOI:10.1038/news.2010.436
9. Gerhardt, I., Liu, Q., Lamas-Linares, A., Skaar, J., Kurtsiefer, C., Makarov, V., "Full-field implementation of a perfect eavesdropper on a quantum cryptography system", *Nature Commun.*, **2** (2011), 349. DOI:10.1038/ncomms 1348
10. Lydersen, L., Wiechers, C., Wittmann, C., Elser, D., Skaar, J., Makarov, V., "Hacking commercial quantum cryptography systems by tailored bright illumination", *Nature Photonics*, **4 (2010)**, 686–689. DOI:10.1038/NPHOTON. 2010.214
11. Gerhardt, I., Liu, Q., Lamas-Linares, A., Skaar, J., Scarani, V., Makarov, V., Kurtsiefer, C., "Experimentally faking the violation of Bell's inequalities", *Phys. Rev. Lett.*, **107** (2011), 170404. DOI:10.1103/PhysRevLett.107. 170404
12. Makarov, V., Skaar, J., "Faked states attack using detector efficiency mismatch on SARG04, phasetime, DPSK, and Ekert protocols", *Quantum Inform. Comput.* **8** (2008), 622–635.
13. Wiechers, C., Lydersen, L., Wittmann, C., Elser, D., Skaar, J., Marquardt, C., Makarov, V., Leuchs, G., "After-gate attack on a quantum cryptosystem", *New J. Phys.*, **13** (2011), 013043. DOI: 10.1088/1367-2630/13/1/013043

14. Lydersen, L., Wiechers, C., Wittmann, C., Elser, D., Skaar, J., Makarov, V., "Thermal blinding of gated detectors in quantum cryptography", *Opt. Express*, **18** (2010), 27938–27954. DOI: 10.1364/OE.18.027938
15. Jain, N., Wittmann, C., Lydersen, L., Wiechers, C., Elser, D., Marquardt, C., Makarov, V., Leuchs, G., "Device calibration impacts security of quantum key distribution", *Phys. Rev. Lett.*, **107** (2011), 110501. DOI:10.1103/ PhysRevLett.107.110501
16. Lydersen, L., Skaar, J., Makarov, V., "Tailored bright illumination attack on distributed-phase-reference protocols", *J. Mod. Opt.*, **58** (2011), 680–685. DOI:10.1080/09500340.2011.565889
17. Lydersen, L., Akhlaghi, M.K., Majedi, A.H., Skaar, J., Makarov, V., "Controlling a superconducting nanowire single-photon detector using tailored bright illumination", *New J. Phys.*, **13** (2011), 113042. DOI:10.1088/1367-2630/13/11/113042
18. Lydersen, L., Makarov, V., Skaar, J., "Comment on 'Resilience of gated avalanche photodiodes against bright illumination attacks in quantum cryptography'", *Appl. Phys. Lett.*, **99** (2011), 196101. DOI:10.1063/1.3658806
19. Sauge, S., Lydersen, L., Anisimov, A., Skaar, J., Makarov, V., "Controlling an actively-quenched single photon detector with bright light", *Opt. Express*, **19** (2011), 23590–23600. DOI: 10.1364/OE.19.023590
20. Lydersen, L., Jain, N., Wittmann, C., Maroy, O., Skaar, J., Marquardt, C., Makarov, V., Leuchs, G., "Superlinear threshold detectors in quantum cryptography", *Phys. Rev. Lett.*, **84** (2011), 032320. DOI:10.1103/PhysRevA.84.032320
21. Lydersen, L., Wiechers, C., Wittmann, C., Elser, D., Skaar, J., Makarov, V., "Avoiding the blinding attack in QKD: REPLY (COMMENT)", *Nature Photonics*, **4** (2010)., 800–801 DOI:10.1038/nphoton.2010.278
22. Makarov, V., "Controlling passively quenched single photon detectors by bright light", *New J. Phys.*, **11** (2009), 065003. DOI:10.1088/1367-2630/11/6/065003
23. Yuen, H.P., "Key generation: Foundation and a new quantum approach", *IEEE J. Sel. Topics Quantum Electr.*, **15** (2009), 1630–1645. DOI:10.1109/JSTQE.2009.2025698
24. Salih, H., Li, Z.H., Al-Amri, M., Zubairy, H., "Protocol for direct counterfactual quantum communication", *Phys. Rev. Lett.* **101** (2013), 170502. DOI:10.1103/PhysRevLett.110.170502
25. Kish, L.B., "Totally secure classical communication utilizing Johnson (-like) noise and Kirchhoff's law", *Phys. Lett. A*, **352** (2006), 178–182. DOI:10.1016/j.physleta.2005.11.062

26. Mingesz, R., Kish, L.B., Gingl, Z., Granqvist, C.G., Wen, H., Peper, F., Eubanks, T., Schmera, G., "Unconditional security by the laws of classical physics", *Metrol. Meas. Syst.* **20** (2013), 3–16. DOI:10.2478/mms-2013-0001
27. Kish, L.B., Abbott, D., Granqvist, C.G., "Critical analysis of the Bennett–Riedel attack on secure cryptographic key distributions via the Kirchhoff-law–Johnson-noise scheme" (2013). Available online at: viXra:1306.0058, arXiv:1306.653.
28. Kish, L.B., "Protection against the man in the middle attack for the Kirchhoff-loop Johnson (-like) -noise cipher and expansion by voltage-based security", *Fluct. Noise Lett.*, **6** (2005), L57–L63. DOI:10.1142/S021947750600 3148
29. Kish, L.B., "Enhanced secure key exchange systems based on the Johnson noise scheme", *Metrol. Meas. Syst.*, **20** (2013), 191–204. DOI:10.2478/mms-2013-0017
30. Mingesz, R., Gingl, Z., Kish, L.B., "Johnson (-like) -noise–Kirchhoff-loop based secure classical communicator characteristics, for ranges of two to two thousand kilometers, via model-line", *Phys. Lett. A*, **372** (2008), 978–984. DOI:10.1016/j.physleta.2007.07.086
31. Saez, Y., Kish, L.B. "Errors and their mitigation at the Kirchhoff-law–Johnson-noise secure key exchange" (2013). Available online at: viXra: 1305.0126; arXiv:1305.4787v1.
32. Kish, L.B., Kwan, C., "Physical uncloneable function hardware keys utilizing Kirchhoff-law–Johnson-noise secure key exchange and noise-based logic" (2013). Available online at: viXra:1305.0068; arXiv:1305.3248.
33. Kish, L.B., Saidi, O., "Unconditionally secure computers, algorithms and hardware", *Fluct. Noise Lett.*, **8** (2008), L95–L98. DOI:10.1142/S0219477508004362
34. Gonzalez, E., Kish, L.B., Balog, R., Enjeti, P., "Information theoretically secure, enhanced Johnson noise based key distribution over the smart grid with switched filters", *PLoS ONE*, **8** (2013), e70206. DOI:10.1371/journal.pone.0070206
35. Kish, L.B., Mingesz, R., "Totally secure classical networks with multipoint telecloning (teleportation) of classical bits through loops with Johnson-like noise", *Fluct. Noise Lett.*, **6** (2006), C9–C21. DOI:10.1142/S021947750600332X
36. Kish, L.B., Peper, F., "Information networks secured by the laws of physics", *IEICE Trans. Commun.*, **E95-B** (2012), 1501–1507.

37. Kish, L.B., Mingesz, R., Gingl, Z., Granqvist, C.G. "Spectra for the product of Gaussian noises", *Metrol. Meas. Syst.*, **19** (2012), 653–658. DOI: 10.2478/v10178-012-0057-0
38. Rice, S.O. "Mathematical analysis of random noise", *Bell System Tech. J.* **23** (1944), 282–332. Available online at: http://archive.org/details/bstj23-3-282
39. Rychlik, I., "On some reliability applications of Rice's formula for the intensity of level crossings", *Extremes*, **3** (2000), 331–348. DOI:10.1023/A:1017942408501

4.4. In the Ring: Response to Charles Bennett and Jess Riedel

4.4.1. At the beginning of April 2013, I received the news from Horace Yuen (who had been informed by Osamu Hirota) that quantum informatics co-founder Charles Bennett (with J. Riedel) published a manuscript in arxiv claiming that KLJN offers zero security against an ideal Eve. Horace recommended that I, in my denial, should focus on Bennett's conditional-information equations (Eqs. 16 and 17 in the paper below). I did that and killed these equations (see Eqs. 21–26 and the whole of Section 2.2 in the paper); however the response also went much beyond just these. The debate was an excellent way to dig deep and not only sharpen, but also extend the analysis of KLJN. A brief and enhanced follow-up-analysis of the unconditional security proof eventually led to an invited paper in a quantum informatics journal [See: L.B. Kish, C.G. Granqvist, *Quantum Information Processing*, **13** (2014), pp. 2213–2219], which we have reproduced in Section 5.1 of this book. Thus this public debate did not only terminate the erroneous 'Kish Cypher' Wikipedia page, on which quantum-fans dominated (see Section 4 of this book) but it also helped to complete many of the unfinished analyses. KLJN came out from the ring highly reinforced.

4.4.2. Derek Abbott was visiting me when the news from Horace arrived. In accordance with his nature, he became very enthusiastic about the high profile attack and the visibility it would give to KLJN; thus I invited him to participate in the response. I dictated to him a quick-list of killer responses we had briefly discussed; they amounted to about a few pages long. However, he stuck upon the no-wave issue of quasi-static electrodynamics because he taught microwave technics at the time; where all the problem solutions were based on waves. I told him that engineering used many unphysical math steps to solve problems, which were usually fine, provided no physical meaning is regarded to the unphysical intermediate math solutions. A good example is the linear response theory where the incoming signal is symbolically replaced by an infinite series of Dirac-pulses, which are completely unphysical but make the math solution easy. Later, this mistake led Derek and his colleagues to their attack published in Nature's

(*Continued*)

(*Continued*)

Scientific Reports (see our denials of all their claims in Sections 4.6.1–4.6.3 of this book). But the lesson was important: KLJN designers could make similar mistakes and produce systems that offer no security! Good examples are Vadim Makarov's successful hacking attacks against commercial quantum communicators.

4.4.3. Soon, I left for Sweden, where life is peaceful — conducive for deep thinking — and worked out the extensive analysis/denial of Bennett's claims and much more with the strong support of Claes Granqvist (see Figure 4.4.1 for some fun). It became one of the longest papers of my carrier.

4.4.4. To help the Reader, there is a table of contents in this paper to help find parts of interest. For experts of security, it may be beneficial to first read the paper in Section 5.1 of this book.

Figure 4.4.1. "*Something is really fishy there*". There were many such moments with Claes Granqvist while detecting erroneous claims and/or arguments in the papers of adversaries or of whole schools of thought. For example, see our extended analysis and denial of virtually all the Bennett–Riedel KLJN claims (our paper is below); the Landauer principle–Szilard engine–Maxwell demon issues described briefly in Section 4.5; and the denial of the Gunn–Allison–Abbott claims with our extended analysis in Section 4.6.

Paper: Critical analysis of the Bennett–Riedel attack on the secure cryptographic key distributions via the Kirchhoff-law–Johnson-noise scheme.[1]

Content

Abstract
Introduction
1.1 Is the security of quantum encryption indeed robust?
1.2 The KLJN secure key exchange system
  1.1.1 *The idealized KLJN scheme and its security*
  1.1.2 *The security of the KLJN scheme is based on the Second Law of Thermodynamics*
  1.1.3 *On active (invasive) attacks and attacks utilizing non-idealities*
  1.1.4 *Foundations of the information theoretic security in practical KLJN schemes*
  1.1.5 *Mathematical proof of the unconditional security of the exchanged key*
  1.1.6 *Optional security addition: cap imposed on q by Alice and Bob*
1.2 Summary of Bennett–Riedel's arguments regarding the KLJN scheme
  1.2.1 *Bennett–Riedel's claim concerning no information transfer in a wire in the no-wave (quasi static) limit* 14
  1.2.2 *Bennett–Riedel's claim that the KLJN system does not offer security*
  1.2.3 *Bennett–Riedel's claim regarding a "thermodynamics-free" key exchange scheme*
  1.2.4 *Bennett–Riedel's wave-transient based attack before the steady state is reached*
  1.2.5 *Bennett–Riedel's passive time-correlation attack in the no-wave limit*
  1.2.6 *Current extraction/injection based active (invasive) attack*

[1]L.B. Kish, D. Abbott, C.G. Granqvist, "Critical analysis of the Bennett-Riedel attack on secure cryptographic key distributions via the Kirchhoff-law-Johnson-noise scheme", *PLoS ONE* **8** (2013) e81810. (Open access).

Discussion and Results

2.1 Physics facts: information, propagation, and wave couplers in the quasi static limit

2.1.1 *The mathematical definition of a wave in physics*

2.1.2 *The quasi-static limit of electrodynamics, and electrical circuitry symbols with lumped elements*

2.1.3 *Signal propagation in the no-wave (quasi-static) limit*

2.1.4 *Further implications of the quasi-static (no-wave) limit: Directional couplers, etc.*

2.2 Refutation of *Bennett–Riedel's* claim about no information transfer in the no-wave limit

2.3 Invalidity of *Bennett–Riedel's* equations and the correct equations

2.3.1 *The wave limit and the Pao-Lo Liu key exchange system*

2.3.2 *Bennett–Riedel's equations for the KLJN scheme in the no-wave (quasi static) limit*

2.4 Proof that *Bennett–Riedel's* key exchanger is unphysical

2.5 Ten ways to crack *Bennett–Riedel's* key exchanger by passive attacks

2.5.1 *Six universal, energy/current-flow-analysis attacks*

2.5.2 *Three transient-damping resistor attacks*

2.5.3 *Wire-resistance Johnson noise attack*

2.5.4 *The above attacks are inefficient against the KLJN system as a result of thermodynamics*

2.6 On transient attacks against the KLJN scheme

2.7 Why *Bennett–Riedel's* passive correlation attack does not work against KLJN

2.8 Why the current extraction/injection active attack does not work against KLJN

2.9 Remarks about potential hacking attacks

Conclusions

Acknowledgements

Abstract: Recently, Bennett and Riedel (BR) (2013)[2] argued that thermodynamics is not essential in the Kirchhoff-law–Johnson-noise (KLJN) classical physical cryptographic exchange method, in an effort to disprove the security of the KLJN scheme. They attempted to demonstrate this by introducing a dissipation-free deterministic key exchange method with two batteries and two switches. In the present paper, we first show that BR's scheme is unphysical and that some elements of its assumptions violate basic protocols of secure communication. All our analyses are based on a technically-unlimited Eve with infinitely accurate and fast measurements limited only by the laws of physics and statistics. For non-ideal situations and at active (invasive) attacks, the uncertainly principle between measurement duration and statistical errors makes it impossible for Eve to extract the key regardless of the accuracy or speed of her measurements. To show that thermodynamics and noise are essential for the security, we crack the BR system with 100% success via passive attacks, in ten different ways, and demonstrate that the same cracking methods do not function for the KLJN scheme that employs Johnson noise to provide security underpinned by the Second Law of Thermodynamics. We also present a critical analysis of some other claims by BR; for example, we prove that their equations for describing zero security do not apply to the KLJN scheme. Finally we give mathematical security proofs for each BR-attack against the KLJN scheme and conclude that the information theoretic (unconditional) security of the KLJN method has not been successfully challenged.

Introduction

Information theoretic (*i.e.*, unconditional) security [1] means that the stated security level—either perfect or imperfect, as in any physical system [2]—holds even for cases when the abilities of an eavesdropper (generally called "Eve") are limited only by the laws of physics. Since 1984, quantum key distribution (QKD) [2] has been claimed to possess unconditional security, and much later, in 2005, an alternative based on classical physics, known as the Kirchhoff-law–Johnson-noise (KLJN) scheme [2], appeared as a competing approach.

Very recently, QKD's co-founder Charles Bennett [3] co-authored a manuscript [4] with Jess Riedel wherein they present an extensive criticism

[2]Bennett, C.H. and Riedel, C.J. (2013). "On the security of key distribution based on Johnson-Nyquist noise". Available online at: http://arxiv.org/abs/1303.7435v1

of the KLJN scheme and deny its security under idealized conditions. Bennett and Riedel (BR) assert that thermodynamics is not essential in the KLJN scheme and argue that it does not provide security. They attempt to prove this claim by showing a dissipation-free deterministic key exchange method with nothing but two batteries and two switches. Moreover, among other statements [4], BR argue that the quasi-stationary (*i.e.*, no-wave) limit of electrodynamics is unsuited for information transfer, thus implying that this (required) assumption [2] for (perfect) security of the KLJN system is unphysical. Our present paper is a detailed critical analysis of the BR scheme. In summary, we show that BR's scheme is unphysical, and we provide further analysis that demonstrates the security of the KJNL scheme.

In this introductory chapter we set the scene for the next chapter, wherein we will fully crack the BR system in various ways and also respond to BR's arguments about the KLJN scheme. We first consider the currently ongoing debates concerning the security of QKD, which is a necessary preamble since BR propound that the security of QKD is robust. Then we briefly outline the KLJN secure key distribution scheme and its main features. Subsequently, we describe the "thermodynamics-free" key exchange system due to BR and the related argumentation in their paper [4].

1.1. *Is the security of quantum encryption indeed robust?*

BR write [4]: "*we emphasize that quantum key distribution has been shown to be robust with imperfect components against very general attacks*". We see this situation very differently and first briefly summarize the currently ongoing debates in the QKD field.

Currently, there is a discussion [5–8] about the fundamental security/non-security of existing QKD schemes. This debate was initiated by Yuen [5,8], who was later joined by Hirota [6] in claiming that the security of existing quantum key distribution schemes is questionable or poor. Recently, Renner [7] entered the discussion to defend the old security claims. It should be noted that Yuen [9] and Zubairy *et al.* [10] have proposed new advanced schemes for non-QKD-based secure quantum communication.

BR's claim that QKD displays "*robust security with imperfect elements*" [4] has been proven incorrect, and QKD has been cracked by utilizing the imperfect nature, such as non-linearity, of necessary building elements. Practical quantum communicators—including several commercial ones—have been fully cracked as shown in numerous recent papers [11–25]. Vadim Makarov, who is one of the leading quantum crypto crackers, stated that "*Our hack gave 100% knowledge of the key, with zero disturbance to the system*" [11]. This statement hits the foundations of quantum encryption schemes, because the often-claimed basis of the security of QKD protocols is the assumption that any eavesdropping activity will disturb the system enough to be detected by the communicating parties (generally referred to as "Alice" and "Bob"). An important aspect of these quantum-based hacking attacks is the extraordinary (100%) success ratio of extracting the "secure" key bits by Eve, which indicates that the security is not only imperfect but simply non-existing against these types of attacks until proper defense strategies or protocol modifications have been added to the scheme in order to restore the information theoretic security they supposedly had before these attacks were known.

In conclusion, and in clear contradiction to BR's claim [4], *quantum key distribution has been found vulnerable to well-designed attacks* for the case of imperfect components.

1.2. *The KLJN secure key exchange system*

The Kirchhoff-law–Johnson-noise key distribution scheme [2,26–39] is a classical statistical physical alternative to QKD, whose security is based on Kirchhoff's Loop Law and the Fluctuation-Dissipation Theorem. More generally, it is founded on the Second Law of Thermodynamics, which certifies that the security of the ideal KLJN scheme is as strong as the impossibility to build a perpetual-motion machine of the second kind. Potential and unique technical applications of the KLJN scheme include non-counterfeitable hardware keys and credit cards via Physical Uncloneable Functions (PUFs) [35]; unconditionally secure hardware, computers and other instruments [35,36]; and unconditionally secure smart grids [37–39]. The short summary of the KLJN scheme given below is based on a previous survey paper [2].

1.1.1. *The idealized KLJN scheme and its security*

The working principle of the KLJN scheme [2,26] is presented in Fig. 1, which shows an idealized configuration without any defense circuitry — such as current-voltage measurement/comparison, filters, *etc* — against invasive and non-ideality attacks. At the beginning of each bit exchange period (BEP), Alice and Bob connect their randomly chosen resistors R_A and R_B, respectively, to the wire line. These resistors are randomly selected by the switches from the set $\{R_L, R_H\}$, $(R_L \neq R_H)$, where the elements represent the low L and high H bit values 0 and 1, respectively.

The Gaussian voltage noise generators — delivering white noise with publicly agreed bandwidth — represent an enhanced thermal (Johnson) noise at a publicly agreed high effective noise-temperature T_{eff} at which their noises are statistically independent from each other, implying that $\langle U_A(t)U_B(t)\rangle = 0$, as well as from the noise during a former BEP. During the first practical implementation of the KLJN scheme, by Mingesz *et al.* [29], the noise-temperature range $8 \times 10^8\text{K} \leq T_{eff} \leq 8 \times 10^{11}\text{K}$ was used, which made the wire temperature insignificant even when the wire resistance was not zero.

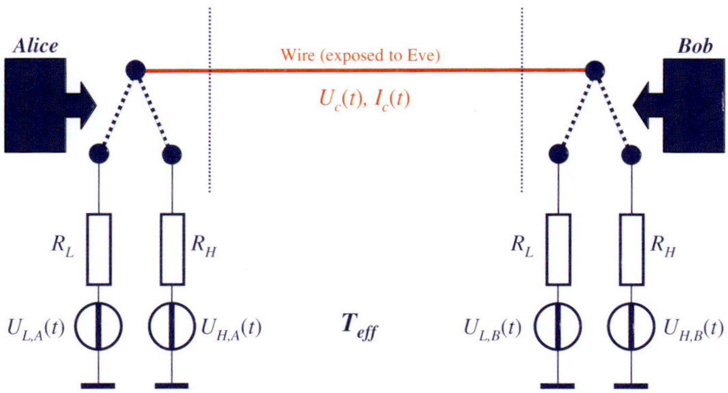

Figure 1. Outline of the core KLJN key exchange system. The communicator parties, Alice and Bob, randomly choose and connect either R_L or R_H to the wire. The (effective) temperature T_{eff} is publicly agreed and kept, and the (enhanced or standard) Johnson noises of the resistors $U_{L,A}(t)$, $U_{L,B}(t)$, $U_{H,A}(t)$, and $U_{H,B}(t)$ are independent and Gaussian. The resulting channel voltage $U_c(t)$ and current $I_c(t)$ are also uncorrelated due to the Second Law of Thermodynamics. Parasitic elements leading to non-ideal features and defense circuitry against active (invasive) attacks and against attacks utilizing non-ideal features are not shown.

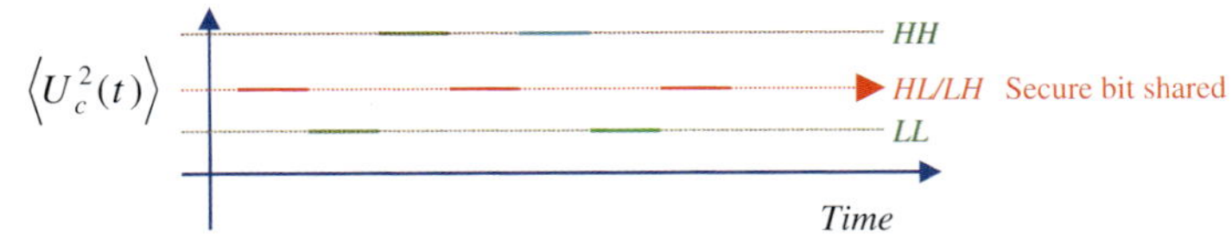

Figure 2. Mean-square voltage (and current) versus time during operation. There are three different levels (dotted lines) depending on the actual bit values; the intermediate value indicates a secure bit exchange period.

Alice and Bob (as well as Eve) can use a measurement of the mean-square voltage and/or current to assess the bit status of the system, as shown in Fig. 2 for the case of voltage. The situations *LH* and *HL* represent secure bit exchange [2,26], because Eve cannot distinguish between them through measurements, and whenever Alice and Bob see the HH/LH situation they know that the other party has the complementary bit value, which means that they infer the full bit arrangement. Eve cannot extract this information, because she does not know any of the bit values. In other words, a secure bit has been generated and shared. The bit situations *LL* and *HH* are insecure, which means that these bits (50% of the executed BEPs) are discarded by Alice and Bob.

According to the Fluctuation-Dissipation Theorem, the power density spectra $S_{u,L}(f)$ and $S_{u,H}(f)$ of the voltages $U_{L,A}(t)$ and $U_{L,B}(t)$, supplied by the voltage generators in R_L and R_H, are given by

$$S_{u,L}(f) = 4kT_{eff} R_L \quad \text{and} \quad S_{u,H}(f) = 4kT_{eff} R_H, \tag{1}$$

respectively. In the case of secure bit exchange (*i.e.*, the *LH* or *HL* situation), the power density spectrum $S(f)$ and the mean-square amplitude $\langle U_{ch}^2 \rangle$ of the channel voltage $U_{ch}(t)$, and the same measures of the channel current $I_{ch}(t)$, are given by

$$\langle U_{c,HL/LH}^2 \rangle = \Delta f \, S_{u,c,HL/LH}(f) = 4kT_{eff} \frac{R_L R_H}{R_L + R_H} \Delta f, \tag{2}$$

and

$$\langle I_{c,HL/LH}^2 \rangle = \Delta f \, S_{i,c,HL/LH}(t) = \frac{4kT_{eff}}{R_L + R_H} \Delta f, \tag{3}$$

respectively, where Δf is the noise bandwidth.

1.1.2. *The security of the KLJN scheme is based on the Second Law of Thermodynamics*

During the *LH* and *HL* cases, linear superposition makes the spectrum given by Eq. (2) represent the sum of the spectra at the two particular situations. Thus one obtains

$$S_{L,u,c}(f) = 4kT_{eff}R_L \left(\frac{R_H}{R_L + R_H}\right)^2 \tag{4}$$

when only the noise generator due to R_L is running and

$$S_{H,u,c}(f) = 4kT_{eff}R_H \left(\frac{R_L}{R_L + R_H}\right)^2 \tag{5}$$

when the only the noise generator due to R_H is running.

If Eve is to identify which end of the wire has R_L or R_H, it is necessary for her to measure and evaluate a physical quantity offering directional information. In the ideal case, the only information of this kind is the direction of the power flow from Alice to Bob (or *vice versa*, depending on the choice of positive current direction). In thermal equilibrium, however, this power must fulfill $P_{A\to B} = \langle U_c(t)\ I_c(t)\rangle = 0$, as required by the Second Law of Thermodynamics. In other words, the ultimate security of the KLJN system against passive attacks is provided by the fact that the power $P_{H\to L}$, by which the noise generator due to resistor R_H is heating resistor R_L, is equal to the power $P_{L\to H}$ by which the noise generator due to resistor R_L is heating resistor R_H [2,26,32]. Thus the fact that the net power flow is governed by $P_{A\to B} = P_{L\to H} - P_{H\to L} = 0$ can easily be shown from Eqs. (4) and (5) for the noise-bandwidth Δf by

$$P_{L\to H} = \frac{S_{L,u,c}(f)}{R_H} = 4kT_{eff}\frac{R_L R_H}{(R_L + R_H)^2}\Delta f \tag{6a}$$

and

$$P_{H\to L} = \frac{S_{H,u,c}(f)}{R_L} = 4kT_{eff}\frac{R_L R_H}{(R_L + R_H)^2}\Delta f \tag{6b}$$

The equality $P_{H\to L} = P_{L\to H}$ is in accordance with the Second Law of Thermodynamics. In other words it is as difficult to crack the ideal KLJN scheme as to build a perpetual motion machine of the second kind [4].

This security proof against passive (listening) attacks holds only for Gaussian noise — *i.e.*, the statistics of thermal noise — which has the well-known property that its power density spectrum or autocorrelation function already provides the maximum achievable information about the noise, and no higher-order distribution functions or other tools, such as higher-order statistics, are able to provide additional information.

The required duration τ of the BEP, at a given bit error probability [34] of the bit exchange between Alice and Bob, is determined by the following arguments: For the *LL* bit status of Alice and Bob, which is not a secure situation, the channel voltage and current satisfy

$$\begin{aligned} \left\langle U_{c,LL}^2 \right\rangle &= \Delta f \, S_{u,c,LL}(f) = 4kT_{eff} \frac{R_L}{2} \Delta f; \\ \left\langle I_{c,LL}^2 \right\rangle &= \Delta f \, S_{i,c,LL}(t) = \frac{2kT_{eff}}{R_L} \Delta f, \end{aligned} \tag{7}$$

while, in the case of the other non-secure situation namely the *HH* bit status, the channel voltage and current satisfy

$$\begin{aligned} \left\langle U_{c,HH}^2 \right\rangle &= \Delta f \, S_{u,c,HH}(f) = 4kT_{eff} \frac{R_H}{2} \Delta f, \\ \left\langle I_{c,HH}^2 \right\rangle &= \Delta f \, S_{i,c,HH}(t) = \frac{2kT_{eff}}{R_H} \Delta f. \end{aligned} \tag{8}$$

During key exchange in this classical way, Alice and Bob must compare the predictions of Eqs. (7) and (8) with the actually measured mean-square channel voltage and current to decide whether the situation is secure (*i.e.*, *LH* or *HL* prevails), while realizing that these mean-square values are different in each of these three situations (*LL*, *LH* or *HL*, and *HH*). If the situation is secure, Alice and Bob will know that the other party has the inverse of his/her bit, which implies that a secure key exchange takes place. Alice and Bob must use sufficiently large statistics to achieve low error probability. Fortunately, the bit error probability decays exponentially with the duration τ of the BEP [34]. Furthermore, a new "intelligent" KLJN protocol [31] can be used, which employs additional circuit calculations by Alice and Bob to reduce the BEP without increasing the error probability.

1.1.3. *On active (invasive) attacks and attacks utilizing non-idealities*

It has been pointed out repeatedly [2,26,28,29,32] that deviations from the earlier shown circuitry and Johnson-like noise — including invasive attacks by Eve, parasitic elements, delay effects, inaccuracies, non-Gaussianity of the noise, *etc* — will cause a potential information leak toward Eve. However it is fortunate that the KLJN system is very simple, which implies that the number of such attacks is strongly limited. The defense methods against the attacks are straight-forward and are generally based on the comparison of instantaneous voltage and current data at the two wire ends via an authenticated communication between Alice and Bob, as indicated in Fig. 3. These attacks [2, 40, 43, 45] are not the subject of the present paper, and we refer to our relevant rebuttals where they have been analyzed [2,32,41,42,44] and where misconceptions and errors have been pointed out and corrected. Our earlier survey paper [2] reviewed various attacks on the KLJN scheme.

It is important to emphasize that Alice and Bob know Eve's best measurement information, because it is given by comparisons of voltage

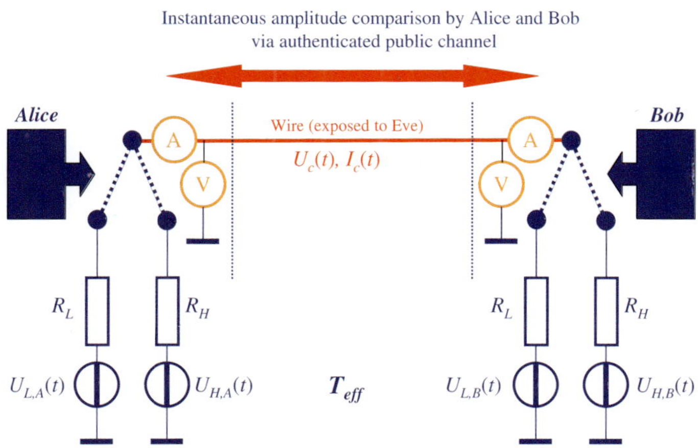

Figure 3. KLJN system minimally armed against invasive (active) attacks, including the man-in-the-middle-attack. Alice and Bob measure the instantaneous channel voltage and current amplitudes and compare them via an authenticated public channel. In this way, they learn all the information Eve can have. Additions to prevent hacking — such as line filters, blinding detectors, *etc* — are not shown. The notation is the same as in Fig. 1.

and current at the two ends of the wire. If Eve uses the best available protocol and the security of a certain bit is compromised, this is known also by Alice and Bob who therefore can decide to discard the bit in order to have a secure key. This is a new and unique situation in cryptography, which raises a number of research questions as mentioned in an earlier paper [32].

Finally, a secure type of privacy amplification [33] — XOR-ing the key bit pairs and producing a new key with this output, which results in half of the original length — is also feasible to enhance the security because of the low bit error probability of KLJN key exchange. The error probability decays exponentially with the increasing duration τ of the BEP [34]. At the experimental demonstration [29] the error probability was 2×10^{-4}.

1.1.4. *Foundations of the information-theoretic security in practical KLJN schemes*

Of course, perfect security of any physical key exchanger exists only under ideal (mathematical) conditions. For example, quantum encryption theoretically can offer perfect security only in the limit of a zero-photon-emission rate [5] (*i.e.*, zero bit exchange rate) and zero detector and channel noise limits, which are unphysical and can never occur in a real system. The KLJN scheme is no exception to this rule [2,26,32]: it offers perfect security only at zero bandwidth or distance as a consequence of transients, cable resistance, capacitance, *etc*. However, just as for claims in favor of QKD, parameters of the KLJN building elements and protocol can be chosen so that the perfect security limit can be approached asymptotically. The general situation in the non-ideal case is that a miniscule DC signal component buried in a much larger Gaussian noise (of fixed variance) must be detected by Eve from small statistics limited by the BEP. This DC signal component is typically the mean value of a finite-time mean-square operation or that of the output component of a cross-correlation operation; further discussion on this issue is given in an earlier paper [31] and in Sec. 2.8. Eve must detect the sign of this small DC component in the large noise. When the parameters approach the ideal situation, the ratio of the DC signal amplitude and the root-mean-square (RMS) amplitude of the noise converges towards zero as a power law decay — typically with exponent -1 or -2 [2,29,42,44] — with regard to the invested resources such as wire volume, current/voltage resolution, BEP duration τ, *etc.*

Here we summarize the foundation of the mathematical analyses below by explaining why and how Eve's information is limited compared to Alice and Bob's information, even though we have not imposed any limitation for Eve's measurement accuracy or her measurement speed anywhere in our analysis. This is a point of frequent misunderstanding and the additional explanation here was triggered by the referee report written by Dr. Bennett.

All our analyses and argumentations are based on a *technically-unlimited eavesdropper* (Eve). Accordingly, Eve's amplifiers and analyzers have infinite amplitude resolution and infinite measurement speed (infinite bandwidth), they are noise-free and absolutely linear. Otherwise, parameters describing characteristic time/frequency cut-off of Eve's measurement apparatus and of the amplitude resolution limit would enter into all of the equations describing Eve's information. Eve's only measurement-technology limitation is of fundamental physical nature: Eve can only very poorly separate the signals originating from the two directions as the result of the Rayleigh scattering at sizes much smaller than the signal wavelength; see Sections 2.1.4 and 2.3.2 below. This fundamental physical process is unavoidable, and Eve must live with the fact that the laws of physics prohibit using an efficient directional coupler. However, the same limitations concerning Rayleigh scattering also apply to Alice and Bob.

Then, the obvious question emerges: *Why does Eve have significantly less information than Alice and Bob*? One should note that this question is relevant only for passive attacks in practice (that is, non-ideal situations with finite distance, finite bandwidth, nonzero cable resistance and capacitance) and active attacks for arbitrary situations, because, for the ideal situation (with zero distance, finite bandwidth, zero cable resistance and capacitance) the Second Law guarantees perfect security. So, how can we summarize the main arguments behind our security proofs against the various attacks presented in this paper, and what is the fundamental mechanism behind the unconditional security shown by our mathematical security proofs?

Is it today's technological limitations of Eve's accuracy? No: it is a *fundamental limit*: the *statistical uncertainty principle* interrelating the finite measurement duration with statistical errors during noise analysis.

Eve can measure with infinite speed, which means infinite bandwidth, but *she will not find anything beyond the noise bandwidth* Δf set by the noise generators and line filters of Alice and Bob (except negligible spurious, stochastic frequency components decaying in an exponential or power-law fashion versus frequency beyond the band limit). According to Shannon's sampling theorem, the maximum sampling frequency giving statistically independent data is $2\Delta f$ thus during the duration τ of the BEP, Eve (just as Alice and Bob) can extract only $s = 2\Delta f\tau$ independent samples of the channel noise even if Eve's measurement apparatus can collect samples with infinite sampling frequency. The quantity s is determined by Alice and Bob because Δf and τ are determined by them and chosen so that the error probability of exchanged bits is miniscule, thus no error correction algorithm is needed. Yuen [5] points out that error correction algorithms provides information to Alice and Bob in QKD, but KLJN can avoid using such. In practical applications one has $s \approx 100$, which is an extraordinarily small number for Eve to safely distinguish the minor differences, 10^{-4} or less [41], in the noise statistics she can extract using non-idealities. For active attacks, Alice and Bob can easily enforce the same small differences in the statistics — for example 14 bits, used in the experimental demonstration, which is equivalent to a difference of less than 10^{-4}. It is virtually impossible to distinguish such a small difference between stochastic signals with the available number s of independent samples. As a result, the mathematical analyses shown in the subsequent sections indicate that Eve's successful guessing probability will be close to 0.5, — i.e., the limit representing zero information — and the accepted measure of security (statistical distance, see below) between Eve's extracted version of the key and a perfectly secure key is exponentially small.

But what about Alice and Bob? Alice and Bob know essential parameters that Eve does not have access to: they know their own resistance value and the exact amplitude of their noise fed into the line. Thus they do not need to utilize the non-ideality-based miniscule differences seen by Eve or the even smaller differences that Eve may generate by active (invasive) attack while staying hidden; they only need to monitor the channel voltage/current and identify which one of the three significantly different levels of the mean-square noise takes place. At $s \approx 100$, the achievable bit exchange has extremely small error probability, such as 10^{-12}, see [34]. Further improvements are offered by the "intelligent" KLJN method, where s can be significantly diminished by Alice and Bob via reducing τ when utilizing the knowledge of their own noise time function, combined with linear network calculations [31] allowed by the classical physical nature of the scheme.

1.1.5. *Mathematical proof of the unconditional security of the exchanged key*

In order to mathematically analyze the security of the shared key, one must compare the probability distribution for successfully guessing each possible key sequence of an N-bit-long key, encompassing 2^N different sequences, with that of the perfect key having uniform distribution. A statistical distance measure, the variational distance Δ [46] between the distributions representing the key guessed by Eve and the distribution representing the ideal (uniform) key is a useful concept. It defined by

$$\Delta(E, I) = \max_{j=1,\ldots,2^N} [P(E_j) - P(I_j)], \tag{9}$$

where E and I represent Eve's extracted key and the perfect key, respectively, and $P(E_j)$ and $P(I_j)$ are the probabilities of correctly guessing the jth version of Eve's key and of the perfect key, respectively. The key exchange has ε-security, as discussed by Hirota [6], if the statistical distance between the distributions representing the key guessed by Eve and the ideal (uniform) keys is less than ε, *i.e.*,

$$\Delta(E, I) \leq \varepsilon \tag{10}$$

for $\varepsilon \geq 0$.

The KLJN scheme provides identically and independently distributed sequences of random variables as key bit values, so that

$$\Delta(E, I) = \max_{j=1,\ldots,2^N} [P(E_j) - P(I_j)] = p^N - 0.5^N, \tag{11}$$

where p is Eve's probability of successfully guessing bits. In non-ideal cases involving an information leak, and when the parameters are sufficiently close to the ideal limit, p can be given as

$$p = 0.5 + q, \tag{12}$$

where $0 < q \ll 0.5$; here $q = 0$ would mean a perfectly secure key. The reason for this behavior is easy to see if one realizes that Eve's small DC signal component offsets the center (mean value) of the probability density burying the large Gaussian noise. The first derivative at the center of the Gaussian density function is zero, implying that its Taylor approximation, to first order, results in a stable value for small changes around the center. In the idealized case (*i.e.*, zero DC signal) Eve's estimation of the mean value of the Gaussian noise would yield $p = 0.5$ (recording a positive sign at 50% of the exchanged key bits and negative sign also at 50% of the cases). In the non-ideal situation, the DC signal and the mean value of noise + signal are positive or negative; hence the flat amplitude distribution within this range makes Eve experience a non-zero q (*cf.*, Eq. 12) which is proportional to the DC signal [47].

As an example, we now consider the case of a non-zero wire resistance [29, 40–42] and assume that capacitive effects are compensated [29] or can be neglected due to the actual bandwidth. More examples will be shown in Chapter 2. For the case of fixed distance and bandwidth, q is proportional to the inverse of the square of wire diameter, *i.e.*, with the inverse of the wire's volume V. In other words

$$q = \vartheta_w V^{-1}, \tag{13}$$

where ϑ_w is a constant valid for a wire-resistance attack. Then, for the case of $Nq \ll 0.5$, one obtains

$$\begin{aligned}\Delta &= (0.5 + q)^N - 0.5^N = 0.5^N[(1 + 2q)^N - 1] \\ &\cong 2Nq0.5^N = 2N\vartheta_w V^{-1}0.5^N\end{aligned} \tag{14}$$

where the last approximation is valid for $q \to 0$. Equation (14) indicates that Δ decays exponentially with increasing value of N and inversely with wire volume V.

For the case of ε-security with $\Delta \leq \varepsilon$, *i.e.*, in the $Nq \ll 0.5$ limit, the required q is given by

$$q(\varepsilon, N) = \frac{\vartheta}{V(\varepsilon, N)} \leq \frac{\varepsilon}{2N} 2^N. \tag{15}$$

During the experimental demonstrations of the KJNL scheme, referred to above [29], it was found that Eve's q equaled 0.025 for secure bit exchange during wire-resistance attacks with a wire resistance being 2% of the loop resistance. For a practical evaluation, let us suppose, that due to additional leaks (transients, cable capacitance, etc.) the actual q is double of that, $q = 0.05$. Due to the $0 < q \ll 0.5$ assumption, in order to use the theory described above, we apply the a privacy amplification described in [33], which keeps the independently and identically distributed nature of the key by XOR-ing pairs of bits in the original key to have a new key with enhanced security and half of the length. Then, repeating this process a second time to obtain the final key with 25% of the original length, the resulting effective q of Eve is $q = 5 \times 10^{-5}$ [33] which allows key lengths N up to the order of 10^4 bits. Equation (11) can then be evaluated with this effective q value. For a 1000-bit-long shared key, it results in $\Delta(E, I)_{1000} = 9.3 \times 10^{-303}$ (*i.e.*, an ε-security with $\varepsilon_{1000} \cong 10^{-302}$); for a 500-bit-long shared key it results in $\Delta(E, I)_{500} = 1.5 \times 10^{-152}$ (*i.e.*, an ε-security with $\varepsilon_{500} \cong 2 \times 10^{-152}$).

Finally, we observe that there are advanced protocols that can enhance the security or limit the required resources in efficient ways while the scaling of q versus the utilized resource (wire volume, *cf.* Eq. 13) shown above in Eqs. (14) and (15) does not change. Below we give a short list of advanced protocols and associated basic security features proposed up to now:

(*a*) Ideal KLJN schemes with passive attacks:

- The Second Law of Thermodynamics and Kirchhoff's Loop Law [2,26].

(*b*) Non-ideal KLJN schemes with passive or active (invasive) attacks:

- Transient protocols involving random-walk from equal resistances [31] and voltage ramping/timing [2,29].

- Selecting the noise bandwidth versus the value of wire resistance and wire capacitance [2,29].
- General defenses that work in any situation including hacking, encompassing comparison of instantaneous voltage and current amplitudes and discarding any bits where they differ or where they provide information to Eve (even it is erroneous). Note, however, that specific protocols apply for different hacking attacks.
- Privacy amplification (XOR-ing key bit pairs [33]).
- Enhanced KLJN protocols, for example the "intelligent" (iKLJN) and "keyed" (KKLJN) methods [31].

1.1.6. *Optional security addition: cap imposed on q by Alice and Bob*

We briefly mention an additional security tool provided by the classical physical nature of the KLJN system; see its detailed description and security analysis elsewhere [48]. The fact that Alice and Bob have a public authenticated communication channel for comparing their instantaneous current and voltage data (together with the channel parameters known by them) allows them to access the measurement information of Eve. Thus they can discard those exchanged bits that give out too much information to Eve. In another word, Alice and Bob can impose a strict upper limit on q in Eqs. (12) and (15). This additional security tool is useful when the available resources of Alice and Bob are insufficient to diminish the q related to the exchanged raw bits and, for some other reasons, they want to avoid privacy amplification to reduce it. This ability of Alice and Bob is another indication that they are in full control of the maximum of statistical information Eve is able to access.

1.2. *Summary of Bennett–Riedel's arguments regarding the KLJN scheme*

BR have presented an extensive analysis [4] that is fundamentally flawed but nevertheless very useful for the purpose of elucidating differences between simplistic or irrelevant model approaches and the physics upon which the KLJN scheme is founded.

An outline of BR's claims reads as follows: It is first stated that the no-wave limit (*i.e.*, quasi-static electrodynamics) is unphysical for signal propagation. Based on this statement, they assert that Eve can separate and measure the "orthogonal" wave components propagating from Alice to Bob and *vice versa*. They also state that the KLJN scheme is deterministic, which means that Eve has a full description of the whole system, including Alice's and Bob's history, if Eve's measurements of the two wave components are limited by nothing but the laws of physics. To support these claims, BR expound that thermodynamics and noise are not essential in the KLJN scheme and that thermodynamics would eradicate determinism as a consequence of fluctuations. Further corroboration of their view is obtained from the construction of a deterministic and thermodynamics-free key exchanger, which looks similar to a KLJN scheme without resistor, and where two of the four noise voltage generators are removed and the remaining ones replaced by batteries with known and identical voltage. Moreover, BR propose a passive correlation-measurement-based attack and an active current-extraction attack against the KLJN scheme.

After briefly describing BR's claims, we critically analyze and refute all of them in Chapter 2, and we also present the physics appropriate for the KLJN scheme.

1.2.1. *Bennett–Riedel's claim concerning no information transfer in a wire in the no-wave (quasi-static) limit*

BR write [4]: "*We believe this no-wave limit is inappropriate and non-physical for analyzing communication protocols (even as a mathematical idealization) because if propagating waves are excluded there is no way for information to get from Alice's side of the circuit to influence Bob's side, or vice versa.*"

Based on this argument, they assert that Eve can separate and measure the "orthogonal" wave components that propagate from Alice to Bob and *vice versa*.

After surveying the relevant physics facts about waves, directional couplers for signal separation and the no-wave limit in Sec. 2.1, we refute the above argument in Sec. 2.2. Furthermore, we show what physics has to say about signal propagation in the no-wave (quasi-static) limit.

1.2.2. *Bennett–Riedel's claim that the KLJN system does not offer security*

BR [4] set up three equations for the KLJN scheme, which invoke the deterministic nature of Maxwell's equations and neglect the stochastic nature of Johnson noise and the secret/random choice of the resistors. With this premise, it is not surprising that they concluded that KLJN does not offer any security. Here we discuss only the first and third of BR's equations — since the second one is redundant — and their main conclusion.

The conditional information $H(F|G)$ represents the remaining uncertainty about the set of data F when the set of data G is known. Now $H(F|G) = 0$ if G completely determines F, whereas $H(F|G) = H(F)$ for the case when G does not provide any information about F. BR's first equation is

$$H(X|Z_A) = H(X|Z_A, Z_B) = H(X|Z, Y), \tag{16}$$

where X is a variable that fully describes the physical quantities on Alice's side of Eve's location during the BEP. These quantities include waves traveling toward Alice and away from her and all of her equipment, as well as noise and memory. The variable Y has the same meaning with regard to Bob. Furthermore, Z_A and Z_B are wave components propagating from Alice and Bob (as observed by Eve), respectively, and $\mathrm{Z} = (Z_A, Z_B)$ represents both wave components. We note that in BR's paper [4] either Z is incorrectly indexed or X and Y must be exchanged.

We first presume that Eq. (16) is valid, which assumes that Z_A and Z_B can be measured separately. This means that the uncertainty about Alice's "full description" X does not change if Eve expands her knowledge of wave Z_A coming from Alice by the knowledge of wave Z_B coming from Bob, and the same remains true even if knowledge of the total description of Bob's data Y is included.

It should be observed that the first equality in Eq. (16) contradicts BR's proposed passive correlation attack [4], which requires knowledge of both Z_A and Z_B and thus implies that $H(X|Z_A) > H(X|Z_A, Z_B)$.

We now introduce the mutual information $I(X; Y)$ of X and Y, which measures how much the knowledge of X or Y tells about the other variable. As a consequence of Eq. (16), and with further argumentation, BR deduce the following equation for the conditional mutual information between X and Y, conditional on Z:

$$I(X;Y|Z) = H(X|Z) - H(X|Z,Y) = 0. \tag{17}$$

This equation, if it is valid, would mean that after measuring the two waves $Z = (Z_A, Z_B)$, Eve's information about X (*i.e.*, Alice's full description) is not increased by learning Y (Bob's full description). Thus after measuring the two waves $Z = (Z_A, Z_B)$, Bob's information about Alice would not be larger than Eve's information about her. The same argumentation would work also in the opposite direction, so that the KLJN system would not offer any security.

We will see below that BR's equations are invalid even in the wave limit; this is a result of multiple reflections as well as of Alice's and Bob's secure reflection coefficients and noises (known only by them) that always guarantee that they know more than Eve.

Most importantly, Eqs. (16) and (17) are entirely unfounded in the no-wave limit because the propagating relaxations Z_A and Z_B (which are not waves) cannot be measured separately; only their sum can be determined.

1.2.3. *Bennett–Riedel's claim regarding a "thermodynamics-free" key exchange scheme*

One of the major claims of BR [4] is that thermodynamics and noise are not essential for security in the KLJN scheme. To prove this, they attempted to construct a deterministic key exchange method with two voltage generators and two switches, as illustrated in Fig. 4. This scheme is in fact already known; it is called the "Orlando system" and was conceived and patented by Davide Antilli in 2005 [49]. Despite its origin, we refer to it as the "BR system" below.

In the idle mode between bit exchange periods, the switches are in position I; thus the wire channel is grounded. At the beginning of the BEP, Alice and Bob randomly choose between the switch positions L or H representing the corresponding bit values, and at the middle of the BEP they change their bit value. If the randomly chosen sequences of bit values

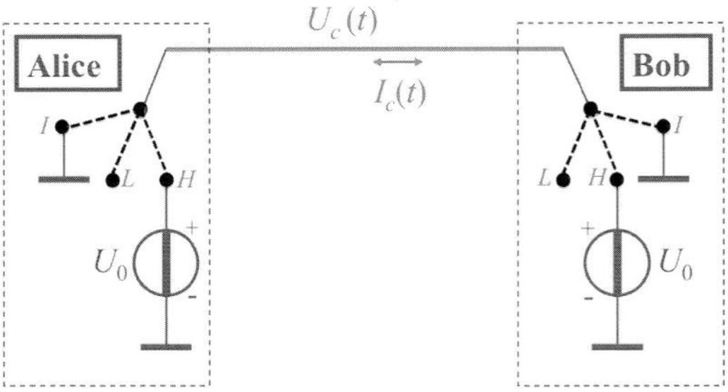

Figure 4. Outline of the (Antilli–)Bennett–Riedel system. The two ends of the wire channel are connected to a three-stage switch with positions I (idle between two bit exchanges), L (low bit value) and H (high bit value). The two DC voltage generators have the voltage U_0, and $U_c(t)$ and $I_c(t)$ are the voltage and current time-functions in the wire. Note that this figure is an improved version of BR's system because they mention the idle situation (necessarily grounded wire) only in the text but do not show it in their figure.

happen to be identical, then the voltage on the wire will be zero for half of the BEP, and these events are disregarded. If the choices by Alice and Bob are complementary, then the voltage is U_0 for the whole BEP.

BR make three statements about the system in Fig. 4, which will be important later: They assert that (*i*) "*The wires and voltage sources are taken to be ideal, with zero thermal noise*" and, as a corollary, that (*ii*) "*Thermodynamics and noise do not play a role.*" Furthermore, they claim (*iii*) that *the BR system is secure in the "no-wave" limit* accomplished in a special way: that Eve waits with her measurements until transients have decayed.

We will see below that statements (*i*) to (*iii*) lead to an unphysical situation, namely that Eve must wait for infinite time before she may start listening. Furthermore, one should note that (*iii*) is an illegal assumption in unconditionally secure communications, because then Eve can only be limited by the laws of physics. Thus statement (*iii*), in itself, would imply only conditional security.

In Secs. 2.1 and 2.2 below we show why BR's scheme is unphysical, and we also crack it fully in a variety of ways while we demonstrate that the KLJN scheme stays unbroken as a consequence of the Second Law of Thermodynamics and of noise.

1.2.4. *Bennett–Riedel's wave-transient-based attack before the steady state is reached*

BR write [4]: *Thus, while the steady state mean square noise voltage in the original KLJN protocol does not allow Eve to distinguish between the LH and HL settings of Alice's and Bob's resistors she can distinguish them using (a) transient waves created by the switching action before the steady state is established.*

For example Bob's resistor affects the phase and amplitude correlations between a right-traveling wave at time t and its left-traveling echo at time $t + \Delta$, where Δ is the transit time from Eve to Bob and back, with the echo vanishing only if the resistor is perfectly impedance matched to his end of the line.

Here it should be noted that BR have not put forward any concrete protocol with a quantitative and testable evaluation scheme. This is unfortunate because, by establishing such a protocol, one can see that, in the no-wave limit, such transients would represent minor information for Eve about Alice's and Bob's status. Even if propagating signal components (not waves) could be measured, the limited information about the noise within a small fraction of its correlation time (and the unknown additive noise and reflection at the other end of the wire) would make the information available to Eve very small. Moreover, even this minuscule information would converge towards zero upon a decrease of the noise bandwidth and/or a reduction of the wire length. The statistical distance between the key guessed by Eve and that of the perfectly secure key (of the same length) will vanish in a fashion similar to the one described by Eqs. (14) and (15). Sections 2.6 and 2.7 below discuss an efficient transient protocol and quantitative analysis.

1.2.5. *Bennett–Riedel's passive time-correlation attack in the no-wave limit*

BR write [4]: *Thus, while the steady state mean square noise voltage in the original KLJN protocol does not allow Eve to distinguish between the LH and HL settings of Alice's and Bob's resistors, she can distinguish them* *using (b) time correlations in the steady-state distribution of traveling waves resulting from the fluctuations that give rise to Johnson-Nyquist noise. For example Bob's resistor affects the phase and amplitude correlations between a right-traveling wave at time t and its left-traveling echo at time $t + \Delta$, where Δ is the transit time from Eve to Bob and back, with the echo vanishing only if the resistor is perfectly impedance matched to his end of the line.*

We will analyze this problem in Secs. 2.6 and 2.7 and give a security proof showing that the statistical distance between the key guessed by Eve and the perfectly secure key will vanish in an exponential fashion versus the length of the key.

1.2.6. *Current extraction/injection based active (invasive) attack*

BR write [4]: ...*she* (Eve) *could still learn the key by an active steady-state attack in which she would place a very high-resistance shunt between her node and ground, and monitor the direction of current flow into it. Of course Alice and Bob could try to detect this weak leakage current also, and abort the protocol if they found it. The result would be an unstable arms race, won by whichever side had the more sensitive ammeter, not the sort of robustness reasonably expected of a practical cryptosystem.*

We observe that this attack is valid only against the BR system because, in the KLJN scheme, the direction of the current flowing into the shunt resistor does not provide any information since its origin is a Gaussian noise process with zero mean and exhibiting perfect symmetry around zero. What BR might want to say for the KLJN scheme is that, by using the shunt resistor in the middle, the change of the RMS current in the wire will be greater in the direction of the lower resistance than in the directions of the higher resistance.

A very small difference in current, such as the one referred to above, results in an extremely poor statistics for Eve, and therefore one of the present authors (LK) has proposed a more efficient attack of the mentioned type in the original paper describing the KLJN scheme [26]: this attack entails a separate noise current generator instead of a shunt resistance as well as an evaluation of the cross-correlations between the injected current and the channel currents at the two sides of the injection. These cross-correlations determine which end of the wire has the low and which one has the high resistance. An attack of this type was disregarded as being inefficient already in the foundation paper for KLJN [26], because Eve would need a very long time to create sufficient statistics to reach a reasonable decision, whereas she only has the short duration of the BEP before the process ends. In Sec. 2.8, we analyze this attack mathematically and give a security proof against it.

Discussion and Results

The flow of analysis and argumentation in this chapter is as follows: First, in Sec. 2.1, we survey well-know facts about the physics related to the no-wave (quasi-static) limit of electrodynamics as well as facts about information transfer in that limit. Then, in Sec. 2.2, we refute BR's claim that there is no information transfer in the quasi-static (no-wave) limit. In Sec. 2.3 we then analyze BR's equations (Eqs. 16 and 17 above) indicating zero security and show that they are invalid for the KLJN scheme not only in the no-wave limit but also in the wave limit, whereas they are indeed valid for BR's thermodynamics-free system. In Sec. 2.4 we demonstrate that BR's thermodynamic-free key exchanger is unphysical because transients will oscillate for infinite time in the wire. Subsequently, in Sec. 2.5, we analyze the real, physical BR system and present ten different ways to fully crack it. We also show there that none of these ways of cracking work against the KLJN scheme, which proves that thermodynamics is essential for the security of KLJN. In Sec. 2.6 we argue that BR are incorrect when they write that the wave-transient attack would crack the KLJN system, and we also find that the statistical distance between the key guessed by Eve and the ideal key exponentially converges zero versus the length of the key. In Secs. 2.7 and 2.8, we demonstrate why BR's passive-correlation attack does not work in the KLJN system and why BR's current-extraction attack fails to change the exponential convergence of the statistical distance to zero. Finally, Sec. 2.9 contains some general remarks about protection against hacking.

2.1. *Physics facts: Information, propagation, and wave couplers in the quasi-static limit*

In Secs. 2.1.1 to 2.1.4 we clarify what is meant by a wave in physics: what are the conditions for the existence of a wave, and what is quasi-static electrodynamics [50] represented by circuit symbols? We also discuss whether electronic circuits are able to transfer signals and information in the quasi-static (no-wave) limit, and we treat the nature of delayed signal propagation in the no-wave limit as well as the inefficiency to separate propagation directions with directional couplers [51]. Here we emphasize it again: the issue is fundamental limitation due to the laws of classical physics.

2.1.1. *The mathematical definition of a wave in physics*

In physics, a wave is defined as a propagating amplitude disturbance $U(x, t)$ that is the solution of the wave equation

$$c^2 \frac{\partial^2 U(x,t)}{\partial x^2} = \frac{\partial^2 U(x,t)}{\partial t^2}, \tag{18}$$

where c is the phase velocity, *i.e.*, the propagation velocity when no dispersion is present. The dynamics of waves is governed by the oscillation of energy between two types, such as the electrical and magnetic field energies. If only one of these types of energy takes part in the propagation — or if the propagation is not based on the bouncing of energy between these two fields — then the propagating field disturbance is not a wave but merely a near-field oscillation with retardation effects.

We now consider a wire with finite size L. The wave equation in Eq. (18) has solutions only for frequencies

$$f \geq f_m = \frac{c}{2L}. \tag{19}$$

In other words, propagating field disturbances with frequency components below the minimum wave frequency f_m do not satisfy the wave equation, and hence they are not waves. We concur with BR that propagation and corresponding time delays (*i.e.*, retardation) are essential notions, but the propagating entities are not waves but field relaxations, and the consequences of this will be outlined below. Thus BR's statements about propagating "orthogonal" wave components that can be separated in the two directions is simply unphysical and leads to incorrect equations and conclusions. Furthermore, when the KLJN scheme operates in the "no-wave limit", this means that the condition

$$f \ll f_m = \frac{c}{2L} \tag{20}$$

applies [2,26], and BR are correct in using the term "quasi-static" to describe this situation. However, in the limit of quasi-static electrodynamics [50] it is incorrect to classify the propagating disturbances as waves; these disturbances are neither the solution of the wave equation nor do their electrical and magnetic fields have wave energy bouncing back and forth between them during propagation.

2.1.2. *The quasi-static limit of electrodynamics, and electrical circuitry symbols with lumped elements*

Quasi-static electrodynamics [50] and Eq. (20) constitute the bases for the operation and associated circuit drawings of any electrical circuit with lumped elements. The physical implication is that — along a line in a circuit drawing and the corresponding wire in the realized circuit, and at a given moment — the instantaneous current and the voltage amplitudes are virtually homogeneous, and retardation effects (including waves) can be neglected. In the absence of these implications, everyday electrical engineering design of circuits with lumped elements would be invalid and impossible.

2.1.3. *Signal propagation in the no-wave (quasi-static) limit*

After the comments above it is obvious that BR's assertion, that without waves in the wire there is no information transfer, is not only unphysical but also in blatant contradiction with everyday experience. No landline phones, no computers or other electrical circuits with lumped elements would be able to function and process information if BR's claim were true! In conclusion, the quasi-static (no-wave) limit [50] is a physically valid working condition for the KLJN system, and it is not unphysical as BR claim.

2.1.4. *Further implications of the quasi-static (no-wave) limit: Directional couplers, etc.*

We now consider *wave-based* directional couplers for extracting and separating signal components in two directions. These couplers simply do not work in the quasi-static limit, and even in the wave limit the cancellation of the irrelevant signal component is strongly frequency dependent because it is determined by the successful destructive interference of wave components in the coupler [51]. Couplers with good directivity are of the size $\lambda_0/4$, where $\lambda_0 = c/f_0$ and f_0 is the frequency for optimal operation. For longer wavelengths (*i.e.*, smaller frequencies), the system is subject to Rayleigh scattering and, accordingly, the separation of intensities decays with a power function scaling according to f^4.

There are also *non-wave-based* directional couplers, which are able to separate signals coming from two directions in the wire. These couplers work with lumped elements, such as transformers or active devices, and can be efficient in a wide frequency range. Their working principle is to cancel the signal of the irrelevant direction by subtracting from the channel voltage another voltage that is induced by the channel current. However, all of these couplers fail with the KLJN key exchanger because, for a proper operation to reveal Alice's voltage spectrum, the designer must know the exact value of her resistor. If instead Bob's resistor value is used, then the resulting signal voltage will be different and signal's spectrum will match Bob's noise spectrum instead. This fact is again a consequence of the Second Law of Thermodynamics, which guarantees that the cross-correlation of the channel voltage and channel current is zero, which leads to statistically independent channel voltage and current as a result of their Gaussian nature. Similarly, measuring the channel voltage $U_c(t)$ and current $I_c(t)$ and creating $U_L^*(t) = U_c(t) \pm I_c(t)R_L$ and $U_H^*(t) = U_c(t) \pm I_c(t)R_H$ would not offer information as a consequence of the independence and the Gaussianity of $U_c(t)$ and $I_c(t)$. According to basic noise calculus, the spectrum of $U_L^*(t)$ and $U_H^*(t)$ would be $4kT_{eff}R_L$ and $4kT_{eff}R_H$, respectively, independently of the sign for the second terms in these sums. In conclusion, non-wave-based directional couplers do not provide useful knowledge for Eve.[3]

2.2. *Refutation of Bennett–Riedel's claim about no information transfer in the no-wave limit*

As already shown in Sec. 2.1.3, there is indeed information transfer in the no-wave limit, and this fact is supported by common experience; *cf.* Eqs. (18) to (20). Therefore, the quasi-static limit is physical in an information processing system.

2.3. *Invalidity of Bennett–Riedel's equations, and the correct equations*

Below, we show that BR's equations are invalid for the KLJN scheme, in the wave limit as well as in the no-wave (quasi-static) limit.

2.3.1. *The wave limit and the Pao-Lo Liu key exchange system*

It is important to note that the default operation of BR's system (*cf.*, Fig. 4) is within the wave limit, which is a consequence of the abrupt switching of the voltage (*cf.*, Sec. 2.4) and the generated high-frequency products. Moreover, in the BR system, no noise unknown by Eve is fed by Alice and Bob into the system. Thus Eqs. (16) and (17) are indeed valid for the BR system (but not for KLJN). As a consequence, the BR system does not offer any security for Alice and Bob, as further discussed in Sec. 2.5.

The wave limit represents an illegal operational condition for the KLJN scheme, and therefore it is unimportant. However there is a software-based protocol working in the wave limit, known as the Pao-Lo Liu key exchange system [52–54], which was inspired by KLJN but does not utilize the Second Law of Thermodynamics. In the Liu protocol, random number samples of infinitesimally low noises (in the ideal situation) at Alice's and Bob's sites

[3]However, Pao-Lo Liu's analysis shows that, in the steady-state regime, even a directional coupler cannot extract the key bit. See: P.-L. Liu, "A key agreement protocol using band-limited random signals and feedback", *IEEE J. Lightwave Tech.* **27** (2009) 5230–5234.

are sent and reflected with random sign of the reflection coefficient. Alice's reflection coefficient, and the noise intensity added by her, is chosen so that, in the steady-state mode of ideal conditions, BR's proposed correlation attack [4] between the incoming and outgoing waves does not yield any information for Eve. The relevant relation for the Liu protocol, in the ideal situation, is

$$H(X|Z_A) = H(X|Z) = H(X) > 0 \tag{21}$$

instead of the zero-security situation, $H(X|Z_A) = H(X|Z) = 0$, implied by BR's considerations and Eq. (17) [4]. Furthermore and surprisingly, Liu's system seems to satisfy

$$I(X; Y|Z) = H(X|Z) - H(X|Z, Y) > 0 \tag{22}$$

in steady-state and at the ideal limit. Liu's system has other weaknesses, though, stemming from the wave limit, *viz.*, the distinct possibility to observe Z_A and Z_B; these flaws lead to problems with transients [53] and vulnerability to non-ideal filters [54]. Neither is Liu's system protected by the Second Law of Thermodynamics or other laws of physics.

Finally, returning to the KLJN scheme but lingering in the wave limit, we have the following comments: If only the waves coming from Alice's direction and denoted Z_A are known, this particular situation provides less information about Alice's total description than the situation when the waves Z_B coming from the direction of Bob are known as well. This is so because Z_A alone offers limited information about the reflection coefficient, and the resistance determining it, at Alice's side. On the other hand, in accordance with BR's passive correlation attack discussed in Sec. 1.2.5 (and also in Sec. 2.7 for the no-wave limit, where it does not work), the cross-correlation of Z_A and Z_B (requiring the wave limit) provides more information about the reflection coefficient at Alice than Z_A does, and thus $H(X|Z_A) > H(X|Z)$. We note, in passing, that BR's attack and its justification contradict their own equation, given in Eq. (16), which claims that adding Z_B to the knowledge of Z_A does not help Eve. The duration of the BEP is limited in the KLJN protocol, and thus the relation $H(X|Z_A) > H(X|Z) > 0$ applies in Eq. (21).

It should be observed that Liu's system [52–54], described above, is slightly different from the KLJN system in the wave limit (which is illegal for KLJN) because, in the Liu protocol, the added and reflected noises are combined at the two ends in such a way that, in the ideal case, the cross-correlation does not yield any information from Eve. Thus Liu's system implies that, in general, the correct relation for the wave limit is $H(X|Z_A) \geq H(X|Z) > 0$.

2.3.2. *Bennett–Riedel's equations for the KLJN scheme in the no-wave (quasi-static) limit*

BR's relations in Eqs. (16) and (17) do not exist for the KLJN system in the quasi-static limit because Z_A and Z_B are not observable separately [51]. Directional couplers that are able to separate such waves would produce outputs corresponding to

$$Z'_A = Z_A + (1 - \kappa) Z_B \tag{23}$$

and

$$Z'_B = Z_B + (1 - \kappa) Z_A \tag{24}$$

with $\kappa \propto 1/f^2$. The largest separation would be at the high cut-off frequency B_{kljn} of the noise bandwidth. As already pointed out in Sec. 2.1.4, this will lead to unconditional ε-security ($\varepsilon \propto B_{kljn}^4$), *i.e.*, results of the same nature as in Eqs. (14) and (15). The resources invested by Alice and Bob are the duration τ of the BEP and the length of the key ($\varepsilon \propto 2^{-N} \tau^{-4}$).

Finally, we set up the correct relations replacing Eqs. (16) and (17): The conditional information terms for the KLJN scheme satisfy

$$H(X) > H(X|U_c, I_c) > H(X|U_c, I_c, Z_A^*) \gg H\big(X|U_c, I_c, Z_A^*, Y\big) > 0, \tag{25}$$

where $U_c(x, t)$, $I_c(x, t)$ are current and voltage amplitudes along the wire in the steady state, where the dependence on x is miniscule and approaches zero for $B_{kljn} \rightarrow 0$, and $Z_A^*(x, t)$ is the initial transient disturbance (not wave) running from Alice toward Bob until Bob's end is reached and Bob's unknown noise is mixed into it. The last conditional information term expresses the fact that Bob, by knowing his own total description, is able to make an almost perfect guess of Alice's description X [31]. However this term is still larger than zero, because errors remain even in this case [31, 34], implying that a small uncertainty is left. Correspondingly, instead of Eq. (17), the correct relations for the conditional mutual information satisfy

$$I\big(X;Y|U_c,I_c,Z_A^*\big)=H\big(X|U_c,I_c,Z_A^*\big)-H\big(X|U_c,I_c,Z_A^*,Y\big)\gg 0. \tag{26}$$

2.4. *Proof that Bennett–Riedel's key exchanger is unphysical*

It is easy to see that BR's key exchanger is unphysical in its present form (*cf.*, Fig. 4). To this end, let us consider how long Eve has to "graciously wait" for the termination of the switching transients before she can measure. This time, in fact, is infinite because the transient will bounce back from the two endpoints of the wire, with the same sign from the open end and with altered sign from the endpoint terminated by the battery.

The observations above serve as a clear proof that, in the absence thermodynamics and the loss/energy dissipation it implies, even BR's key exchanger cannot function, and this holds true also if we permit violations of the basic rules of security — *viz.*, that Eve is allowed to measure whenever she can and wants — and instead force Eve to wait until the transients decay, which takes infinitely long time.

In conclusion, BR's scheme is unphysical, and one must realize that there are losses in "real" physical systems and that the related energy dissipation is controlled by thermodynamics.

2.5. *Ten ways to crack Bennett–Riedel's key exchanger by passive attacks*

Below, we show ten ways to crack BR's thermodynamics-free system with 100% success rate, and we furthermore point out that the same cracking methods do not work with the KLJN scheme, which is a consequence of thermodynamics and noise.

2.5.1. *Six universal energy/current-flow-analysis attacks*

To circumvent the problem of waves, Alice and Bob use proper voltage envelopes to prevent wave-modes (high-frequency components belonging to the wave limit). In practice, the wave modes should be kept negligible. Another alternative is that Alice and Bob use filters. One should note that convergence requires some loss, which is unavoidable for any real physical system.

Six universal energy/current-flow-analysis attacks are based on the fact that any wire has a geometrical capacitance, and to charge the wire one needs a current flow, energy flow and power flow. Measurement of voltage and current, and determination their product, gives the power flow and its direction as shown in Fig. 5. This power flow is the quasi-static analogue of the Poynting vector in electromagnetics.

The power flow vector is given by

$$\vec{P}(t) = U_c \vec{I}_c(t), \tag{27}$$

and the energy flow vector is its integral over the BEP according to

$$\vec{E} = \int_0^{\tau} \vec{P}(t)dt. \tag{28}$$

Eve's situation is fully characterized by the direction of the current vector $\vec{I}_c(t)$, the mean power flow vector $\langle \vec{P}(t) \rangle$, and the energy flow vector $\vec{E}$. The magnitudes of the $\langle \vec{P}(x,t) \rangle$, $\vec{E}(x)$ and $\langle \vec{I}(x,t) \rangle$ vectors with regard to

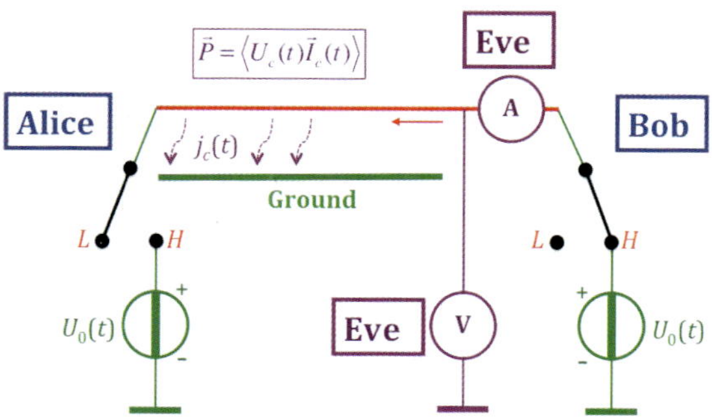

Figure 5. Universal energy-flow-analysis attack against BR's scheme in the no-wave limit. The no-wave limit is provided by the specific time-function of the voltage $U_0(t)$. The capacitive current density $j_c(t)$ toward the ground is spatially homogeneous along the wire, which leads to a maximum channel current amplitude $\vec{I}_c(t)$ power flow vector and energy flow vector at the closed end, and zero at the open end. The direction of these vectors during the charge-up period is pointing toward the open end.

location also fully inform Eve and compromise security. The further away from the connected voltage source these location-dependent quantities are evaluated, the less are their values, and they are zero at the open end. The directions of these vectors during the charge-up period are toward the open end.

In conclusion, the direction and the location-dependence of the three measurable quantities offer six ways to fully crack the key in the BR system.

2.5.2. *Three transient-damping resistor attacks*

To make the system physical and stop the transient after one return, Alice and Bob may use damping resistors to match the wave resistance of the wire, as shown in Fig. 6. These resistors will cause a continuous noise current flowing into the geometrical capacitance of the wire. There are then three more ways to utilize thermodynamics to crack this system during the steady state.

The noise current is correlated with the time derivative of the channel voltage and can be written

$$\vec{Y} = \left\langle \frac{dU_c(t)}{dt} \vec{I}_c(t) \right\rangle . \tag{29}$$

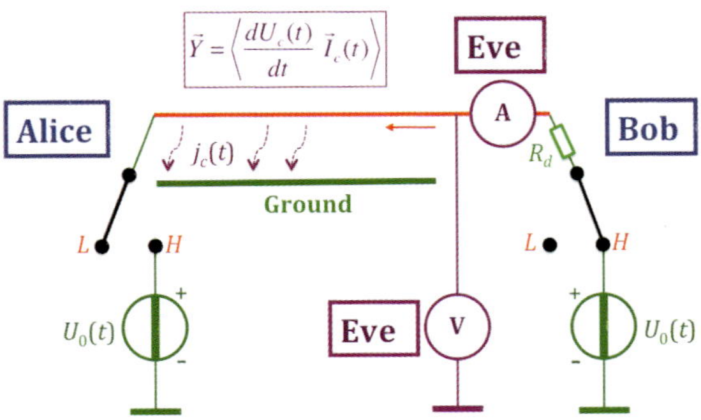

Figure 6. Transient-damping-resistor version of BR's scheme, and capacitive noise current attack. The direction and the location-dependent value of the cross-correlation vector $\vec{Y}$ for the time-derivative of the channel voltage and the current vector provide two ways to crack the key, while the location-dependence of the RMS channel current offers a third way.

Both the sign of the cross-correlation vector $\vec{Y}$ and its value with regard to location fully inform Eve about the situation; their absolute values are zero at the free end of the wire and maximum at the closed end.

A third way to crack the key is given by the location-dependence of the RMS channel current, which is zero at the open end and maximum at the closed end.

2.5.3. *Wire-resistance Johnson-noise attack*

Any wire will have non-zero resistance, and thus it produces Johnson noise. Eve can simply measure the voltage noise between the wire and the ground at the two ends of the wire, as indicated in Fig. 7.

The free end of the wire will have a voltage noise spectrum given by

$$S_u = 4kTR_w, \tag{30}$$

while the connected end shows zero noise. Consequently Eve can fully crack the system. One may note that this attack can be avoided if the connected end of the wire has a large additive noise to conceal the noise given by Eq. (30), but then the former attacks utilizing the current, power flow and energy flow vectors will still crack the system even in the steady state.

2.5.4. *The above attacks are inefficient against the KLJN system as a result of thermodynamics*

It is easy to understand how thermodynamics and noise, fed by the two communicating parties, protect the KLJN scheme against the above attacks.

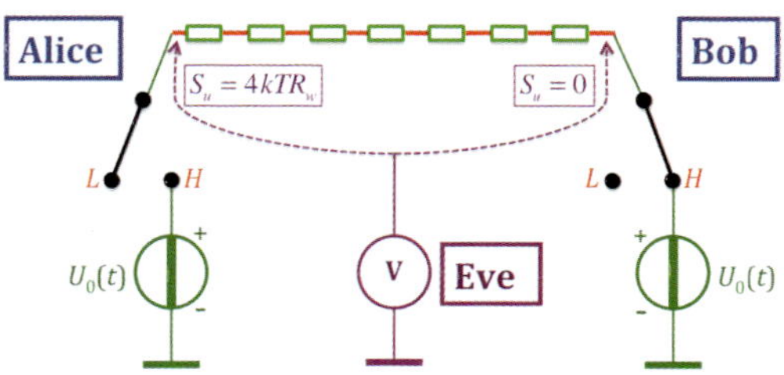

Figure 7. BR's scheme with finite wire resistance and Johnson noise attack. With a wire resistance of R_w and in the steady-state mode, Eve will measure a zero-power density spectrum S_u at the closed end of the wire and $S_u = 4kTR_w$ at the open end.

The resistors used by Alice and Bob make the system thermodynamic and produce Johnson noise. The noise voltages are much larger than the parasitic Johnson noise of the wire, because the wire resistance must be small (maximum 1 to 2% of $R_L + R_H$). Similarly the noise bandwidth is chosen so that the capacitive currents are negligible compared to the channel current.

The implication of the considerations above is that, when the above described attacks are used against the KLJN scheme, Eve's measurement will be a small DC signal buried in a large noise. This leads to relations similar to those shown for the wire resistance voltage drop in Eqs. (12) to (15), and the information theoretic security will be almost perfect. In the case of analogous attacks against the BR case, on the other hand, there is no other type of noise to bury Eve's signal. The rectified noise voltages and currents, and the cross-correlation results, are all uni-polar noises for which either the polarity of this quantity provides the result or, when its size matters, the size compared to zero. For example, the Johnson noise of the wire should be evaluated only at its two ends and should be zero at one end and non-zero at the other. Neither statistics nor averaging is needed to crack the BR system with these attacks; the result is virtually instantaneous and within the correlation time of the noise.

2.6. *On transient attacks against the KLJN scheme*

This attack is different from other attacks in the literature and in this paper in the sense that, in the no-wave (quasi-static) limit where KLJN operates, no concrete realization has ever been proposed with a measurement and evaluation protocol. Therefore, at the moment, this attack is only hypothetical but is brought up here for the sake of completeness and debate.

Researchers working with the KLJN scheme have realized from the very beginning that transients pose vulnerabilities, and various schemes have been proposed to reduce the potential information leak; they include ramping up/down of the noise, starting from zero noise amplitude (and velocity), and adiabatic random walking of Alice's and Bob's resistance [31]. The efficiency of any attack is strongly limited; firstly, this is due to the quasi-static condition that is manifested by the noise voltage envelope and filters and, secondly, it is due to the unknown resistances and noises at the two ends of the wire. As soon as the front of the propagation (not wave) of the band-limited noise reaches the other end, new information about the particular noise is strongly reduced. Even in the case of a no-transient protocol, Eve has effectively only a very small sample of a noise whose duration is much shorter than its correlation time.

First we describe the so far best-known transient protocol, which is based on random-walk resistances [31]. Alice and Bob arbitrarily choose R_L or R_H as their R_A and R_B, and they use continuously variable resistors — such as potentiometers, MOSFETs, *etc* — to execute the key exchange. If noise generators are employed to enhance the noise temperature, then their band-limited white noise spectra also need to be variable in a synchronized fashion so that the noise temperature stays constant at the publicly agreed value T_{eff}. Furthermore, suppose that the noise bandwidth in the KLJN scheme is secured by line filters at Alice's and Bob's ends. At the beginning of the KLJN clock period, both Alice and Bob start with

$$R_A(0) = R_B(0) = \frac{R_L + R_H}{2} \tag{31}$$

and they stay at this value until the noises equilibrate in the wire. Thus no informative transients can be observed just after connecting the resistors to the line, because the bit values have not yet been realized. Then Alice and Bob execute independent, adiabatically slow continuum-time random walks with their resistor values (in a fashion synchronized with the spectral parameter of their noise generators). The random walks are performed so slowly that — from a thermodynamic point of view — the system is changing in the adiabatic limit; thus there is almost thermal equilibrium in the wire during the whole random-walk process.

There is a publicly pre-agreed time t_r to execute these independent random walks. If Alice and Bob reach their randomly preselected values R_A and R_B within this time period, then they stop the random walk and stay at this value. After the time period t_r they restart the measurements in the regular fashion. This procedure virtually removes the transient effects and the information leak they may cause.

If, by the end of the time t_r, the random walk of Alice or Bob (or both) does not reach the randomly preselected resistance value, then he/she (or both) submit a cancellation signal via an authenticated channel, and the bit exchange process is immediately aborted; then a new independent KLJN-clock-period starts in the way described above.

Concerning security, the production of spurious frequency products is proportional to the RMS speed v_{rms} of the random walk and, if a concrete attack is implemented, it is reasonable to assume that it satisfies

$$q = \vartheta_{tr} v_{rms}, \tag{32}$$

where ϑ_{tr} is a constant relevant for the transient attack against this scheme (*cf.* Eqs. 12 and 13). The above assumption leads to unconditional ε-security ($\varepsilon \propto v_{rms}$) with results of the same nature as those given in Eqs. (12) and (15) and with statistical distance

$$\Delta = (0.5 + q)^N - 0.5^N \cong 2Nq\, 0.5^N = 2N\vartheta_{tr} v_{rms} 0.5^N. \tag{33}$$

If q is not small enough, it can further be reduced by the privacy amplification [33] because of the high fidelity of the KLJN scheme.

Here the resource used to approach the perfect security is the duration τ of the BEP, because it is inversely proportional to v_{rms} when the random walk time is dominating. In other words, at fixed key length the "price" of increasing the security is a reduction of the speed of key exchange, and $\varepsilon \propto 0.5^N \tau^{-1} N$ can again be arbitrarily small.

2.7. *Why Bennett–Riedel's passive correlation attack does not work against KLJN*

Directional couplers have limited bandwidth, work in the wave limit and — given that their directivity is good — have a size $\lambda_0/4$ (*cf.* Sec. 2.1.4). For much longer wavelengths — *i.e.*, smaller frequencies, as in the KLJN scheme — the system displays Rayleigh scattering and accordingly (*cf.* Eqs. 23 and 24) the passive correlation attack results in a correlation coefficient with power function scaling according to f^4. These conditions lead to unconditional ε-security ($\varepsilon \propto f^4 0.5^N$) with results that again are of the same nature as those in Eqs. (12) and (15) and with statistical distance

$$\Delta = (0.5 + q)^N - 0.5^N \cong 2Nq\, 0.5^N = 2N\vartheta_{cr} B_{kljn}^4 0.5^N, \tag{34}$$

where ϑ_{cr} is a constant defined as in Eq. (13). This value can further be enhanced by privacy amplification [33], if needed.

Here the resource used to approach perfect security is the duration τ, ($\tau \propto 1/B_{kljn}$) of the BEP, because it is inversely proportional to the highest frequency in the noise-bandwidth. In other words, at fixed key length the cost of increasing the security is a reduction of the speed of key exchange, and $\varepsilon \propto 0.5^N \tau^{-4} N$ can again be arbitrarily small.

2.8. *Why the current extraction/injection active attack does not work against KLJN*

BR [4] propose an active (invasive) attack wherein Eve connects a grounded resistor to the line in order to extract some current and also monitors the current direction in the wire. BR's verbatim statement was reproduced in Sec. 1.2.6 above.

We fully agree with the above assessment when it refers to the BR system. However this attack is inefficient against the KLJN scheme, and this fact was pointed out already in the original paper describing the KLJN scheme [26]. In fact, this latter paper proposes a technically more efficient attack of the same nature: that Eve injects a stochastic current at the middle and monitors the cross-correlation of this current with the channel currents in the two directions; the correlation coefficient will be greater in the direction of the smaller resistance. This attack was later pointed out also by Reiner Plaga and Horace Yuen in private communications. Alice and Bob monitor the channel currents at the two ends and compare their instantaneous amplitudes via an authenticated public channel. If the currents differ, the bit exchange event is terminated and that bit is discarded.

In this section we show a mathematical proof that the *uncertainly principle between measurement duration and statistical errors* makes it impossible for Eve to crack the key and the unconditional security remains even against this type of attack. The usual argument to justify the attacks referred to above is that Eve may use miniscule current amplitudes, which are below the detection limit of the comparisons by Alice and Bob. This argument does not work, however, because Alice and Bob can design their current resolution so that Eve, by implementing this attack, cannot extract enough information. Mathematically, the channel current at Alice's side of Eve is

$$I_{cA}(t) = I_c(t) - \gamma I_E(t), \tag{35}$$

and at Bob's side of Eve it is

$$I_{cB}(t) = I_c(t) + (1-\gamma) I_E(t), \tag{36}$$

where $I_E(t)$ is Eve's injected current and $(1-\gamma)/\gamma = R_A/R_B$. The cross-correlations with Eve's current during the BEP are

$$\begin{aligned} \rho_A &= \left\langle [I_c(t) - \gamma I_E(t)] I_E(t) \right\rangle_\tau = \left\langle I_c(t) I_E(t) \right\rangle_\tau - \gamma \left\langle I_E^2(t) \right\rangle_\tau \\ &= U_{cE\tau}(t) - \gamma \left\langle I_E^2(t) \right\rangle - \gamma U_{EE\tau}(t) \end{aligned} \tag{37}$$

$$\begin{aligned} \rho_B &= \left\langle [I_c(t) + (1-\gamma) I_E(t)] I_E(t) \right\rangle_\tau \\ &= \left\langle I_c(t) I_E(t) \right\rangle_\tau + (1-\gamma) \left\langle I_E^2(t) \right\rangle_\tau \\ &= U_{cE\tau}(t) + (1-\gamma) \left\langle I_E^2(t) \right\rangle + (1-\gamma) U_{EE\tau}(t), \end{aligned} \tag{38}$$

where $\langle\ \rangle_\tau$ stands for finite-time (τ) average, U for noise components, and $\langle\ \rangle$ for the exact average (requiring infinite time). The dominant terms at the right-hand side of Eqs. (37) and (38) are the noise terms of the cross-correlations between Eve's current and the channel current, with mean-square amplitudes scaling as τ^{-1}. The RMS amplitude Eve's correlation signal (after finite-time averaging) is negligible compared to that of the channel current, and its DC component

$$\rho_{DC} = \sigma \left\langle I_E^2(t) \right\rangle, \tag{39}$$

where $\sigma < 1$. The last noise terms at the right-hand side of Eqs. (37) and (38) are negligible compared to the first noise terms. The detection problem is again the same as the one encountered at the wire-resistance-attack: a small DC component (of the second term) in a large noise (dominated by the first term). Thus, as inferred from Eqs. (12) and (13), q will again satisfy

$$q = \vartheta_{ci} \sigma \left\langle I_E^2(t) \right\rangle, \tag{40}$$

where ϑ_{ci} is a constant relevant for this current injection/extraction attack at fixed τ (note that ϑ_{ci} is inversely proportional to τ). Again one reaches *unconditional* ε*-security* ($\varepsilon \propto 0.5^N \sigma$), with results of the same nature as those in Eqs. (12) to (15) and with statistical (variational) distance

$$\Delta = (0.5 + q)^N - 0.5^N \cong 2Nq0.5^N = 2N\vartheta_{ci}\sigma \left\langle I_E^2(t) \right\rangle 0.5^N. \qquad (41)$$

As a practical example, let us consider 9 bits accuracy of current/voltage comparison defense by Alice and Bob, which means $\sqrt{\langle I_E^2(t)\rangle}/\sqrt{\langle I_c^2(t)\rangle} < 2 \times 10^{-3}$ implying that Eve's relative DC component will have the same maximum. For a similar value of asymmetry during the experimental demonstration [29], the resulting q was 0.025. In that case, a two-step XOR-type privacy amplification [33], described in Section 1.1.4 will achieve $q < 5 \times 10^{-5}$, which allows key lengths N up to the order of 10^4 bits. Similarly to the situation in Section 1.1.4, for a 1000-bit-long shared key, it gives $\Delta(E, I)_{1000} = 9.3 \times 10^{-303}$ (*i.e.*, an ε-security with $\varepsilon_{1000} \cong 10^{-302}$); for a 500-bit-long shared key it results in $\Delta(E, I)_{500} = 1.5 \times 10^{-152}$, *i.e.*, an ε-security level with $\varepsilon_{500} \cong 2 \times 10^{-152}$.

2.9. *Remarks about potential hacking attacks*

Mathematical models of physical systems and their building elements are always approximate, and security proofs can only be given for these model systems. Particularly dangerous are the elements that are directly exposed to Eve. Thus a commercial secure key exchanger must be carefully designed with considering all the foreseeable hacking attacks.

For example, a real KLJN scheme must be armed with extra circuitry and protocol steps against Makarov-style blinding attacks [11], circulator-based attacks [32], and other unexplored possibilities such as out-of-frequency-range probing attacks, *etc*.

Conclusions

We showed that thermodynamics, noise, and the Second Law of Thermodynamics — *i.e.*, the impossibility to construct a perpetual motion machine of the second kind — are essential for the security of the classical physical key exchanger in the KLJN scheme. Furthermore we supplied mathematical security proofs for each attack proposed by Bennett and

Riedel [4]. Our results indicate that the security of the KLJN system has not been successfully challenged by them.

We also showed that the Bennett–Riedel scheme is unphysical and we cracked it with 100% success by passive attacks, in ten different ways. It was found that the same cracking methods do not function for the KLJN scheme. Some other claims by BR we subjected to critical analysis as well; for example, we proved that their equations for describing zero security do not apply for the KLJN scheme.

It is important to emphasize that all our analyses have assumed a *technically-unlimited Eve* with infinitely accurate and fast measurements. For non-ideal situations and at active (invasive) attacks, the *uncertainly principle between measurement duration and statistical errors* makes it impossible for Eve to extract the key regardless of the accuracy or speed of her measurements.

Acknowledgements

LK is grateful to Horace Yuen, Vadim Makarov, Renato Renner and Vincent Poor for helpful discussions, consultations and critical remarks about relevant security measures, and to Robert Mingesz, Henning Dekant, John Norton and Gabor Schmera for related discussions. Furthermore, we are indebted to Charles Bennett and Jess Riedel for their work, cited in Ref. [4], which has given us an opportunity to sharpen our arguments in favor of the Kirchhoff-law–Johnson-noise (KLJN) scheme.

References

1. Liang, Y., Poor. H.V., Shamai, S. (2008). "Information theoretic security", *Foundations Trends Commun. Inform. Theory* **5**, 355–580. DOI: 10.1561/0100000036.
2. Mingesz R, Kish LB, Gingl Z, Granqvist CG, Wen H, Peper F, Eubanks T, Schmera G (2013) Unconditional security by the laws of classical physics. *Metrology & Measurement Systems* **XX**, 3–16. DOI: 10.2478/mms-2013-0001. Available online at: http://www.degruyter.com/view/j/mms.2013.20.issue-1/mms-2013-0001/mms-2013-0001.xml
3. Bennett, C.H., Brassard, G. (1984) *Proc. Int. Conf. on Computers, Signals, and Signal Processing, Bangalore, India*. pp. 175–179.

4. Bennett, C.H., Riedel, C.J. (2013). "On the security of key distribution based on Johnson-Nyquist noise". http://arxiv.org/abs/1303.7435
5. Yuen, H.P. (2012) "On the foundations of quantum key distribution — Reply to Renner and beyond". arXiv:1210.2804.
6. Hirota, O. (2012). "Incompleteness and limit of quantum key distribution theory". arXiv:1208.2106v2.
7. Renner, R. (2012) "Reply to recent scepticism about the foundations of quantum cryptography". arXiv:1209.2423v.1.
8. Yuen, H.P. (2012). "Security significance of the trace distance criterion in quantum key distribution". arXiv:1109.2675v3.
9. Yuen, H.P. (2009). "Key generation: Foundation and a new quantum approach", *IEEE J. Selected Topics in Quantum Electronics* **15**, 1630.
10. Salih, H., Li, Z.H., Al-Amri, M., Zubairy, M.S. (2013). "Protocol for direct counterfactual quantum communication", *Phys. Rev. Lett.*, **110**, 170502.
11. Merali, Z. (29 August 2009). "Hackers blind quantum cryptographers", *Nature News*, DOI:10.1038/news.2010.436.
12. Gerhardt, I., Liu, Q., Lamas-Linares, A., Skaar, J., Kurtsiefer, C., Makarov, V. (2011). "Full-field implementation of a perfect eavesdropper on a quantum cryptography system", *Nature Commun.* **2**; article number 349. DOI: 10.1038/ncomms1348.
13. Lydersen, L., Wiechers, C., Wittmann, C., Elser, D., Skaar, J., Makarov, V. (2010). "Hacking commercial quantum cryptography systems by tailored bright illumination", *Nature Photonics* **4**, 686–689. DOI: 10.1038/NPHOTON.2010.214.
14. Gerhardt, I., Liu, Q., Lamas-Linares, A., Skaar, J., Scarani, V., Makarov, V., Kurtsiefer, C. (2011). "Experimentally faking the violation of Bell's inequalities", *Phys. Rev. Lett.*, **107**, 170404. DOI: 10.1103/PhysRevLett.107.170404.
15. Makarov, V., Skaar, J. (2008), "Faked states attack using detector efficiency mismatch on SARG04, phase-time, DPSK, and Ekert protocols", *Quantum Inform. Comp.*, **8**, 622–635.
16. Wiechers, C., Lydersen, L., Wittmann, C., Elser, D., Skaar, J., Marquardt, C., Makarov, V., Leuchs, G. (2011). "After-gate attack on a quantum cryptosystem". *New J. Phys.*, **13**, 013043. DOI: 10.1088/1367-2630/13/1/013043.
17. Lydersen, L., Wiechers, C., Wittmann, C., Elser, D., Skaar, J., Makarov, V. (2010). "Thermal blinding of gated detectors in quantum cryptography", *Opt. Express*, **18**, 27938–27954. DOI: 10.1364/OE.18.027938.

18. Jain, N., Wittmann, C., Lydersen, L., Wiechers, C., Elser, D., Marquardt, C., Makarov, V., Leuchs, G. (2011). "Device calibration impacts security of quantum key distribution". *Phys. Rev. Lett.*, **107**. 110501. DOI: 10.1103/PhysRevLett.107.110501.
19. Lydersen, L., Skaar, J., Makarov, V. (2011). "Tailored bright illumination attack on distributed-phase-reference protocols", *J. Mod. Opt.*, **58**, 680–685. DOI: 10.1080/09500340.2011.565889.
20. Lydersen, L., Akhlaghi, M.K., Majedi, A.H., Skaar, J., Makarov, V. (2011). "Controlling a superconducting nanowire single-photon detector using tailored bright illumination". *New J. Phys.*, **13**, 113042. DOI: 10.1088/1367-2630/13/11/113042.
21. Lydersen, L., Makarov, V., Skaar, J. (2011). "Comment on 'Resilience of gated avalanche photodiodes against bright illumination attacks in quantum cryptography'", *Appl. Phys. Lett.*, **99**. 196101. DOI: 10.1063/1.3658806.
22. Sauge, S., Lydersen, L., Anisimov, A., Skaar, J., Makarov, V. (2011). "Controlling an actively-quenched single photon detector with bright light", *Opt. Express*, **19**, 23590–23600.
23. Lydersen, L., Jain, N., Wittmann, C., Maroy, O., Skaar, J., Marquardt, C., Makarov, V., Leuchs, G. (2011). "Superlinear threshold detectors in quantum cryptography", *Phys. Rev. Lett.*, **84**, 032320. DOI: 10.1103/PhysRevA.84.032320.
24. Lydersen, L., Wiechers, C., Wittmann, C., Elser, D., Skaar, J., Makarov, V. (2010). "Avoiding the blinding attack in QKD: Reply (Comment)", *Nature Photonics*, **4**. 801–801. DOI: 10.1038/nphoton.2010.278.
25. Makarov, V. (2009). "Controlling passively quenched single photon detectors by bright light", *New J. Phys.*, **11**, 065003. DOI: 10.1088/1367-2630/11/6/065003.
26. Kish, L.B. (2006). "Totally secure classical communication utilizing Johnson(-like) noise and Kirchhoff's law", *Phys. Lett. A*, **352**, 178–182.
27. Cho, A. (2005). "Simple noise may stymie spies without quantum weirdness", *Science*, **309**, 2148. Available online at: http://www.ece.tamu.edu/~noise/news_files/science_secure.pdf.
28. Kish, L.B. (2006). "Protection against the man-in-the-middle-attack for the Kirchhoff-loop-Johnson(-like)-noise cipher and expansion by voltage-based security", *Fluct. Noise Lett.*, **6**, L57–L63. http://arxiv.org/abs/physics/0512177.

29. Mingesz, R., Gingl, Z., Kish, L.B. (2008). "Johnson(-like)-noise-Kirchhoff-loop based secure classical communicator characteristics, for ranges of two to two thousand kilometers, via model-line", *Phys. Lett. A*, **372**, 978–984.
30. Palmer, D.J. (2007). "Noise encryption keeps spooks out of the loop", *New Scientist*, iss. 2605 p.32; http://www.newscientist.com/article/mg19426055.300-noise-keeps-spooks-out-of-the-loop.html.
31. Kish, L.B. (2013). "Enhanced secure key exchange systems based on the Johnson-noise scheme", *Metrology & Measurement Systems*, **XX**, 191–204. Open access: http://www.degruyter.com/view/j/mms.2013.20.issue-2/mms-2013-0017/mms-2013-0017.xml?format=INT
32. Kish, L.B., Horvath, T. (2009). "Notes on recent approaches concerning the Kirchhoff-law-Johnson-noise-based secure key exchange", *Phys. Lett. A*, **373**, 901–904.
33. Horvath, T., Kish, L.B., Scheuer, J. (2011). "Effective privacy amplification for secure classical communications", *Europhys. Lett.*, **94**, 28002. http://arxiv.org/abs/1101.4264.
34. Saez, Y., Kish, L.B., Mingesz, R., Gingl, Z., Granqvist, C.G. (2013). "Current and voltage based bit errors and their combined mitigation for the Kirchhoff-law-Johnson-noise secure key exchange". http://arxiv.org/abs/1309.2179. http://vixra.org/abs/1308.0113.
35. Kish, L.B., Kwan, C. (2013). "Physical uncloneable function hardware keys utilizing Kirchhoff-law-Johnson-noise secure key exchange and noise-based logic". http://vixra.org/abs/1305.0068; http://arxiv.org/abs/1305.3248
36. Kish, L.B., Saidi, O. (2008). "Unconditionally secure computers, algorithms and hardware", *Fluct. Noise Lett.*, **8**, L95–L98.
37. Gonzalez, E., Kish, L.B., Balog, R., Enjeti, P. (2013). "Information theoretically secure, enhanced Johnson noise based key distribution over the smart grid with switched filters". http://vixra.org/abs/1303.0094; http://arxiv.org/abs/1303.3262.
38. Kish, L.B., Mingesz, R. (2006). "Totally secure classical networks with multipoint telecloning (teleportation) of classical bits through loops with Johnson-like noise", *Fluct. Noise Lett.*, **6**, C9–C21.
39. Kish, L.B., Peper, F. (2012). "Information networks secured by the laws of physics", *IEICE Trans. Commun.*, **E95-B**, 1501–1507.
40. Scheuer, J., Yariv, A. (2006). "A classical key-distribution system based on Johnson (like) noise — How secure?", *Phys. Lett. A*, **359**, 737–740.

41. Kish, L.B., Scheuer, J. (2010). "Noise in the wire: The real impact of wire resistance for the Johnson(-like) noise based secure communicator", *Phys. Lett. A*, **374**, 2140–2142.
42. Kish, L.B. (2006). "Response to Scheuer–Yariv: 'A classical key-distribution system based on Johnson (like) noise — How secure?'", *Phys. Lett. A*, **359**, 741–744.
43. Hao, F. (2006). "Kish's key exchange scheme is insecure", *IEE Proc. Inform. Soc.*, **153**, 141–142.
44. Kish, L.B. (2006). "Response to Feng Hao's paper 'Kish's key exchange scheme is insecure'", *Fluct. Noise Lett.* **6**, C37–C41.
45. Liu, P.L. (2009). "A new look at the classical key exchange system based on amplified Johnson noise", *Phys. Lett. A*. **373**, 901–904.
46. Arora, S., Barak, B. (2009). *Computational Complexity*. Cambridge University Press, Cambridge.
47. Kish, L.B. (2006). "Thermal noise driven computing", *Appl. Phys. Lett.* **89**, 144104.
48. Kish, L.B., Granqvist, C.G. (2013). "On the security of the Kirchhoff-law-Johnson-noise (KLJN) communicator". http://arxiv.org/abs/1309.4112.
49. Antilli, D. (2005). "System and method for the propagation of deterministic and non-deterministic values by means of electrical conductors". Patent publication number EP1952573 (A2). See: http://www.google.com/patents/EP1952573A2?cl=en
50. Pauli, W. (2000). *Electrodynamics*. Dover Publications, New York.
51. Matthaei, G.L., Young, L., Jones, E.M.T. (1964). *Microwave Filters, Impedance-Matching Networks, and Coupling Structures*. McGraw-Hill, New York.
52. Liu, P.L. (2009). "A key agreement protocol using band-limited random signals and feedback", *IEEE J. Lightwave Technol.*, **27**, 5230–5234.
53. Liu, P.L. (2009). "Security risk during the transient in a key exchange protocol using random signals and feedback", *Phys. Lett. A*, **373**, 3207–3211.
54. Kish, L.L., Zhang, B., Kish, L.B. (2010). "Cracking the Liu key exchange protocol in its most secure state with Lorentzian spectra", *Fluct. Noise Lett.*,**9**, 37–45.

4.5. In the Ring: Non-validity of the Landauer Principle, and Leo Szilard's Mistake

4.5.1. The problem of heat dissipation at control and memory erasure is directly related to Sections 2.1, 2.2 and 4.1. This paper was also important from a historical point of view because this one was on hold in arXiv for the longest duration even though it had already been published as an open access paper earlier on. Even though arXiv was designed to be a repository of electronic pre-prints, and to automatically publish submissions, it is not always such a smooth process. After KLJN was featured by *MIT Technology Review*, my endorsement rights on arXiv were removed; and I did not receive any explanatory notification for it. Most of my submissions thereafter, were put on hold and heavily delayed — and when they were finally published, the stated submission dates were incorrect. Recently, the situation has improved much especially when I submit to computer science. However, my endorsement rights have still yet to be restored; even my students have been awarded their own rights due to our joint papers.

Figure 4.5.1. John D. Norton. (Courtesy of J.D. Norton). He is full professor and center director at the University of Pittsburg. John is one of the most prominent deniers of Landauer's principle. When he saw my struggles with arXiv he sent an encouraging note to me (dated December 8, 2014; see following quote).

"Sorry to see the struggles continue. If it is any consolation, my experience in my field is that the only sure recipe for passing the gatekeepers is to write a paper that, in virtually all the content, simply repeats what everyone thinks already. You then

Figure 4.5.2. Carl-Gustaf Ribbing. He was Emeritus Professor in Solid State Physics and Optics at Uppsala University, Sweden. He died in 2015. Carl, a very open-minded scientist, was intensively involved in the critical analysis and related denial of global warming claims on a strictly scientific basis. In his and his colleagues' view, climate and temperature are naturally varying and the human contribution to that via greenhouse gases is below the measurable limit. Even its sign, positive or negative temperature contribution, is an open question. He was also interested in the heralded statements in modern science and technology, and I often told him that *Landauer theorem is the global warming in physical informatics*.

> *add the smallest epsilon of novelty that you can. The smaller you make epsilon, while keeping it non-zero, the greater your chance of passing. The trouble is that you are actually saying something new that shakes things up."*
>
> — *John N.*

Paper: Demons: Maxwell's demon, Szilard's engine and Landauer's erasure–dissipation[1,2]

Abstract. This talk addressed the following questions in the public debate at HoTPI: (*i*) energy dissipation limits of switches, memories and control;

[1] Kish, L.B., Granqvist, C.G., Khatri, S.P., Wen, H. (2014). "Critical Remarks on Landauer's principle of erasure-dissipation", *Int. J. Mod. Phys. Conf. Ser.*, **33**, 1460364 (open access). Available online at: http://vixra.org/abs/1408 and http://arxiv.org/abs/1412.2166.

[2] Perhaps, our deepest recent related work showing that information change and energy dissipation are unrelated: L.B. Kish, C.G. Granqvist, S.P. Khatri, F. Peper, "Response to "Comment on 'Zero and negative energy dissipation at information-theoretic erasure'", *J. Computational Electronics* **15** (2015) 335–339. http://arxiv.org/abs/1507.08906

(*ii*) whether reversible computers are possible, or does their concept violate thermodynamics; (*iii*) Szilard's engine, Maxwell's demon and Landauer's principle: corrections to their exposition in the literature; (*iv*) whether Landauer's erasure–dissipation principle is valid, if the same energy dissipation holds for writing information, or if it is invalid; and (*v*) whether (non-secure) erasure of memories, or the writing of the same amount of information, dissipates most heat.

Keywords: Energy requirement of control.

In the subject of energy requirement for information processing (including erasure), and the related field of thermal demons (engines with single thermal bath utilizing measurement information), there is a large body of papers which miss some of the key aspects. Here we briefly comment on several of these issues.

1. Energy requirement for changing a single bit value

Brillouin [1], Alicki [2] and Kish [3–5] have arrived at similar results which show, in independent ways, that the lower limit of energy requirement for changing a bit value in a binary device is given as

$$E_{\min} \approx kT \ln\left(\frac{1}{\varepsilon}\right), \tag{1}$$

where k is Boltzmann's constant, T is temperature and ε is error probability of bit operation. Strictly speaking, Eq. (1) is valid only for observation times up to the correlation time of the stochastic thermal excitation (thermal noise) in the system. For longer observation time windows t_w, Eq. (1) was generalized [4,5] to read as

$$E_{\min} \approx kT \ln\left(\frac{1}{\varepsilon}\frac{t_w}{\tau}\right), \tag{2}$$

where τ is time and the $\varepsilon \to 0$ limit is assumed.

Figure 1 shows a double-well potential model of bit states and two-stage switches. A bit error is generated as the thermal noise reaches beyond the energy barrier. When the bit state is changed during information processing, the particle must be pushed over the barrier into the other well. The invested energy cannot be saved, because any attempt to do so will result in more

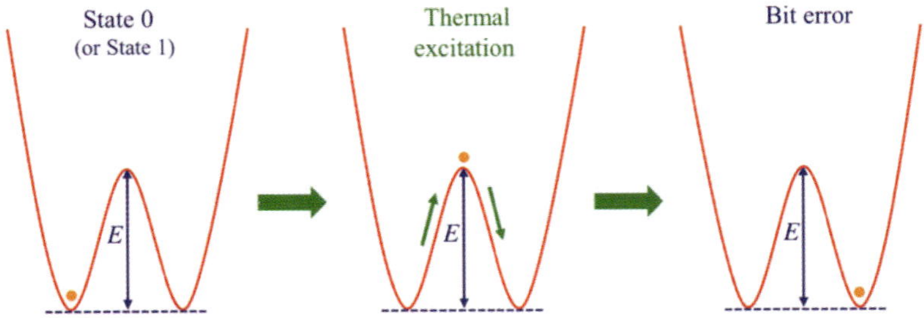

Figure 1. Double-well potential model of bit states and two-stage switches. A bit error is generated when the thermal noise reaches beyond the energy barrier.

energy dissipation than the original one as a consequence of the need of extra switches and control steps and/or timing [4,5].

2. Brillouin's negentropy-of-information principle *versus* Landauer's principle

Brillouin's negentropy principle of information [6], published in 1953, claims that in order to *change* a bit of information one has to dissipate an energy of at least

$$E_{\min} \approx kT\ln(2). \tag{3}$$

This result agrees with Eq. (1) for $\varepsilon = 0.5$, *i.e.*, when the binary system has zero bits of information. Thus Eq. (3) is always valid, whereas Eqs. (1) and (2) pose more strict/physical lower limits.

Landauer's principle [7], which was developed in steps after 1961, claims that an energy of at least

$$E_{\min} \approx kT\ln(2) \tag{4}$$

must be dissipated in the memory in order to *erase* a bit of information. While this result looks formally identical to Brillouin's assertion above, Landauer's result is incorrect when the bit state does not change during erasure.

We note that there is an extensive and valuable literature — for example by Porod, Ferry, Norton and their coworkers [8–13] — showing that Landauer's principle is unphysical, and we refer the reader to these works after a study of the present short note.

3. Is a recent "experimental proof" of Landauer's principle valid?

Recent experiments by Bérut *et al.* [7] to prove Landauer's Theorem use advanced tools and are interesting, but the conceptual basis for this proof, and for its interpretation, show characteristic features of poor science. Here are a few examples:

(*i*) *What was the potential well?* An optical field of focused laser light (an optical tweezer) was used in a fluid. This is a highly non-equilibrium situation, and a *strongly driven system in steady state*. Rigorously speaking, this system does not even have a temperature! And an equivalent temperature can be anything, even attaining "negative" values (as for population-inversion in lasers).

(*ii*) *Is relocating the particle between the two wells indeed a single-bit erasure?* Why it is not the *writing* of single-bit information? Nothing would then change in the experiments and their results!

(*iii*) The studied system is a single-bit memory with friction force (loss) *during a change of the state*, it is not damping of excess energy after the bit change is completed, as in friction-free potential wells. What happens if one reduces the friction coefficient (viscosity) and forces the particle over the same path and with the same speed? No such test was done; neither was the distance between the two wells varied. One can easily imagine that friction losses would become even smaller than Landauer's energy limit.

(*iv*) The *energy dissipation of keeping up and modifying the potential barrier* during the process, which is of the order of 10^{20} kT, was neglected. It could be noted that Landauer committed a similar error in his famous analysis [7]. This neglected energy — *i.e.*, the modification of potential barriers, or control of gate potentials for MOSFETs — is the dominant fundamental energy dissipation in today's computers [14].

4. Missed control energy in classical and quantum demons

Surprisingly, Landauer's Theorem has been put into seemingly inherent connection with thermodynamic demons such as Maxwell's demon, Szilard engines and Quantum demons. We are unable to cite all of the vast literature on this topic, instead we refer to a few examples [15–17] that point clearly to the source of this error.

Equations (1) and (2) make it obvious that whenever a single switching or decision operation takes place during the engine cycle, then energy exceeding $kT\ln(2)$ is dissipated during *each* of these operations. This is extremely important, because the energy the demon is supposed to produce is exactly $kT\ln(2)$ (Szilard engine), or at least have a similar value (Maxwell's demon). Scrutinizing one of the cited papers by Bennett [16], all or almost all operations of a single demon cycle involve such single-bit operations, which means that the control steps in themselves generate many times greater heat than the one given by the $kT\ln(2)$ limit. Words such as "insert", "observe", "if", "go to", "attach", (letting to) "expand", "remove", "transform", *etc.*, signify such a minimum single-bit control operation in the mentioned paper [16]. Specific operations are exemplified below from [16], where L/R means left/right and M1 ... L3 are designations used by Bennett [16] (we added the Italic typeface):

M1. *Insert* partition [L]
M2. *Observe* the particle's chamber [L] or [R]
M3. *If* memory bit = R, *go to* R1 [R]
M4. *If* memory bit = L, *go to* L1 [L]
R1. *Attach* pulleys so right chamber can expand [R]
R2. *Expand*, doing isothermal work W [R]
R3. *Remove* pulleys [R]
R4. *Transform* known memory bit from R to L [L]
R5. *Go to* M1 [L] L1. *Attach* pulleys so left chamber can expand [L]
L2. *Expand*, doing isothermal work W [L]
L3. *Remove* pulleys [L] L4. Go to M1 [L]

Even Szilard missed the crucial aspect of control energy in his famous paper [18] because he used a two-stage lever to control a gear. The control of this gear consumes all of the energy the Szilard engine was supposed to produce. *Thus the Szilard-engine-puzzle about the Second Law and an intelligent being is non-existent from the beginning.* In hindsight, Szilard's error might be considered fortunate because it led to an expansion of the research field. In the case of Szilard's engine, it is possible to fix the mentioned flaw [5], but the question as to the role of an intelligent being remains obsolete.

Finally, we note that Renner [17] and others claim that control energies can be saved by using a large number of parallel demons mechanically coupled to each other. This belief is incorrect since demons are fundamentally random so that they require independent control units and related energy dissipation. Conservation of control energy can be done in periodic engines, where this effect can be executed both in space (parallel-coupled engines) and time (resonators with high Q-factor) [19], but it is not possible for demons.

5. Lower limit of energy requirement for erasing a large memory

Finally, we address the problem of non-secure bit erasure in computers. Do computers execute erasure when they discard information?

The answer is "no" since, in practice, they do not reset the memory bits but just change the address of the boundary of the free part of memory as illustrated in Figure 3. The number of bits in the address scales as $\log_2 N$, where N is the size of the whole memory, and hence (in accordance with Eq. 2) the energy dissipation is of the order of

$$E_{\text{er}} = kT \ln\left(\frac{1}{\varepsilon}\frac{t_w}{\tau}\right) \log_2 N. \tag{5}$$

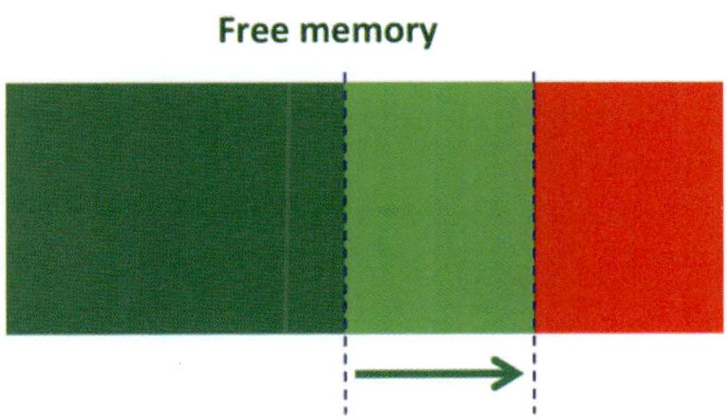

Figure 2. Simplified one-dimensional illustration of discarding information in an idealized computer memory. Green signifies space free for writing, and Red denotes occupied space. The address of the boundary of the free memory is moved along the arrow to discard information and increase the free-memory part.

At a fixed error rate and observation time window, the energy dissipation of erasure scales as

$$E_{\text{er}} \propto \log_2 N, \tag{6}$$

which is a much more optimal situation than for the Landuer limit of $kT \ln(2)$ energy dissipation per bit. Thus real memories typically work in the "sub-Landauer" regime during erasure, which again indicates that Landauer's principle is invalid.

References

1. Brillouin, L. (1962). *Science and Information Theory*. Academic: New York.
2. Alicki, R. (2013). "Stability versus reversibility in information processing", *Int. J. of Mod. Phys.: Conf. Series*, **33**, 1460353.
3. Kish, L.B. (2004). "Moore's Law and the energy requirement of computing versus performance", *IEE Proc. — Circ. Dev. Syst.*, **151**, 190–194.
4. Kish, L.B., Granqvist, C.G. (2012). "Energy requirement of control: Comments on Szilard's engine and Maxwell's demon", *EPL*, **98**, 68001.
5. Kish, L.B., Granqvist, C.G. (2012). "Electrical Maxwell demon and Szilard engine utilizing Johnson noise, measurement, logic and control", *PLoS ONE*, **7**, e46800.
6. Brillouin, L. (1953). "The negentropy principle of information", *J. Appl. Phys.*, **24**, 1152.
7. Bérut, A., Arakelyan, A., Petrosyan, A., Ciliberto, S., Dillenschneider, R., Lutz, E. (2012). "Experimental verification of Landauer's principle linking information and thermodynamics", *Nature*, **483**, 187–189.
8. Porod, W. (1988). "Energy requirements in communication — Comment", *Appl. Phys. Lett.*, **52**, 2191–2191.
9. Porod, W., Grondin, R.O., Ferry, D.K. (1984). "Dissipation in computation", *Phys. Rev. Lett.*, **52**, 232–235.
10. Porod, W., Grondin, R.O., Ferry, D.K., Porodm G. (1984). "Dissipation in computation — Reply, *Phys. Rev. Lett.*, **52**, 1206–1206.
11. Norton, J.D. (2011). "Waiting for Landauer", *Stud. Hist. Philos. Mod. Phys.*, **42**, 184–198.
12. Norton, J.D. (2005). "J.D. Eaters of the lotus: Landauer's principle and the return of Maxwell's demon", *Stud. Hist. Philos. Mod. Phys.*, **36**, 375–411.
13. Norton, J.D. (2013). "The end of the thermodynamics of computation: A no go result", *Philos. Sci.*, **80**, 1182–1192.

14. Kish, L.B. (2002). "End of Moore's law: Thermal (noise) death of integration in micro and nano electronics", *Phys. Lett. A*, **305**, 144–149.
15. Bennett, C.H. (1987), "Demons, engines and the second law", *Sci. Am.*, **257**, 108–116.
16. Bennett, C.H. (2003). "Notes on Landauer's principle, reversible computation, and Maxwell's demon", *Studies Hist. Philos. Modern Phys.*, **34**, 501–510.
17. Renner, R. (2014). "Landauer's Principle revisited". Talk presented at Hot Topics of Physics Informatics (HoTPI-2013) at Hunan University, China, on November 12, 2013, unpublished.
18. Szilard, L. (1929). "On the decrease of entropy in a thermodynamic system by the intervention of intelligent beings", *Z. Phys.*, **53**, 840.
19. Kish, L.B. (2011). "Thermal noise engines", *Chaos Solitons Fractals*, **44**, 114–121.

4.6. The Analyses Triggered by the Gunn–Allison–Abbott Attack

4.6.1. Derek Abbott, believing in the existence of waves in KLJN, gave the job of cracking KLJN using directional wave dividers to his talented engineering PhD student, Lachlan Gunn, and involved his wise engineering associate, Andrew Allison. At the Hot Topic of Physical Informatics (HoTPI) 2013 meeting in Changsha, China, they told me they had succeeded. I thought that was highly unlikely because, first of all, there are no waves in KLJN; secondly, in Pao-Lo Liu's KLJN emulation attempt, though there seemed to be waves that could be observed in both directions, it still seemed to be secure. My answer was: "Great, maybe there is indeed a successful means of cracking the KLJN system; and, if it is so, I will be happy to develop defense against it; however waves cannot exist at KLJN conditions in the cable". I told them that, if they published their findings without mentioning waves, I would support their publication, because the claim of the existence of waves would violate the laws of physics. They did not listen. And though their paper was initially turned down by *Physical Review Letters*, it was later accepted by Nature's *Scientific Reports* [Gunn, L.J., Allison, A., Abbott, D. (2014). "A directional wave measurement attack against the Kish

(*Continued*)

Figure 4.6.1. Derek Abbott in "superposition of interests". (Courtesy of D. Abbott).

(Continued)

key distribution system", *Scientific Reports*, **4**, 6461]. In the next three sections, we nullify their arguments and experimental results. The experiments reflect a poor KLJN design and their failure is a good lesson to prospective KLJN developers, too.

4.6.2. The paper in Section 4.6.1 proves, among others, that no waves exist at KLJN's steady-state conditions; otherwise perpetual motion machines could be built.

4.6.3. The paper in Section 4.6.2 shows the correct analysis of the cracking scheme, and indicates that the attack is less efficient than the (Bergou)–Scheuer–Yariv one. It also lists a number of potential experimental artifacts.

4.6.4. The paper in Section 4.6.3 deals with the big design error in how Gunn *et al.* modeled the cable attenuation. The system was simply not a KLJN system. One should note that this paper was possible due to the *honesty and courtesy of Lachlan Gunn* who gave us some of the unpublished details of their experiments after publication.

4.6.1. Waves or only handwaving? What does physics say?[1]

4.6.5. For physicists, it is usually an evident claim that, in a string or cable of finite length L, the longest wavelength is $2L$, and that waves with lower frequency than this limit (wavelengths longer than $2L$) are forbidden states. Acoustics, musical instruments, harmonic analysis, the quantum theory of materials, etc. are based on these facts. However, for many electrical engineers this is not obvious because of microwave theories, where waves are often used in calculations as a mathematical trick.

4.6.6. We clarified the matter from various angles in the paper below. It is important to keep in mind that the question of security is not important here, and the next paper (Section 4.5.2) will handle that in detail. The key question here is: What does physics say about low-frequency waves in a short cable?

4.6.7. Our opponents say that claim in Note 4.6.5 is true only in shorted or open-end cables. That is an incorrect answer, because adding a damping to an oscillator system never introduces new oscillatory modes (poles/zeros), and then Planck's formula from the black-body radiation would not hold.

4.6.8. We point out below that the existence of waves in the forbidden low-frequency limit would allow the construction of perpetual motion machines of the second kind — that is the violation of the Second Law of thermodynamics.

[1] L.B. Kish, S.P. Chen, C.G. Granqvist, J. Smulko, "Waves in a short cable at low frequencies, or just hand-waving? What does physics say?", invited talk at the 23rd International Conference on Noise and Fluctuations (ICNF 2015), Xian, China, June 2–5, 2015. DOI: 10.1109/ICNF.2015.7288604. http://arxiv.org/abs/1505.02749 (April 16, 2015).

Figure 4.6.2. Hsien-Pu (Barry) Chen. He was a PhD student for the simulation of cable related effects in the KLJN system. Barry has a background in circuit/chip design, and joined our group at the same time when the project of answering the Gunn–Allison–Abbott (GAA) attack started. He simulated the data for Figure 3 and Table 1 for the paper in this section. The same results were also used in the conceptual base of the next paper (see Section 4.6.2). During his PhD research, his goal was to develop a full-scale real-time simulation system to test cable related information leaks and develop related tools of defense against hackers.

Paper: Do electromagnetic waves exist in a short cable at low frequencies? What does physics say?[2]

Abstract. We refute a physical model, recently proposed by Gunn, Allison and Abbott (GAA) [http://arxiv.org/pdf/1402.2709v2.pdf], to utilize electromagnetic waves for eavesdropping on the Kirchhoff-law–Johnson-noise (KLJN) secure key distribution. Their model, and its theoretical underpinnings, is found to be fundamentally flawed because their assumption of electromagnetic waves violates not only the wave equation but also the Second Law of Thermodynamics, the Principle of Detailed Balance, Boltzmann's Energy Equipartition Theorem, and Planck's formula by implying infinitely strong blackbody radiation. We deduce the correct mathematical model of the GAA scheme, which is based on impedances at the quasi-static limit. Mathematical analysis and simulation

[2]Chen, H.P., Kish, L.B., Granqvist, C.G., Schmera, G. (2014). *Fluctuation and Noise Lett.* **13**, 1450016 (1–13); http://arxiv.org/abs/1404.4664 ; http://vixra.org/abs/1403.0964.

results confirm our approach and prove that GAA's experimental interpretation is incorrect too.

Keywords: Waves; KLJN secure key exchange; physics in engineering education.

1. Introduction

Recently Gunn, Allison and Abbott (GAA) [1] proposed a new, delay-based attack against the Kirchhoff-law–Johnson-noise (KLJN) secure key distribution scheme [2–5]. GAA claim — contradicting earlier statements most recently expounded in work by Kish, Abbott and Granqvist (KAG) [4] — that waves exist in a *finite-size* cable at *arbitrarily low frequencies.*

The theoretical basis of GAA's assertion [1] is the fact that, whereas wave-guides usually have a low-frequency cut-off for wave modes versus the *diameter* of the wave-guide, no such cut-off exists for transversal electromagnetic (TEM) wave modes in the case of *infinitely* long wave-guides. GAA write that, because coaxial cables have TEM wave modes, there is no frequency cut-off of the wave-based component of the electrical transport down to zero frequency. As a consequence of their presumption, GAA use the d'Alembert solution [1]

$$U(t,x) = U_+\left(t - \frac{x}{v}\right) + U_-\left(t - \frac{x}{v}\right) \tag{1}$$

for propagating lossless fluctuations — which may or may not be waves — in a linear medium to model the propagation of voltage in the cable used for key exchange in the KLJN scheme, where U_+ and U_- are voltage components of waves propagating to the right and left along the x axis, and v is propagation velocity.

The experimental support of GAA's claim is that they have measured the voltage between the ends of a short coaxial cable at low-frequency sinusoidal voltage drive with an impedance-matched load at the other end and, at first sight, have found that their results confirmed some of the implications of Eq. (1), as further elaborated in Section 3 below.

We have analyzed GAA's statements [1] and found most of them invalid. Specifically, our findings and conclusions are the following:

(*i*) In cables, wave modes with wavelengths greater than twice their length are forbidden states, meaning that such modes do not exist;

consequently there are no waves in cables in the frequency range pertinent to the KLJN scheme.

(*ii*) Instead, time-dependent propagation processes are non-wave type *retarded potentials* in a distributed impedance system; one of the implications of this is that Eq. (1) does not hold.

(*iii*) GAA's interpretation of their own "wave-verification" experiments [1] is invalid.

(*iv*) A correct derivation of the problem posed by GAA allows a proof that their attack does not work, and an identification of the source of errors in their "attack-demonstration" experiments.

(*v*) We prove that directional coupler attacks under KLJN conditions can never offer more information leak than the old wire resistance attacks. It should be noted that directional couplers in KLJN can never separate Alice's and Bob's contribution.

(*vi*) Finally we are able to formulate a general proof that any future (not yet existing) advanced attack type, that may be able even to *separate* the contribution of Alice and Bob, can never crack the KLJN scheme. Instead such an attack would lead to more advanced KLJN schemes that would not only be faster than earlier versions but would also be immune against these types of attacks.

The results delineated in (*i*)–(*vi*) will be derived and presented in three separate publications. The present paper is focused on items (*i*)–(*iii*), and moreover points out that the erroneous wave picture at low frequencies, referred to above, seems to be a common misconception in electrical engineering education.

2. Refutation of GAA's theory and experimental interpretation

As mentioned above, GAA's attack [1] on the KLJN scheme employs waves and related delays in a cable to extract information. While attempts to utilize time delays in cables for information purposes are to be encouraged, the asserted use of waves, which do not exist, is a fundamental flaw that invalidates GAA's basic considerations, proposed experiments, and interpretation of these experiments.

The wavelengths corresponding to the frequency range of concern in the KLJN scheme are much longer than the physical extent of the cable, and we earlier referred to that range the "no-wave" or "quasi-static" limit [2–5]. As remarked above, GAA argue that TEM wave modes do not exhibit any low-frequency cut-off. It is true that TEM wave modes in a wave-guide do not have a low-frequency cut-off *versus the diameter* of the wave-guide, but this argument is irrelevant because wave modes do have a cut-off *versus the length* of the cable. This does not imply that the electrical transport itself has a cut-off; it solely means that, when wave modes are forbidden, electrical transport takes place via non-wave phenomena — such as drift and relaxation — which constitute the form of transport in the quasi-static region of electrodynamics.

In this section, we use three separate fields of physics and mathematics to prove, in six different ways, that GAA's assumption of the existence of waves in the short cable within the frequency range pertinent to the KLJN scheme violates the following equations and laws of physics (our proofs are given subsequently):

(a) the wave equation,
(b) essential dynamical behavior of magnetic and electric energy in electromagnetic wave modes,
(c) Boltzmann's Energy Equipartition Theorem of Thermodynamics,
(d) the Principle of Detailed Balance of Thermodynamics,
(e) the Second Law of Thermodynamics, and
(f) theory and experiments for blackbody radiation.

2.1. *Violation of the wave equation*

It was recently pointed out by KAG [4] (including one of the proponents of the GAA model) that the wave equation precludes the existence of waves in the frequency range of concern for the KLNJ scheme. Next we provide more details about this fact and first illustrate the distributed inductance–capacitance–resistance (LCR) model of the cable in the KLJN scheme in Figure 1.

For the sake of simplicity but without losing generality, we discuss the case of a lossless cable. The main conclusion about the lack of wave modes is general because the inclusion of damping terms in harmonic differential

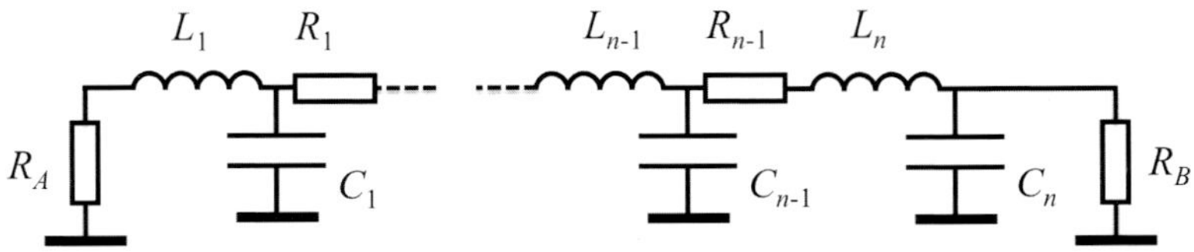

Figure 1. Outline of the pertinent part of the KLJN scheme with a distributed LCR model of a long and leakage-free cable. When the cable losses can be neglected, one may omit the R_i resistors representing the distributed resistance of the cable. Alice's and Bob's resistors, denoted R_A and R_B, respectively, are randomly selected from the set $\{R_L, R_H\}$ with $(R_L \neq R_H)$ at the beginning of each bit-exchange period. These resistors, with associated serial generators (not shown), emulate thermal noise with high noise temperature and strongly limited bandwidth.

equations *can never produce new eigen-frequencies*; they can only modify them and their bandwidth.

The wave equations of voltage $U(x, t)$ and current $I(x, t)$ in lossless cables are

$$\frac{\partial^2 U(x,t)}{\partial x^2} = \frac{1}{v_c^2}\frac{\partial^2 U(x,t)}{\partial t^2}, \tag{2}$$

$$\frac{\partial^2 I(x,t)}{\partial x^2} = \frac{1}{v_c^2}\frac{\partial^2 I(x,t)}{\partial t^2}, \tag{3}$$

where the phase propagation velocity of waves in the cable is

$$v_c = \sqrt{\frac{1}{L_u C_u}}. \tag{4}$$

Here L_u and C_u are inductance and capacitance "densities" of the cable (with units of H/m and F/m), *i.e.*, the unit-length (one-meter) cable inductance and capacitance.

The general classical-physical solutions of these equations in infinite ideal cables are superpositions of waves, with arbitrary frequency, propagating in positive and negative directions in accordance with the d'Alembert solution in Eq. (1). However, in a cable with finite length D, the frequency-space of solutions is quantized to discrete values so that integer multiples of the half-wavelength fit in the cable. Thus the wavelength λ_{max} of the wave with the lowest frequency f_{min} can be written as

$$\lambda_{max} = 2D, \quad f_{min} = \frac{v_c}{2D}. \tag{5}$$

Frequencies below f_{min}, down to zero frequency, constitute a forbidden band of wave states.

The KLJN key exchanger operation strictly requires for security that its frequency range satisfies the quasi-static condition, *i.e.*,

$$f \ll f_{min}. \tag{6}$$

Thus the wave-based scheme and considerations of GAA for eavesdropping violate not only the wave equations in Eqs. (2) and (3), and its d'Alembert solution in Eq. (1), but also the related other fields of classical and quantum physics of waves, because such non-existent solutions are forbidden states.

We note, in passing, that if the wave-based treatment by GAA [1] were correct, we would not have quantization of atomic electron shells, a forbidden band (energy gap) would not exist in solid state physics, semiconductor devices would not work, and even chemistry would be non-existent or at least very different.

2.2. *Violation of the energy dynamics within wave modes*

It is easy to visualize the pertinent physical situation even without the wave equation. While the energy of the wave mode is constant, any electromagnetic wave signifies propagation and oscillatory "bouncing" of energy between magnetic and electric energy forms. To have a physical wave, both aspects are essential, and the propagation must take place via "bouncing" of the energy between its two forms. For example, an L–C resonator has its "bouncing" phenomenon between the electric and magnetic energy forms but lacks propagation, and hence this oscillation is not a wave.

Consider now a short cable satisfying Relation 6, which is driven by an ac voltage generator $U(t)$ at one end and closed with a resistance R_w given by

$$R_w = \sqrt{\frac{L_u}{C_u}} \tag{7}$$

at the other end. Then the cable has equal amounts of energy in the electric and magnetic forms, as discussed below, and both of these energies are oscillating. However, these energies are not "bouncing" between the electric and magnetic forms but oscillate between these forms and the generator. This fact is obvious from the considerations that follow next.

Quasi-static conditions (Relation 6) imply that the time-dependent voltage and current are spatially homogeneous along the wire, *i.e.*,

$$U(x,t) \simeq U(t), \tag{8}$$

$$I(x,t) \simeq I(t). \tag{9}$$

Thus the electrical and magnetic energies, denoted $E_e(t)$ and $E_m(t)$, respectively, in the cable can be written

$$E_e(t) = \frac{1}{2} C_c U^2(t), \tag{10}$$

$$E_m(t) = \frac{1}{2} L_c I^2(t) = \frac{1}{2} L_c \frac{U^2(t)}{R_w^2}, \tag{11}$$

where $C_c = DC_u$ and $L_c = DL_u$ are the capacitance and inductance of the whole cable, respectively, and the last term of Eq. (11) originates from the well-known fact that a cable of arbitrary length, closed with the wave impedance (resistance) R_w, has an input impedance of exactly R_w. Thus

$$\frac{E_e(t)}{E_m(t)} = \frac{C_c}{L_c} R_w^2 = \frac{C_u}{L_u} R_w^2 = 1 \tag{12}$$

at each instant of time. Obviously, no "bouncing" phenomenon takes place between the electric and magnetic forms of energy because the energy is bouncing between the generator and the magnetic and electric fields, not between each other, as would be expected for a wave.

Now suppose that R_A and R_B, rather than R_w, terminate the cable ends. The only significant change that will occur in the quasi-static limit is that the ratio of electrical and magnetic energies will be different from unity, which further goes against the wave hypothesis.

In conclusion, even the simple picture of wave modes given above proves that waves do not exist in the cable in the quasi-static limit.

2.3. *Violation of thermodynamics*

Simple but fundamental thermodynamic considerations also prove that waves cannot exist in the cable in the quasi-static frequency range.

2.3.1. *Violation of the Energy Equipartition Theorem, the principle of Detailed Balance and the Second Law of Thermodynamics*

According to Boltzmann's *Energy Equipartition Theorem* [6] for thermal equilibrium at temperature T, each electromagnetic wave mode has a mean thermal energy equal to kT, where k is Boltzmann's constant. Half of this mean energy is electrical and the other half is magnetic. For N different wave modes in the system, both the electrical and magnetic fields carry a total energy equal to $NkT/2$.

It is easy to see that the assumption that, in a *hypothetical* wave system, these wave energies are less than the above given values violates not only the *Energy Equipartition Theorem* but also the *Principle of Detailed Balance* [6] and the *Second Law of Thermodynamics*: Coupling this hypothetical wave system to a regular one would hit the *Detailed Energy Balance* of equilibrium between wave modes in the hypothetical and regular systems because it would yield an energy flow toward the hypothetical one. This energy flow could then be utilized for perpetual motion machines of the second kind, *i.e.*, violate the *Second Law of Thermodynamics*.

2.3.2. *Violation of Planck's Law and experimental facts for blackbody radiation*

Planck [7] deduced his law of thermal radiation, for simplicity, from the properties of a box with black walls, *i.e.*, internal walls with unity absorptivity and emissivity. In thermal equilibrium, thermal radiation within an infinitely large box with walls of arbitrary absorptivity, emissivity and color has a power spectral intensity for each polarization given by

$$I(f) = \frac{4\pi h f^3}{c^2} \frac{1}{e^{hf/kT} - 1}, \qquad (13)$$

where h is Planck's constant. The derivation of this formula is based on counting existing wave modes in the frequency range $f > f_{min}$, where the minimum frequency is obtained by the same frequency quantization as we claim to exist in a cable. GAA's notions (as well as common electrical engineering education) imply that in closing the finite-size cable by the wave impedance (resistance) R_w at its two ends, all of the lower frequencies, at $f < f_{min}$, will also be available for wave modes as a consequence of the unity absorptivity and emissivity of the impedance match. However, one must realize that a cable closed by the wave impedance (resistance) R_w at its two ends is a one-dimensional realization of Planck's box with perfectly absorbing (black) walls. Allowing the $f < f_{min}$ frequency range for wave modes results in a non-quantized continuum distribution of wave modes, which yields an infinite number$N$ of wave modes in any finite frequency band. Such a situation in the finite-size cable results in an infinite amount of thermal energy NkT in any finite frequency band within the classical-physical frequency range $f \ll kT/h$. This situation is not a problem in a cable or box with infinite size. However, in a finite-size box, infinite thermal energy in the sum of wave modes would naturally result in infinite intensity of blackbody radiation.

Furthermore, the situation is similar for the KLJN scheme in which the cable ends are terminated by R_A and R_B rather than R_w. According to Planck's results, discussed above, the thermal radiation field in the closed box and cable does not depend on the absorptivity and emissivity of the walls — *i.e.*, on the cable termination resistances — since otherwise Planck's Law would be invalid and his formula violate the Second Law of Thermodynamics. Therefore the above argumentation regarding infinite energies and infinite thermal radiation holds for arbitrary termination and wall color.

If there is a loss in the cable and the cable is terminated by R_A and R_B, the same situation holds provided R_A and R_B, and the cable, have the same (noise) temperature because the system is still in thermal equilibrium.

In conclusion, GAA's claims violate the *Planck formula* and *experimental facts about blackbody radiation.*

2.3.3. *Proof that the total energy of the cable is less than the energy required for a single wave mode*

To further support the above conclusions we show that in the quasi-static frequency limit, wherein the KLJN scheme operates, the total energy is not sufficient even to supply a single wave mode, and consequently waves cannot exist without violating several principles and one law of thermodynamics.

As an example, we mathematically analyze how the no-wave situation manifests itself in a lossless cable, which is closed at both ends by resistors equal to the wave impedance R_w at the temperature T. These conditions are not necessary but serve to simplify the calculations. Suppose that, in accordance with the KLJN conditions, the Johnson noise of the resistors has an upper cut-off frequency at f_c so that

$$f_c \ll f_{min} = \frac{v_c}{2D} \tag{14}$$

hold in order to satisfy Relation 6. Thus the spatial homogeneity of current and voltage, Eqs. (8) and (9), is valid for the wire, and it is straightforward to calculate the electric and magnetic energies of thermal origin as shown next.

According to the Johnson–Nyquist formula [8], the thermal electrical energy in the cable capacitance is

$$E_{e,th} = \frac{1}{2}C_c\left\langle U^2(t)\right\rangle = \frac{1}{2}C_c\int_0^{f_c}\frac{4kTR_w/2}{1+f^2/f_{0C}^2}df$$

$$\cong kTC_cR_wf_c = \frac{kT}{2}\frac{f_c}{f_{min}} \ll \frac{kT}{2} \tag{15}$$

and the thermal magnetic energy in the cable inductance is

$$E_{m,th} = \frac{1}{2}L_c\left\langle I^2(t)\right\rangle = \frac{1}{2}L_c\int_0^{f_c}\frac{4kT/\left(2R_w\right)}{1+f^2/f_{0L}^2}df$$

$$\cong kT\frac{L_c}{R_w}f_c = \frac{kT}{2}\frac{f_c}{f_{min}} \ll \frac{kT}{2}, \tag{16}$$

where the characteristic frequencies of the Lorentzian spectra in Eqs. (15) and (16) are defined as

$$f_{0C} = \frac{1}{2\pi C_cR_w/2} = \frac{1}{\pi DC_u}\sqrt{\frac{C_u}{L_u}} = \frac{1}{\pi D}\sqrt{\frac{1}{L_uC_u}} = \frac{1}{\pi}\frac{v_c}{D} = \frac{2}{\pi}f_{min} \tag{17}$$

and

$$f_{0L} = \frac{2R_w}{2\pi L_c} = \frac{1}{\pi L_c}\sqrt{\frac{L_u}{C_u}} = \frac{1}{\pi D}\sqrt{\frac{1}{L_u C_u}} = \frac{1}{\pi}\frac{v_c}{D} = \frac{2}{\pi} f_{min}. \quad (18)$$

Similar calculations can be carried out for the general case in which the cable ends are not terminated by R_w but with different resistance values R_A and R_B. Specifically, the parallel resultant resistance R_A and R_B (instead of $R_w/2$) enters in the left-hand side of Eqs. (15) and (17), and the serial (sum) loop resistance $R_A + R_B$ (instead of $2R_w$) enters in the left-hand side of Eqs. (16) and (18), while the final inequalities shown by Eqs. (15) and (16) remain.

If there is a loss in the cable, and the cable is terminated by R_A and R_B, the same situation holds provided R_A and R_B, as well as the cable, have the same (noise) temperature, because the system is still in thermal equilibrium. If the cable is cooler then there is an energy flow out of the cable, which further strengthens the inequalities at the right hand sides of Eqs. (15) and (16).

In conclusion, Eqs. (15) and (16) prove that for a short cable and within the frequency range of interest for the KLJN scheme, the sum of electrical and magnetic energies in all of the hypothetical "wave modes" is much less than the energy needed for a single wave mode in thermal equilibrium.

3. Correct treatment of cable delays in the frequency range for the KLJN scheme

3.1. *General considerations*

We showed above that wave modes cannot exist in the cable at the KLJN condition $f \ll f_{min}$. A number of questions then arise naturally, such as (*i*) what type of system is the cable under these conditions, (*ii*) what is the nature of the propagating fluctuations caused by Alice's and Bob's noise generators, and (*iii*) are there any other implications of the KLJN condition?

To answer these questions, we first note that the system under consideration is not a waveguide, as implied by GAA [1], but a *distributed impedance network* in the *quasi-static limit*. Secondly, the propagating fluctuations are not waves but phase-shifted voltages and currents; in the language of physics they are related to *retarded potentials* of non-wave solutions, and in electrical engineering vocabulary they are *spatio-temporal fluctuations in an impedance network.*

The general implications of the KLJN conditions are very pervasive, as elaborated and discussed in Section 2 above. The specific consequences for the KLJN scheme are that the mathematical and physical framework used by GAA [1] is invalid and that the same applies to their experimental analysis, as shown in the present paper and as will be further discussed in forthcoming articles.

When the frequency converges towards zero, the impact of the inductance and capacitance of the cable on the cable current and voltage also rapidly diminish. However, the voltage drop over the cable is determined by its serial resistance R_c and inductance L_c, because the capacitive shunt currents approach zero. Thus the first-order approximation of the cable impedance is

$$Z_c \cong R_c + j2\pi f L_c. \tag{19}$$

For simplicity, we analyze the situation wherein the cable loss (resistance) is negligible so that

$$Z_c \cong j2\pi f L_c. \tag{20}$$

The corresponding phase delay of the voltage at Bob's end, compared to that of Alice's end, is

$$\varphi_{AB} = -2\pi f L_c / R_B \tag{21}$$

when the voltage is generated by Alice. This phase delay corresponds to a *frequency-independent time delay* according to

$$\tau_{AB} = L_c / R_B, \tag{22}$$

which at first glance seems to suggest that we are dealing with waves and that the d'Alembert equation holds, as stated by GAA. However one must realize that this *time delay depends on the load resistance* R_B at the other end of the cable, which implies that the *time delay and measured phase velocity in the two directions are different* due to the condition $R_A \neq R_B$ during secure bit exchange [2–5], *i.e.*, under circumstances such that GAA's method [1] is supposed to function. To illustrate this dichotomy, we evaluate the phase delay for voltage propagation in the opposite direction, *i.e.*, when the voltage is generated by Bob. Now one finds

$$\varphi_{BA} = 2\pi f L_c / R_A \tag{23}$$

and

$$\tau_{BA} = L_c / R_A. \tag{24}$$

Thus the d'Alembert equation, applied by GAA [1] to prove the existence of waves, cannot be used in the present situation.

3.2. *Simulation based on a circuit model for the cable*

To corroborate the theoretical considerations above, we used Linear Technology's LTspice-IV cable simulator software to analyze the experimental situation in GAA's work [1] and confirmed all of their stated results. The simulations proved that a coaxial cable with parameters and conditions similar to those employed by GAA [1] can be modeled with the lumped impedance circuitry shown in Fig. 2, where parts (a) and (b) represent a cable with and without loss, respectively.

Figure 3 shows results of our simulation addressing the experimental data in GAA's article [1]. The conditions are the same as those of GAA [1] and reported in *their Figure* 5, but our simulation uses a practical cable model and simple impedance representations (see Fig. 2) which fit the cable data to a high degree of accuracy. The practical lossy cable and the simple

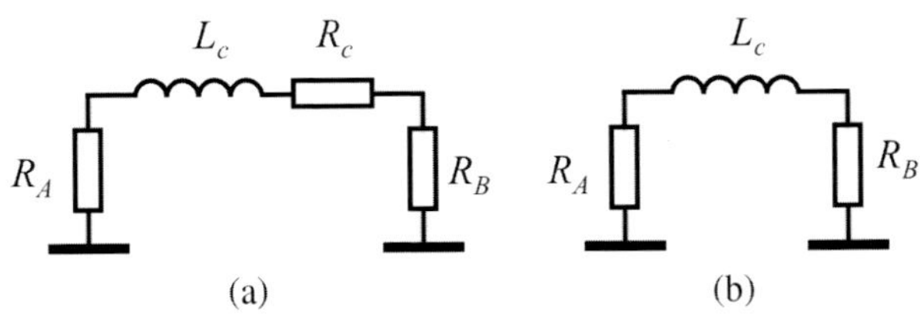

Figure 2. Lumped impedance-components-based model of a cable at low frequencies for analyzing voltage drop along the cable and phase shift in the limit $f \ll f_{min}$. Part (a) represents a cable with loss (cable inductance and resistance are designated L_c and R_c, respectively), and part (b) represents a lossless cable.

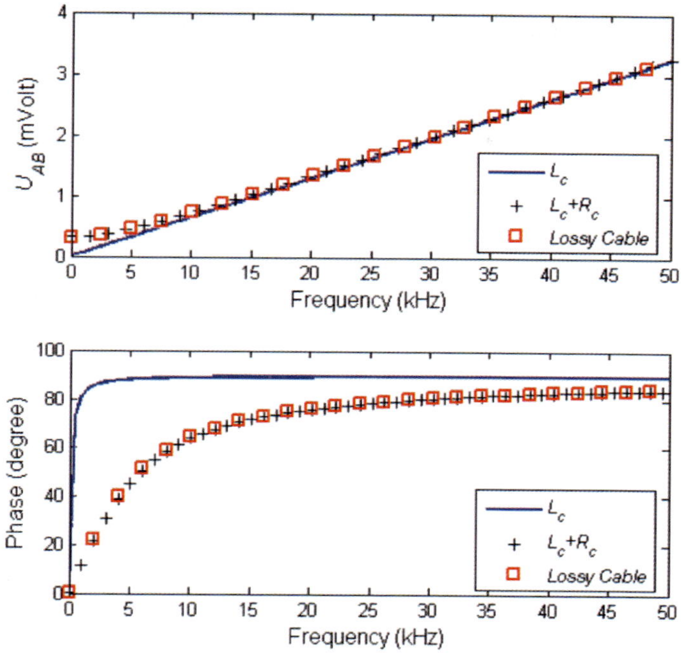

Figure 3. Comparison of simulated data based on impedance models, using LTspice, with those of a real (lossy) cable. Alice's and Bob's resistors, denoted R_A and R_B, have the resistance R_W, and Alice drives the cable with a sinusoidal voltage generator (1 V) via Bob's resistor. The cable is characterized by length $D = 1.5\,\text{m}$, $L_c = 1.03\,\mu\text{H}$, $R_c = 0.0315\,\Omega$, and $C_c = 150\,\text{pF}$. Upper panel shows voltage drop U_{AB} over the cable between Alice's and Bob's ends and lower panel shows phase shift of U_{AB} compared to that of the voltage at Alice's end. Squares signify simulations of a lossy cable (model RG58), crosses represent data obtained by use of the lumped-parameters-model in Fig. 2a, and solid line was derived from an inductance model devised to simulate a lossless cable (Fig. 2b). These results are in full agreement with the experimental data shown in *Figure 5 of GAA's work* [1].

impedance model in Figure 2(a) give identical results, while results for the lossless cable (corresponding to data compensated for loss in GAA's work [1]) are nicely represented by the simple inductance model in Figure 2(b). Moreover, it is obvious that the cable inductance L_c produces a voltage drop that is the time-derivative of the current, which is determined predominantly by the resistances in the loop. Thus the voltage drop for the lossless cable is the time derivative of Alice's generator voltage, and this experimental finding by GAA [1] to "support" the d'Alembert equation is simply an inductor-type voltage response and it has nothing to do with waves.

Table 1. Simulated equivalent phase velocity calculated from phase shifts between the two ends of the cable versus driving frequency and load resistance (resistance of termination at the other end). The dependence on the resistance is in violation of the d'Alembert equation, for example, in the KLJN system during secure key exchange where the terminal resistances are different. Cable parameters are given in the caption for Fig. 3.

	1 kHz	5 kHz
10 Ω	3.99998×10^7 m/s	4.00018×10^7 m/s
2 Ω	7.99996×10^7 m/s	8.00038×10^7 m/s
50 Ω	1.99999×10^8 m/s	2.00007×10^8 m/s
1 k Ω	3.99993×10^9 m/s	4.00011×10^9 m/s
10 k Ω	3.99946×10^{10} m/s	4.00041×10^{10} m/s

To subject Eqs. (22) and (24) to a final test, we evaluated the phase delay and corresponding time delay toward Bob when the resistor at Bob's end was varied and the cable was lossless. Data are shown in Table 1 and verify the correctness of our Eq. (22) to an accuracy of five digits. GAA's "propagation velocity" toward Bob is practically independent of frequency but depends on Bob's resistor. During secure bit exchange, the "propagation" times toward Alice and Bob are different. This fact verifies our conclusion that GAA's use of the d'Alembert equation, as the base of their mathematical considerations, is indeed incorrect.

One should observe that, for the cases of 1 kΩ and 10 kΩ, GAA's "phase velocity" is greater than the speed of light. This is acceptable and happens often with the phase velocity of oscillations in a *driven impedance system in the steady-state*; however, it is prohibited in the d'Alembert equation as a consequence of the theory of special relativity [9]. Similar effects can happen in wave-based systems with reflecting boundary conditions in stationary mode after the waves fill the system; however, in wave-based systems the phase velocity would be the same for the left and right directions.

In conclusion, the proper KLJN scheme is a simple impedance circuitry with related phase shifts where the corresponding time shifts are asymmetrical during secure bit exchange.

4. Additional notes

One of the reviewers raised an interesting question: In the light of the results of the present paper, how should one see the situation when a small part of a wave-based system is observed or when that part has different properties? Are there waves in those parts? Such a situation is common in physical systems and the answer is straightforward. We show two examples to elucidate the issue.

(a) First consider that a small part of a large wave-based system is observed, so small that Eq. (6) is satisfied. The answer to the question above is then simple: the part is a non-separable component of the wave system, and the observer views a small spatial fraction of the whole wave field. The energy of the observed wave mode is not the energy within this small part but the integral of related wave energy over the large system. However, if we separate the small system from the large one and terminate its ends, wave solutions would not exist at these frequencies.

(b) Another special example is a quarter-wavelength ($\lambda/4$) cable. If we terminate one end and drive the other end with a generator, the latter end acts as a distributed impedance transformer without wave modes. However if, instead of a generator, the same end is connected to a long cable with the same R_w, it will belong to the relevant wave mode that extends over the system and contains a small part of the energy of that wave mode.

5. Conclusions

The efforts of GAA [1] to utilize time delays in cables to crack the KLJN scheme represent an interesting and novel approach and, as such, deserve attention. However, it should be mentioned that Liu [10] earlier used a similar technique, but with unphysical conditions for the simulations [11]. As shown in considerable detail above, GAA's efforts can be irrevocably refuted. Further attempts to crack the KLJN scheme are welcome, but it is essential that they be founded on correct physical models.

Finally, we hope that the fact that waves cannot exist below f_{min} in a short cable, as discussed in Section 2, will eventually enrich the physical underpinnings in electrical engineering courses and textbooks.

Acknowledgements

Discussions with Peter Hrasko, Robert Nevels, Lazar Kish, Krzysztof Michalski, Mehrdad Ehsani, Muneer Mohammad, Dave Ferry and Mark Dykman are appreciated. We are grateful to Lachlan Gunn for showing us the results in [1] and to Derek Abbott for discussions.

References

1. Gunn, L.J., Allison, A., Abbott, D. (2014). "A directional coupler attack against the Kish key distribution system", manuscript http://arxiv.org/abs/ 1402.2709 versions 1 and 2.
2. Kish, L.B., Granqvist, C.G. (2014). "On the security of the Kirchhoff-law–Johnson-noise (KLJN) communicator", *Quantum Information Processing* 13 (2014) 2213–2219.
3. Mingesz, R., Kish, L.B., Gingl, Z., Granqvist, C.G., Wen, H., Peper, F., Eubanks, T., Schmera, G. (2013). "Unconditional security by the laws of classical physics", *Metrology & Measurement Systems*, **XX**, 3–16. Open access: http://www. degruyter.com/view/j/mms.2013.20.issue-1/mms-2013-0001/mms-2013-0001.xml
4. Kish, L.B., Abbott, D., Granqvist, C.G. (2013). "Critical analysis of the Bennett–Riedel attack on secure cryptographic key distributions via the Kirchhoff-law–Johnson-noise scheme", *PLoS ONE*, **8**, e81810. Open access: http://www.plosone.org/article/info%3Adoi%2F10.1371%2Fjournal.pone. 0081810
5. Abbott, D., Schmera, G. (2013). "Secure communications using the KLJN scheme", *Scholarpedia*, **8**, 31157.
6. Boltzmann, L. (1964). *Lectures on Gas Theory*, University of California Press, Berkeley, CA, USA.
7. Planck, M., Morton, M. (1914). *The Theory of Heat Radiation*, P. Blakiston's Son & Co, Philadelphia, PA, USA.
8. Nyquist, H. (1928) "Thermal agitation of electric charge in conductors", *Physical Rev*iew, **32**, 10–113.
9. Einstein, A. (1905). "Zur elektrodynamik bewegter Körper", *Annalen der Physik*, **17**, 891–921.
10. Liu, P.-L. (2009). "A new look at the classical key exchange system based on amplified Johnson noise", *Physics Letters A*, **373**, 901–904.
11. Kish, L.B., Horvath, T. (2009). "Notes on recent approaches concerning the Kirchhoff-law–Johnson-noise-based secure key exchange", *Physics Letters A*, **373**, 2858–2868.

4.6.2. The correct analysis and the failure of the Gunn-Allison-Abbott attack

Figure 4.6.3. Lachlan Gunn and Derek Abbott, on November 10, 2013, at the foot of the Mao Zedong statue in Changsha, China, during the inaugural Hot Topics in Physical Informatics (HoTPI) conference. Right after this image was taken, Lachlan told me about his "wave-based" scheme and experiments, and I warned him that there are no waves present. Lachlan, who is the PhD student of Derek, is a creative applied mathematician and a very hard worker.

Paper: On the "cracking" scheme in the paper "a directional coupler attack against the Kish key distribution system" by Gunn, Allison and Abbott[1]

Abstract. Recently, Gunn, Allison and Abbott (GAA) [http://arxiv.org/pdf/1402.2709v2.pdf] proposed a new scheme to utilize electromagnetic waves for eavesdropping on the Kirchhoff-law–Johnson-noise (KLJN) secure key distribution. We proved in a former paper [*Fluct. Noise Lett.*, **13** (2014), 1450016] that GAA's mathematical model is unphysical. Here we analyze GAA's cracking scheme and show that, in the case of a loss-free cable, it provides less eavesdropping information than in the earlier (Bergou)–Scheuer–Yariv mean-square-based attack [Kish, L.B., Scheuer, J. (2010). *Phys. Lett. A*, **374**, 2140–2142], while it offers no information in the case of a lossy cable. We also investigate GAA's claim to be experimentally capable of distinguishing — using statistics over a few correlation times only — the distributions of two Gaussian noises with a relative variance difference of less than 10–8. Normally such distinctions would require hundreds of millions of correlation times to be observable. We

[1]Chen, H.P., Kish, L.B., Granqvist, C.G., Schmera, G. (2014). *Metrology and Measurement Systems*, **21**, 389–400. http://arxiv.org/abs/1405.2034 ; http://vixra.org/abs/1404.0081.

identify several potential experimental artifacts as results of poor KLJN design, which can lead to GAA's assertions: deterministic currents due to spurious harmonic components caused by ground loops, DC offset, aliasing, non-Gaussian features including non-linearities and other non-idealities in generators, and the time-derivative nature of GAA's scheme which tends to enhance all of these artifacts.

Keywords: KLJN key exchange; information theoretic security; unconditional security.

1. Introduction

Recently Gunn, Allison and Abbott (GAA) [1] proposed a new scheme to utilize electromagnetic waves for eavesdropping on the Kirchhoff-law–Johnson-noise (KLJN) secure key distribution. In a former paper [2], we proved that claims concerning electromagnetic waves are unphysical in GAA's attack, since the quasi-static limit holds for the KLJN scheme and implies that physical waves do not exist in its wire channel. An assumption of wave modes in a short cable, and in the low-frequency limit, in fact violates a number of laws of physics, including the Second Law of Thermodynamics. One aspect related to these mistakes is that, in electrical engineer jargon, all oscillating and propagating time functions are called waves, whereas in physics the corresponding retarded potentials can be wave-type or non-wave-type. Physical waves involve two dual energy forms that regenerate each other during propagation; these forms can involve electrical and magnetic fields, or deal with kinetic and potential energy as in the case of elastic waves. Non-wave-type retarded potentials in the quasi-static regime, however, have negligible crosstalk between the two energy forms, and energy exchange takes place between them and generators [2]; this latter situation pertains to the KLJN scheme. We note in passing that, while there are no physical waves in the KLJN system, the propagation delay of the non-wave-type retarded potentials may still provide information for Eve, and therefore a correct analysis is essential.

In the steady-state driving case, the correct analysis [2] shows that the starting d'Alembert equation

$$U(t, x) = U_+\left(t - \frac{x}{v}\right) + U_-\left(t + \frac{x}{v}\right), \qquad (1)$$

which is the foundation of GAA's approach, is invalid because the system under study cannot be described with a single phase velocity [2], but these velocities are directionally dependent during secure key exchange. Here U_+ and U_- are voltage components of waves propagating to the right and left along the x-axis, and v is a single propagation velocity. GAA used Eq. (1) to deduce the equations

$$\frac{dU}{dt} + v\frac{dU}{dx} = 2\frac{dU_+}{dt} \tag{2}$$

and

$$\frac{dU}{dt} - v\frac{dU}{dx} = 2\frac{dU_-}{dt} \tag{3}$$

as a basis of their "directional coupler" attack. Their claim [1] is that the quantities at the left-hand side of Eqs. (2) and (3) are measurable so that the time derivatives at the right-hand side of the equations can be calculated and used for eavesdropping.

Before analyzing the experimental claims and potential artifacts, we take a closer look at the mathematics of Eqs. (1–3).

2. Mathematical analysis of GAA's scheme

In this section, we present a correct analysis of GAA's scheme and show that Eve's eavesdropped information is always less within the GAA scheme than within the old mean-square attack based on the comparison of two end-voltages [3], unless there are flaws in the realization of the KLJN key exchanger.

We assume in the rest of the paper that the bit-value arrangement between Alice and Bob is mixed, *i.e.*, one of them connects the large resistance to the cable and the other uses the small resistance. This situation indicates not only a secure key exchange event but also that different phase velocities must be used for the two directions in Eq. (1) during steady-state conditions (see also related theory and verifications by simulation in our former paper [2]).

2.1. *General considerations*

Even for waves, Eq. (1) is not suitable for steady-state excitations [4] and the second term violates causality. However there is a way to modify this equation under steady-state conditions in the case of KLJN by using direction-dependent phase velocities [2] of retarded potentials. Furthermore causality is ascertained by setting

$$U(t, x) = U_+\left(t - \frac{x}{v_+}\right) + U_-\left(t - \frac{D - x}{v_-}\right), \tag{4}$$

where $x = 0$ and $x = D$ pertain to the left-hand and right-hand ends of the cable, respectively, and D is cable length. The phase velocities are

$$v_+ = \frac{DR_B}{L_c} \quad \text{and} \quad v_- = \frac{DR_A}{L_c}, \tag{5}$$

where R_A and R_B are Alice's and Bob's resistances, L_c is the inductance of the cable and it is assumed that Alice is at $x = 0$ and Bob is at $x = D$.

It is important to realize that, according to Eqs. (4) and (5), Eve must know Alice's and Bob's resistor values in order to have the correct input for the GAA experiments. This implies that Eve's one-bit uncertainty persists, which is the *indication* of security. A *proof* of security will be given below in Eq. (21).

Our earlier work [2] proved that, in the quasi-static frequency limit pertinent to the KLJN scheme, the exact distributed-impedance rendition of the cable shown in Fig. 1 leads to the simplified serial impedance models in Figs. 2a and 2b because the capacitive currents converge towards zero in the limit of low frequencies. Figure 2a is a first-order approximation of the real situation while Fig. 2b models a situation wherein the cable is lossless or the voltage drop on the resistive component is negligible compared to that of the inductive component in the dominant frequency range of the quasi-static regime.

Figure 3 provides a starting point for our security analysis of the KLJN scheme. U_A and U_B are voltages of the (thermal) noise voltage generators, and R_A and R_B are Alice's and Bob's resistors, respectively. Here U_1 and U_2 are voltages at the two ends of the cable, and Z_c denotes cable impedance. Furthermore we set $U_{12} = U_1 - U_2$.

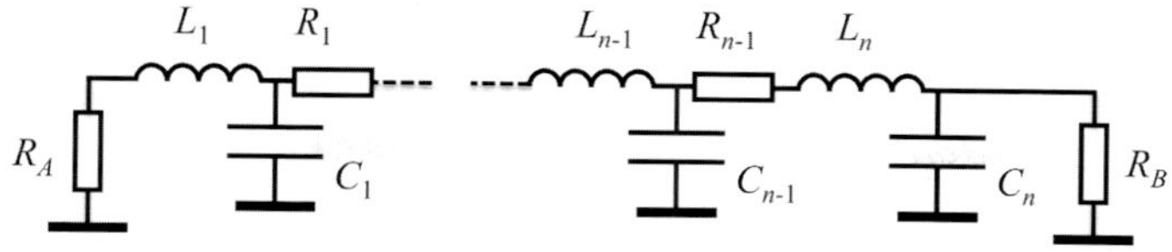

Figure 1. Outline of the pertinent part of the KLJN scheme with a distributed LCR model of a long and leakage-free cable [2]. When the cable losses can be neglected, one may omit the R_i resistors representing the distributed resistance of the cable. Alice's and Bob's resistors — denoted R_A and R_B, respectively — are randomly selected from the set $\{R_L, R_H\}$with $(R_L \neq R_H)$ at the beginning of each bit-exchange period. These resistors, with associated serial generators (not shown), emulate thermal noise with high noise temperature and strongly limited bandwidth.

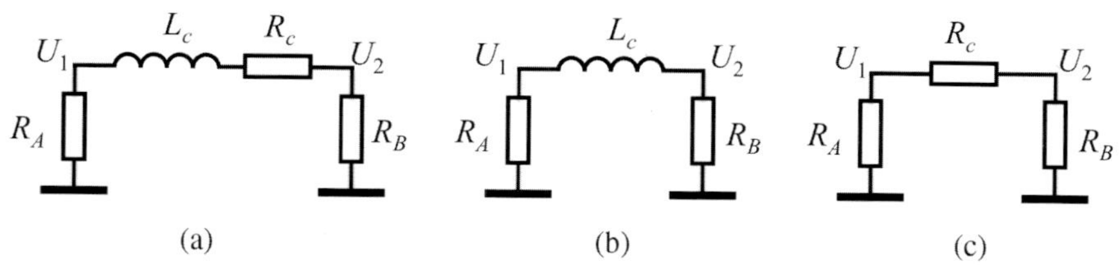

Figure 2. Lumped impedance-components-based model of a short cable at low frequencies for analyzing voltage drop along the cable and phase shift in the quasi static limit [2]. Part (a) represents a cable with loss (cable inductance and resistance are designated L_c and R_c, respectively), and part (b) represents a lossless cable. Part (c) is used to determine the voltage drop in the asymptotic case where loss dominates the cable impedance (this case is not practical and is used only for the sake of completeness).

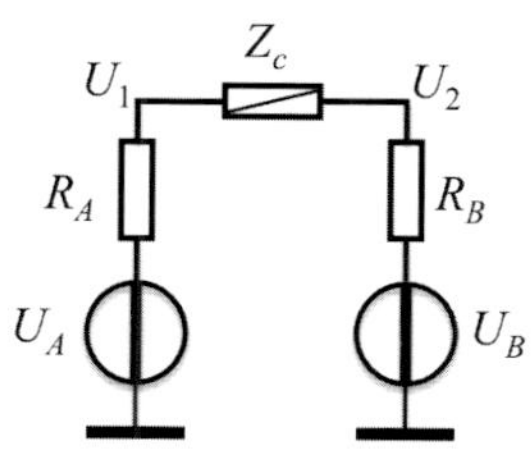

Figure 3. Circuit for impedance-based analysis of GAA's attack, as described in the main text.

In order to accomplish security, the voltage drop on the cable must be kept small in comparison with other voltages [3], *i.e.*,

$$U_1 \cong U_2 \equiv U. \tag{6}$$

We first consider the earlier mentioned wire-resistance-based attack [3, 5], wherein miniscule differences between the mean-square voltages $\langle U_1^2 \rangle$ and $\langle U_2^2 \rangle$ served as information leak toward Eve. In the experimental demonstration, [5], chossing the number of observed correlation times N_{oc} during bit-exchange to 50, the wire resistance to 200 Ω, $R_A = 2$ kΩ and $R_B = 9$kΩ, resulted in Eve's successful guessing probability $p = 0.525$, which means that the relative information leak was 0.19% of the exchanged key bits. This implies that two-stage privacy amplification would be needed [6] to reduce this leak to a level below the desired one of 10^{-8}. For this type of attack, p scales as [7]

$$p = 0.5 + \theta \frac{|Z_c|^2}{R_A R_B} \tag{7}$$

at fixed N_{oc}, where θ is a constant which depends only on N_{oc}.

Turning now to GAA's experiments [1], $|Z_c|^2$ is about 10^5 times smaller than before while their R_A and R_B have similar values (1 kΩ and 10 kΩ, respectively). With the same N_{oc} as above and using the old method [3] together with GAA's parameters, Eve's probability of successful bit guessing would be [6]

$$p \approx 0.5000002, \tag{8}$$

which is better than even the value of $p$needed to secure the upper limit of 10^{-8} for the relative information leak ($p = 0.5006$ [6]).

In stark contradiction to the results above, GAA assert that by using their standard statistical method they measure

$$p \approx 1 \tag{9}$$

at the given conditions. This is an extraordinarily strong claim which — if correct — would mean that Eve can perform a nearly-deterministic guess not only about the bit-states but also of the exact time dependence of Alice's and Bob's noise voltages.

2.2. *Case 1: Lossless short cable with very small impedance*

In the case of a *lossless cable* with very small impedance, we assume that Rel. 6 holds but that U_{12} remains measurable. Suppose now that Eve employs Eq. (2) to extract information. Using the proper velocity and the measurable quantities at the left-hand side of Eq. (2a), we get

$$\frac{dU}{dt} + v_+ \frac{dU}{dx} = 2\frac{dU_x}{dt}. \tag{10}$$

The resulting voltage $U_x(t)$ in the right-hand side of Eq. (2a) needs clarification at this point. To this end we Fourier transform Eq. (6) and obtain

$$j\omega U(\omega) + v_+ \frac{dU(\omega)}{dx} = 2\,j\omega U_x(\omega), \tag{11}$$

so that

$$2U_x(\omega) = U(\omega) + \frac{v_+}{j\omega}\frac{dU(\omega)}{dx} \tag{12}$$

or

$$2U_x(\omega) = U(\omega) + \frac{DR_B}{j\omega L_c}\frac{dU(\omega)}{dx} = U(\omega) + \frac{DR_B}{j\omega L_c}\frac{U_{12}(\omega)}{D}, \tag{13}$$

where $U_{12}(\omega) = U_1(\omega) - U_2(\omega)$. Using Ohm's law for impedance, we have

$$2U_x(\omega) = U(\omega) + R_B \frac{U_{1,2}(\omega)}{j\omega L_c} = U(\omega) + R_B I(\omega). \tag{14}$$

A relation

$$U(\omega) \cong U_1(\omega) \cong U_2(\omega), \tag{15}$$

which is the Fourier transform of Eq. (8), holds when the cable impedance is very small (*cf.* Fig. 3) so that

$$2U_x(\omega) = U_2(\omega) + R_B I(\omega) = U_B(\omega). \tag{16}$$

After inverse Fourier transformation and substituting the voltages back into Eq. (6), it is found that the corrected Eq. (2a) reads as

$$\frac{dU}{dt} + v_+ \frac{dU}{dx} = \frac{dU_B(t)}{dt}. \tag{17}$$

Similar considerations for GAA's other equation, with the opposite sign of the second term, lead to

$$\frac{dU}{dt} - v_{-}\frac{dU}{dx} = \frac{dU_A(t)}{dt}. \tag{18}$$

The right-hand sides of Eqs. (17) and (18) give the time derivative of the voltages of Alice's and Bob's generators provided Eve uses the *correct guess* and consequently substitutes the *correct resistances* in these equations. This result proves that GAA do not have a directional coupler but something else, which can be called a "separator" and is able to extract the voltage amplitudes of Alice's and Bob's generators (without the voltage-division caused by the resistor at the other end). Such a tool would be even better for Eve, but it works only if the correct phase velocity is assumed. The phase velocity in the steady state is determined by the unknown resistor terminating the cable toward the propagation direction [2], and therefore Eve must correctly guess the value of the resistor at that end in order to obtain the correct voltage.

The most important question is this: what happens if Eve assumes the *wrong resistor value* at Bob's side, *i.e.*, if Eve assumes Alice's resistor value? Obviously, the resulting voltage $U_x(t)$ will then be a weighted superposition of the voltages seen by Alice and Bob. However, the real question concerns the statistical properties associated with Eve's choice. Can Eve utilize these properties to extract information?

The answer to the questions is simple, and we first observe that the voltage $U_x(t)$ will be a Gaussian noise [8–10], because a linear combination of Gaussians results in a Gaussian as a consequence of the Central Limit Theorem [11]. Thus the real question regards the variance of the voltage. Its calculation is straightforward, and Eq. (16) becomes

$$2U_x(\omega) = U_2(\omega) + R_A I(\omega) \cong U_1(\omega) + R_A I(\omega). \tag{19}$$

Here it is important to realize that the cable voltage and cable current are orthogonal — *i.e.*, uncorrelated — in order to ensure zero net power flow and satisfy the Second Law of Thermodynamics [7,12–15], so that

$$\langle U(t)I(t)\rangle = 0. \tag{20}$$

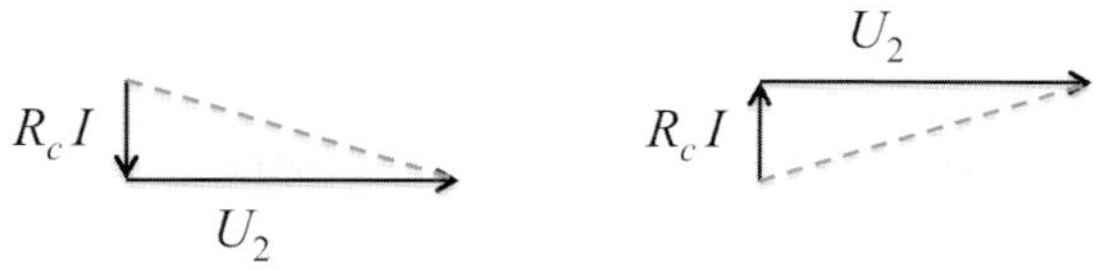

Figure 4. Illustration showing that added orthogonal noise voltages produce the same RMS voltage and mean-square voltage even if the sign of the current is flipped to the opposite value. The resulting time-dependent voltages will be different although their RMS and mean-square and RMS amplitudes remain the same.

Thus Pythagoras' Rule gives that the variance (mean-square) of the sum at the right-hand side of Eq. (19) is invariant to changing the plus sign to a minus sign, which is given a pictorial rendition in Fig. 4. It follows that

$$\langle U_x^2(t)\rangle = \langle U_1^2(t)\rangle + R_A^2\langle I^2(t)\rangle = \langle U_1^2(t)\rangle - R_A^2\langle I^2(t)\rangle, \qquad (21)$$

which is exactly the variance of Alice's noise voltage in accordance with Kirchhoff's law (*cf.* Fig. 3). It should be remarked that GAA used time derivatives, but this does not change the situation of orthogonality.

The mean square voltage always corresponds to that of the noise source of the assumed resistor, which means that Eve gets what she assumes instead of learning about the true bit-situation provided Rel. 6 holds. Thus the only role of the inductance of the lossless cable is to detect the current in the wire, so that Eve's one-bit inaccuracy remains which proves the security of the key exchange against this attack. It should also be noted that "separators" of the above mentioned kind can easily be realized by directly measuring the current and using Ohm's law with guessed resistance values to determine the voltages at Alice's and Bob's ends. A further discussion of the latter issue was given elsewhere [7], where the separators were described and referred to as "impedance-based directional couplers", and where it was pointed out that they are useless for Eve. Thus the obtained mean-square voltages satisfy the *supposed* resistance value, and Eve cannot extract any information by using this system provided Rel. 6 holds. Section 2.4 below elucidates the role of the approximation leading to Rel. 6.

2.3. *Case 2: Short cable dominated by loss*

For a *lossy cable*, the voltage drop over the resistor makes even the modified d'Alembert-equation-approach in Eq. (4) invalid, even if the correct phase velocity is used. Equations (12) and (13) become

$$2U_x(\omega) = U(\omega) + \frac{v_+}{j\omega}\frac{U_{12}(\omega)}{D}$$
$$= U(\omega) + \frac{DR_B}{j\omega L_c}\frac{U_{12}(\omega)}{D} = U(\omega) + \frac{R_B}{L_c}\frac{U_{12}(\omega)}{j\omega}, \qquad (22)$$

and inverse Fourier transformation yields

$$2U_x(t) = U(t) + \frac{R_B}{L_c}\int U_{12}(t)dt. \qquad (23)$$

The obtained U_x does not have any meaning or information for Eve, because $U(t)$ and the integral of $U_{12}(t)$ — which is proportional to the time integral of the current — are orthogonal even if the current and $U(t)$ have some small correlation due to loss.

2.4. *Conclusion of sections 2.2 and 2.3*

As shown above Eve cannot extract any information, neither in the lossless nor in the lossy cable, provided Rel. 6 holds. However, this relation is only approximate, because there is a non-zero difference U_{12} between U_1 and U_2. This small difference causes a small offset between the related results which indeed is information for Eve. But this offset is the very same as that utilized directly in the old mean-square-comparison based wire-resistance-attack method [3] without the extra noise components shown above. The extra noise components weaken Eve's information, and therefore the conclusion is straightforward: GAA's method always provides less information than the old wire-resistance-attack [3].

3. Experiments: What could go wrong?[2]

Many things could go wrong in GAA's experiments claimed to prove the validity of their attack against the KLJN scheme. Here we try to identify the

[2]This time we could only guess what did go wrong because we did not know that GAA used a voltage divider (attenuator) to simulate lossy cables, which breaks the KLJN loop into two loops. These are not KLJN and not secure. This fact shows how poorly documentad their Science Report paper was. For a full analysis, see Section 4.6.3.

most probable deficiencies but presume that conceptual errors concerning the experiments are not present. From the many possibilities, we select only a few and only those directly related to the realization of the KLJN scheme but not to the measurement set-up as such.

3.1. *The experimental claim*

GAA [1] used a standard statistical method to compare distributions of extracted voltage components and to identify the bit (resistor) arrangement at the two ends of the wire in the KLJN scheme. They asserted that they were able to identify the resistor arrangement within a very short time in the case of lossy cables.

Let us now estimate the observable relative difference of the mean-square voltages at the two cable ends in GAA's experiment: The resistors were 1 kΩ and 10 kΩ, and the cable length was 1.5 m and 2 m. GAA did not specify their cable parameters, but a 1.5-meter-long cable taken to be, as a reasonable assumption, a copper wire with a cross section of 1 mm^2 yields a cable resistance of 0.07 Ω.

As seen above, the old wire-resistance-attack [3] gives an upper limit for the extracted information. A mean-square operation is an efficient estimator for Gaussian processes [8], and therefore other statistical methods cannot offer much advantage. Using the result in an earlier paper [3] for the measurable relative mean-square voltage difference, we obtain

$$\Delta_{rel}^2 \approx \frac{\left|\langle U_1^2(t)\rangle - \langle U_2^2(t)\rangle\right|}{\langle U_1^2(t)\rangle} \approx \frac{\left|\langle U_1^2(t)\rangle - \langle U_2^2(t)\rangle\right|}{\langle U_2^2(t)\rangle}$$
$$\approx \frac{R_c^2}{R_A R_B} = \frac{0.07}{10^3 10^4} = 7 \times 10^{-9}, \qquad (24)$$

showing that the imbalance of the mean-square voltages of the two Gaussian noises is less than 10^{-8}. GAA's claim to identify which one of these distributions is the narrower by sampling a few correlation times is untenable, and normally hundreds of millions of correlation times would be required for a reasonably low error probability.

The question then arises as to what GAA did measure and how they obtained their surprising results?

3.2. *Non-Gaussianity*

According to the security proofs in earlier work [9, 10], it is a strict mathematical requirement for the security of the KLJN scheme to have Gaussian processes, which means that the time derivatives also must be Gaussians. GAA did not specify their waveform generator, and thus the degree of Gaussianity remains unclear.

It is important to notice that most commercial noise generators use algorithms and filtering to approach Gaussians. Due to the Central Limit Theorem [11], time-integration shifts the statistics of noises toward Gaussians whereas time derivatives, which were used by GAA, strongly amplify non-Gaussian components.

Thus one strong candidate for causing the poor performance of the KLJN system in GAA's study [1] is non-Gaussianity of time derivatives.

3.3. *Aliasing effects, non-linearity, and spurious noise components*

Aliasing effects (which cause high-frequency non-Gaussian noises), non-linearity and other types of spurious signals in the generator are other strong candidates for destroying Gaussianity. Again, a time derivative will severely emphasize these weaknesses.

3.4. *Deterministic currents in the loop*

A low-frequency or DC current component may exist in the cable of the KLJN scheme and may be caused by a ground loop (leading to a 50/60Hz sinusoidal current) or DC offset. The voltage drop originating from such parasitic currents will introduce a location-dependent bias into the key distributions and quickly uncover the nature of the resistors at the two ends of the wire, as illustrated in Fig. 5.

However, Eve does not need to use GAA's method [1] to elucidate the resistor values: She can simply measure and compare the DC or 50/60 Hz voltage components of the strongly correlated voltage noises at the two ends of the wire and extract the key or its inverse. Figure 6 shows, as an example, computer simulations of two strongly correlated noises with a small DC shift. In this particular case, a single-time measurement is able to identify

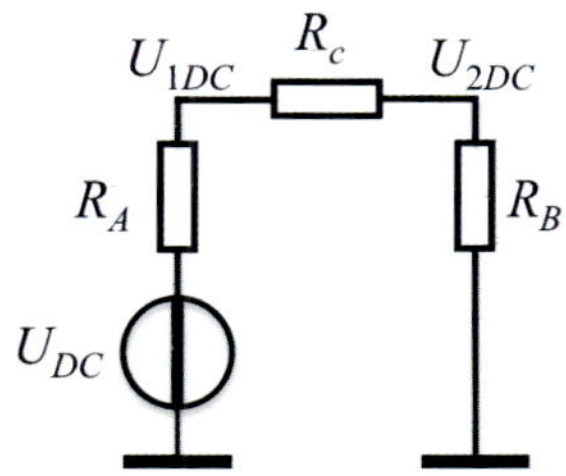

Figure 5. Schematic circuit for illustrating parasitic DC and low-frequency artifacts caused by ground loops. For the sake of simplicity, the parasitic source is assumed to exist solely at Alice's side. Only the parasitic voltage generator is shown, because its impact is additive to the voltages caused by other circuit deficiencies. The parasitic DC or low-frequency components U_{1DC} and U_{2DC} of U_1 and U_2, respectively, are sensitive to the location of the low/high resistor at Alice's and Bob's side.

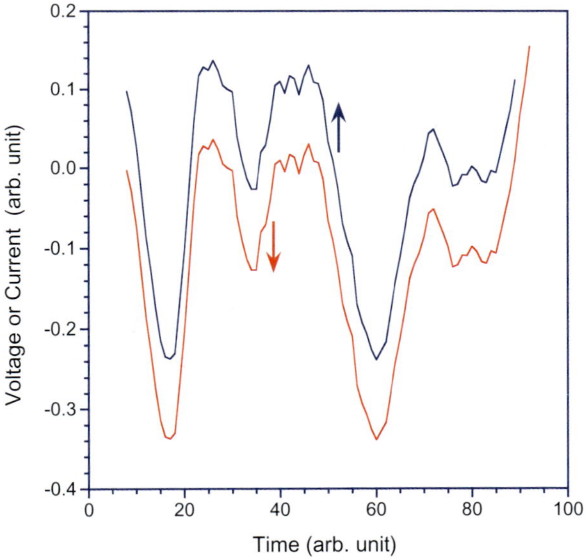

Figure 6. Computer-generated illustration of how a DC shift can distinguish between two strongly correlated noises by comparison at a single moment of time. The arrows indicate the directions of shift.

the DC voltage shift and uncover the key or its inverse. If the DC shift is greater than the stochastic difference between the time functions, then a single-time measurement is sufficient to distinguish the two noises and the bit-situations in the KLJN scheme. Concerning a 50/60 Hz parasitic signal in the loop, the period duration is about a hundred times greater than the correlation time of the noise with 5 kHz bandwidth used by GAA and thus, during the few correlation times used by GAA, this disturbance behaves practically as a parasitic DC shift.

For the situation illustrated above, GAA's result, that Eve's successful guessing probability is progressively enhanced by increasing cable loss, is obvious. A lossless cable is represented by an inductance (*cf.* Fig. 2b) which produces a voltage drop proportional to the time derivative of the current. This means zero DC voltage and zero shift between the distribution functions due to a parasitic DC current in the loop. However the situation is changed when cable loss is present, and then a DC voltage shift will occur in accordance with Ohm's law as a result of the cable resistance R_C. This effect will be strongly enhanced by the time derivation of the channel voltage in GAA's scheme because the voltage drop in the cable is not time-derivated.

4. Conclusion

We have shown that GAA's approach [1] is invalid and that their experimental results must be caused by artifacts. Nevertheless, a correct interpretation of GAA's results is very enlightening because it shows clearly that nonlinearities, non-Gaussianity (even a weak one), aliasing effects and parasitic currents constitute very dangerous potential non-idealities in a practical KLJN system. The removal of such effects is straightforward, however, and can be accomplished by careful circuit design, filters, etc., while ignoring these effects can lead to cracking of the key. Furthermore, a well-defended KLJN system can execute spectral and statistical analysis on the noise in the cable and, together with a proper computer model of the cable, ascertain that effects due to parasitic currents are not present and thereby assure safe results. These types of checks are possible because the KLJN system is a classical physical one, and classical physics permits continuous monitoring of signals and parameters of the channel without destroying this information, which is a situation very unlike that in a quantum system. Thus the robustness of the KLJN state is offered by classical physics and is essential for the security of the key exchange within the KLJN scheme.

Acknowledgements

Discussions with Janusz Smulko and Robert Nevels are appreciated.

References

1. Gunn, L.J., Allison, A., Abbott, D. (2014). "A directional coupler attack against the Kish key distribution system". Manuscript: http://arxiv.org/abs/1402.2709 versions 1 and 2.
2. Chen, H.-P., Kish, L.B., Granqvist, C.G., Schmera, G. (2014). "Do electromagnetic waves exist in a short cable at low frequencies? What does physics say?", *Fluct. Noise Lett.*, **13**, 1450016 (1–13), http://arxiv.org/abs/1404.4664, http://vixra.org/abs/1403.0964.
3. Kish, L.B., Scheuer, J. (2010). "Noise in the wire: The real impact of wire resistance for the Johnson (-like) noise based secure communicator", *Phys. Lett. A*, **374**, 2140–2142.
4. Nevels, R.D. This flaw of GAA's [1], based on misusing Eq. (1) for steady-state signals instead of for pulses, was first pointed out by Dr. Nevels in private communications (March 2014).
5. Mingesz, R., Kish, L.B., Gingl, Z. (2008). "Johnson(-like)-noise–Kirchhoff-loop based secure classical communicator characteristics, for ranges of two to two thousand kilometers, via model-line", *Phys. Lett. A*, **372**, 978–984.
6. Horvath, T., Kish, L.B., Scheuer, J. (2011). "Effective privacy amplification for secure classical communications", *EPL* (formerly *Europhys. Lett.*), **94**, 28002 (1–6).
7. Kish, L.B., Abbott, D., Granqvist, C.G. (2013). "Critical analysis of the Bennett–Riedel attack on secure cryptographic key distributions via the Kirchhoff-law–Johnson-noise scheme", *PLoS ONE*, **8**, e81810 (1–15).
8. Smulko, J. (2014). "Performance analysis of the "intelligent" Kirchhoff's-law–Johnson-noise secure key exchange", *Fluct. Noise Lett.*, **13**, 1450024.
9. Gingl, Z., Mingesz, R. (2014). "Noise properties in the ideal Kirchhoff-law–Johnson-noise secure communication system", *PLoS ONE*, **9**, e96109 (1–4). doi:10.1371/journal.pone.0096109.
10. Mingesz, R., Vadai, G., Gingl, Z. (2014). "What kind of noise guarantees security for the Kirchhoff-loop–Johnson-noise key exchange?", *Fluct. Noise Lett.*, **13**, 1450021. http://arxiv.org/abs/1405.1196.
11. Rényi, A. (2007). *Probability Theory*. Dover Publ., Mineola, NY, USA.
12. Mingesz, R., Kish, L.B., Gingl, Z., Granqvist, C.G., Wen, H., Peper, F., Eubanks, T., Schmera, G. (2013). "Unconditional security by the laws of

classical physics", *Metrol. Measurem. Syst.*, **XX**, 3–16. Available online at: http://www.degruyter.com/view/j/mms.2013.20.issue-1/mms-2013-0001/mms-2013-0001.xml.

13. Kish, L.B. (2013). "Enhanced secure key exchange systems based on the Johnson-noise scheme", *Metrol. Measurem. Syst.*, **XX**, 191–204. Available online at: http://www.degruyter.com/view/j/mms.2013.20.issue-2/mms-2013-0017/mms-2013-0017.xml.
14. Kish, L.B., Horvath, T. (2009). "Notes on recent approaches concerning the Kirchhoff-law–Johnson-noise-based secure key exchange", *Phys. Lett. A*. **373**, 2858–2868.
15. Kish, L.B. (2006). "Totally secure classical communication utilizing Johnson(-like) noise and Kirchhoff's law", *Phys. Lett. A*, **352**, 178–182.

4.6.3. The experimental attenuator artifact in the Gunn–Allison–Abbott attack

4.6.9. In October 2014, I spent two weeks in Szeged, Hungary, as part of my sabbatical research leave. With Zoltan Gingl and Robert Mingesz (who wanted to reproduce Lachlan Gunn's experiments) we started a conversation with Lachlan Gunn to learn the details of the experiments that had not been included in their paper in *Scientific Report* (Nature). We learned important facts, such as how the "noise" reported was not, in actuality, a noise; but a periodic signal pattern, because a very short noise pattern was repeated during making of the statistics. Then I asked Lachlan how they modeled the attenuation in the cable, which was very short; that was when I learned that a commercial microwave attenuator had been used — such devices create two Kirchhoff loops instead of one, a state that is completely illegal in KLJN! I rushed back to the hotel and analyzed the circuitry and the system behavior. It was an absolute failure — a totally cracked system; nothing to do with KLJN. I checked the statistical aspects with Janusz Smulko, before quickly writing the paper below. We did not want to lose any time and promptly submitted it to *Fluctuation and Noise Letters* to get our findings out swiftly.

4.6.10. At the same time, I proposed the names of Lachlan and Robert to the organizers of the ICNF noise conference series for invited talks and a related public debate; however, they only accepted Robert — thus, we must still wait to see a public debate between these "experimentalists".

Paper: Analysis of an attenuator artifact in an experimental attack by Gunn–Allison–Abbott against the Kirchhoff-law–Johnson-noise (KLJN) secure key exchange system[1,2]

Abstract. A recent paper by Gunn–Allison–Abbott (GAA) [L.J. Gunn *et al.* (2014). *Scientific Reports*, **4**, 6461] argued that the Kirchhoff-law–Johnson-noise (KLJN) secure key exchange system could experience a severe information leak. Here we refute their results and demonstrate that GAA's arguments ensue from a serious design flaw in their system. Specifically, an attenuator broke the single Kirchhoff-loop into two coupled loops, which is an incorrect operation since the single loop is essential for the security in the KLJN system, and hence GAA's asserted information leak is trivial. Another consequence is that a fully defended KLJN system would not be able to function due to its built-in current-comparison defense against active (invasive) attacks. In this paper we crack GAA's scheme via an elementary current comparison attack which yields negligible error probability for Eve even without averaging over the correlation time of the noise.

Keywords: KLJN secure key exchange system; directional coupler; unconditional security; experimental artifacts.

1. Introduction

Very recently, Gunn–Allison–Abbott (GAA) published a new type of attack [1] — which has been criticized in earlier papers of ours [2–4] — against the Kirchhoff-law–Johnson-noise (KLJN) secure key distribution system [5–11] and asserted, on the grounds of experiments as well as simulation, that an extraordinarily large information leak could occur with cable losses of 0.1 to 1 dB. In particular, GAA found that at a cable loss of 1 dB, and within a fraction of the correlation time of the noise, Eve can extract the key bit with an error probability of around 0.1, which means that her

[1]Kish, L.B., Gingl, Z., Mingesz, R., Vadai, G., Smulko, J., Granqvist, C.G. (2015). *Fluctuations and Noise Lett.*, **14**, 1550011; http://arxiv.org/abs/1411.0818; http://vixra.org/abs/1410.0122.

[2]This paper of ours would have been unnecessary if the criticized paper [1] were properly documented. Then the reviewers would have refused the paper or request to remove the experimental errors.

probability of successfully guessing the key bit is $p \approx 0.9$. If this claim were correct it would imply that Eve could separate Alice's and Bob's noises of very different intensities. Even though the validity of the mathematical claim concerning the unconditional security [5] of KLJN would remain intact, GAA's assertions would — if correct — imply that applications of the KLJN scheme would be limited to intra-instrument or inter-chip security.

At face value GAA's claims may seem compelling, but efforts at the Department of Technical Informatics at the University of Szeged in Hungary to reproduce GAA's experiment led to personal communication between GAA and us. Important unpublished details of GAA's experiments were then disclosed, including the display of a serious design deficiency caused by the break-up the Kirchhoff loop in the KLJN system by an attenuator in order to provide the desired loss. This deficiency in GAA's experiments led to their above-mentioned claims, but these claims are flawed and clearly founded on experimental artifacts. It should be emphasized that a complete (*i.e.*, fully defended) KLJN scheme would not have been able to function at all under GAA's conditions, which is a consequence of KLJN's current comparison alarm [13], which would shot down the communication within a fraction of the single bit exchange duration.

Even though the design flaw in GAA's experiment is obvious once their experimental conditions were revealed, it is instructive to demonstrate its seriousness. In this paper we first briefly summarize a number of theoretical arguments why GAA's attack cannot function, and we then analyze their experimental artifact and expose its information leak.

2. Why the GAA attack cannot work

We refer to our earlier work on the physical impossibility of low-frequency waves in short cables [3] — which is not directly related to security — and our detailed analysis of GAA's scheme [2], as well as to general arguments [4] about the impossibility of directional couplers. This prior work leads to the following analysis of the problem at hand:

(a) GAA's attack [1] aims at creating a directional coupler to separate the noise components generated by Alice and propagating toward Bob, and generated by Bob and propagating toward Alice, in the KLJN scheme.

(b) The GAA model assumes reflections and propagation at the electromagnetic phase velocity c_p in the cable, which is justified only for waves. However, this is not a serious problem since linear response theory allows one to divide the low-frequency signal into short spikes for which the wave equation does work, to study the response to these spikes separately (including reflected waves propagating at c_p), and then to sum up these responses.

(c) The wave based d'Alembert representation used by GAA [1] is well known and works for short transient signals. However, a directional coupler will provide the required output only as long as mixing of forward-going and reflected signals does not occur.

(d) For low-frequency signals, *i.e.*, in the quasi-stationary limit, reflections create a mixture of signals injected at the two ends of the cable. This leads to an *effective phase velocity* c_{pe}, which is proportional to the resistance terminating the end toward which the wave propagates, as demonstrated by computer simulations in our earlier work [3]. It should be noted that c_{pe} equals c_p when the terminating resistor is equal to the wave impedance (which is 50 Ω in GAA's work [1,3]).

(e) However, GAA's method [1] requires knowledge of c_{pe} in both directions, which in its turn necessitates information on Alice's and Bob's resistance values. Consequently Eve cannot separate Alice's and Bob's signals unless she knows their magnitudes [2].

(f) Our earlier work [2] analyzed feature (e) mathematically and showed that if Eve, while attempting to extract Bob's noise, assumed the correct value of c_{pe} for wave propagation toward Bob she could indeed distinguish between Alice's and Bob's signals. However, if Eve assumed the wrong c_{pe} value, *i.e.*, the value for wave propagation toward Alice, then she would extract a non-existent noise that had the *same mean-square value as Alice's noise*. In other words, Eve gets what she assumes: if she assumes a termination with the *high* resistance then her evaluated noise will agree with that assumption, but if she assumes a termination with the *low* resistance then her evaluated noise will agree with that assumption instead. Therefore Eve's one-bit uncertainty persists within the GAA method.

(g) GAA's theoretical analysis and computer simulation results are in accordance with the above facts in (f), because both of them show that Eve *cannot extract any information from a lossless cable* [1]. Furthermore, GAA's computer simulations for the smallest loss of 0.01 dB (see Fig. 3 in their work [1]) were in accordance with earlier experimental tests [10] of the KLJN system, which showed similar losses and thus a minor information leak that, however, could be remedied by simple privacy amplification [12].

(h) The *only* situations for which GAA were able to extract information were the set-ups with *losses*. This fact indicates that propagation effects were not the cause of the measured information leak but some other phenomenon. Unfortunately GAA did not show any cable-length-dependent information leak, which might have been used to test the real role of propagation delay effects. According to the information we received from GAA, there remained a significant loss, corresponding to 0.1 dB cable loss, in their system even at the experimental situation indicated as "zero loss", which is not immediately obvious from their paper [1]. Therefore GAA imply by "zero loss" that no additional attenuator was used to increase the losses.

(i) GAA mention [1] that our recent defense method against wire resistance based attacks [14] protects also against the GAA attack. However, this defense method — originally developed against the Second Law attack — boosts the noise temperature at the lower resistance end [14]. The effectiveness of this defense method against GAA's attack is another indication of the irrelevance of the propagation effects in GAA's results.

In conclusion, both our own [3] and GAA's [1] theoretical analyses agree in that no information leak exists in the loss-free case whereas propagation delays are present. In the next section we will show that GAA's experimental results, with attenuators to produce their required loss, contain a severe artifact which would have prohibited the functioning of a complete KLJN system as a consequence of its current-comparison defense against active attacks [13]. We show that Eve can extract the information by elementary measurements with very low error rate, virtually immediately, even without using GAA's statistical tool [1].

3. Attenuator artifact in the GAA experiments, and its analysis

The attenuator is a symmetrized voltage divider which provides not only attenuation but also 50 Ω input impedance when the far end of the cable is terminated by 50 Ω. Figure 1 shows the 1 dB attenuator (with approximate values) and its incorporation into the KLJN loop. It is obvious that the shunt resistor R_2 breaks the originally single Kirchhoff loop into two loops with a common side. This is an incorrect realization of the KLJN system, because security has been guaranteed only for a single loop involving Alice and Bob. The violation inherent in GAA's circuit is very significant, because the value of R_2 (500 Ω) is 20 times smaller than that of Bob's resistor R_B (10 kΩ) and two times smaller than Alice's resistor R_A (1 kΩ). These data prove that GAA's experiments, in the cases with higher losses and information leak, were not conducted on the KLJN system but on something else.

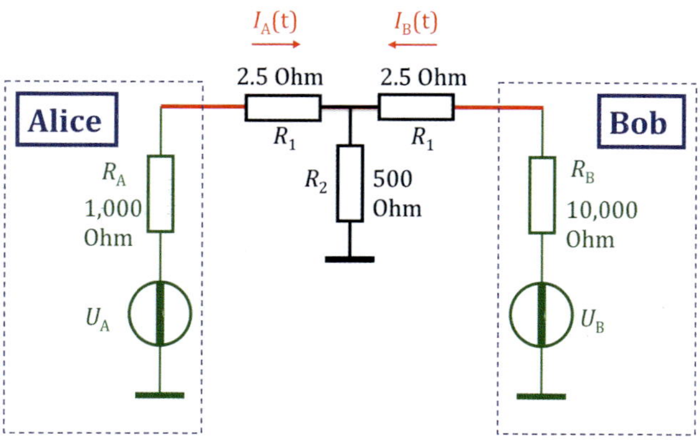

Figure 1. Circuit diagram for the 1 dB attenuator applied by GAA to crack the KLJN scheme. *U*, *R*, *I* and *t* denote voltage, resistance, current and time, respectively. GAA's 0.1 dB attenuator had the same structure with appropriate resistances.

We show next that the time-dependent currents $I_A(t)$ and $I_B(t)$ of Alice and Bob, respectively, instead of being equal, as required for security, are strongly unbalanced, and Alice's mean-square current is about five times larger than Bob's. Straightforward circuit noise analysis provides the mean-square currents of Alice and Bob according to

$$\left\langle I_A^2(t)\right\rangle = S_{iA}(f)B_{kljn} \simeq \frac{4kT_{eff}R_A}{[R_A + R_B R_2/(R_B + R_2)]^2}B_{kljn} + \left(\frac{R_A^{-1}}{R_A^{-1} + R_2^{-1}}\right)^2 \frac{4kT_{eff}R_B}{[R_B + R_A R_2/(R_A + R_2)]^2}B_{kljn} \quad (1)$$

and

$$\left\langle I_B^2(t)\right\rangle = S_{iB}(f)B_{kljn} \simeq \frac{4kT_{eff}R_B}{[R_B + R_A R_2/(R_A + R_2)]^2}B_{kljn} + \left(\frac{R_B^{-1}}{R_B^{-1} + R_2^{-1}}\right)^2 \frac{4kT_{eff}R_A}{[R_A + R_B R_2/(R_B + R_2)]^2}B_{kljn}\,, \quad (2)$$

where it is assumed that the resistors R_1 are much smaller than the total resistances of the loops; $S_{iA}(f)$ and $S_{iB}(f)$ are (white) noise spectra for Alice's and Bob's currents; and B_{kljn} and T_{eff} are noise bandwidth and effective noise temperature of the generators, respectively. Substituting the practical values used in GAA's experiments, we obtain that

$$\frac{\left\langle I_A^2(t)\right\rangle}{\left\langle I_B^2(t)\right\rangle} = 4.95, \quad (3)$$

which means that Alice's mean-square current is about five times stronger than Bob's. This extraordinary difference means that even the simplest comparison methods can extract information so that GAA's complex statistical tool [1] is unnecessary. To illustrate this fact, we show below that a simple current comparison, without making statistics or averaging, is sufficient to create an efficient attack.

Eve's task, in order to extract the bit, is to guess which mean-square current is the larger one: Alice's or Bob's. According to Eq. (3) this task is equivalent to guessing, from a few measurement samples, if a current $I_1(t)$

with unit mean-square value

$$\langle I_1^2(t)\rangle = 1, \tag{4}$$

or another current $I_2(t)$ with mean-square value

$$\langle I_2^2(t)\rangle = 4.95, \tag{5}$$

is the larger. Without loss of generality, we use this example and assume that the values of the measurement noise currents are already normalized in accordance with Eqs. (2) and (3), which is straightforward because the magnitudes of $\langle I_A^2(t)\rangle$ and $\langle I_B^2(t)\rangle$ are theoretically known by Eve since all the resistors and the effective temperature are public knowledge [4–6].

The simplest protocol is to perform a single measurement on the currents and compare their square with the threshold 4.95. If one of the currents squared is greater than the threshold and the other is smaller, then Eve concludes that the first one is $I_2(t)$.

We obtain the following probabilities about the behavior of the square of a *single measurement value* of the current:

$$\left.\begin{aligned} P\big(I_1^2(t) < 4.95\big) &= F_1(4.95) = 0.974 \\ P\big(I_1^2(t) > 4.95\big) &= 1 - F_1(4.95) = 0.026 \\ P\big(I_2^2(t) < 4.95\big) &= F_2(4.95) = F_1(1) = 0.68 \\ P\big(I_2^2(t) > 4.95\big) &= 1 - F_2(4.95) = 1 - F_1(1) = 0.32 \end{aligned}\right\} \tag{6}$$

where F_1 and F_2 are the chi-squared distributions of $I_1^2(t)$ and $I_2^2(t)$, respectively, with one degree of freedom.

The probability of successful (but not necessarily error-free) guessing, *i.e.*, $I_1^2(t) < 4.95$ and $I_2^2(t) > 4.95$, is

$$P_s = 0.974 * 0.32 = 0.31. \tag{7}$$

In this case, Eve's guess is $\langle I_1^2(t)\rangle < \langle I_2^2(t)\rangle$. The error probability of this guess is

$$P_\varepsilon = P\big(I_1^2 > 4.95\big)P\big(I_2^2 < 4.95\big) = 0.026 * 0.68 = 0.018, \tag{8}$$

which is less than 2%, thus indicating over 98% fidelity of Eve's successful guessing.

It is obvious that the probability of "there is no answer" — *i.e*, when both measured values are below the threshold or when both of them are above the threshold — is given as $0.974 * 0.68 + 0.026 * 0.32 \approx 0.67$. Thus on the average three measurements are needed to get an answer which will have over 98% fidelity. Three independent measurements can be done in three correlation times and yield Eve's 1.8% error probability. Very interestingly, GAA's computer simulations gave the same error probability in three correlation times by using their demanding statistical method as our elementary non-optimized method provides, as apparent from Fig. 3 in their work [1]. (Note that the threshold is at the mean-square value of I_2, and much greater than the mean-square value of I_1, and therefore a successful measurement is expected to happen with near-unity probability during the correlation time of the noise, because the square of a Gaussian noise typically goes through virtually all values between zero and its mean-square value during the correlation time.)

The *ad hoc* and non-optimized protocol described above, and its small error probability, serve as an illustration of the astronomically large information leak that is caused by the attenuator artifact in GAA's work [1].

4. Conclusion[3]

A recent paper by Gunn–Allison–Abbott [1] claimed that the KLJN secure key exchange system — referred to by them as the "Kish key distribution system" — could display a devastating information leak. The present paper refuted their results and showed that GAA's arguments arise from a serious deficiency in the design of their system. Specifically, GAA used an attenuator which broke the single Kirchhoff-loop, which is an essential feature for the security in the KLJN system. Therefore GAA's alleged information leak is trivial. We also cracked GAA's scheme via an elementary current-comparison attack, which yielded negligible error probability for Eve in a short time of the order of the correlation time of the noise.

---

[3]In the light of the 3 rebuttals, Sections 4.6.1–4.6.3, the cumulative conclusion is that the GAA paper [1] has *failed at all levels: concept, theory and experiments*. However it reflects ona harc work and its impact is still *positive* because it may protect others to get into these traps. Science often develops via recognized mistakes.

Acknowledgements

We are grateful to Lachlan Gunn for sharing unpublished details on GAA's experimental realization of the KLJN scheme and thereby helping us to identify the design artifact that yielded the major information leak in the attack described in GAA's paper.

LBK's visit of the University of Szeged (October 6–17, 2014) was supported by Texas A&M University. RM's, ZG's and GV's research was partially supported by the European Union and the State of Hungary, co-financed by the European Social Fund in the framework of TÁMOP 4.2.4. A/2-11-1-2012-0001 'National Excellence Program'. JS's research was partially supported by the Statutory Funds of Gdansk University of Technology, Faculty of Electronics Telecommunications and Informatics for JS.

References

1. Gunn, L.J., Allison, A., Abbott, D. (2014). "A directional wave measurement attack against the Kish key distribution system", *Scientific Reports*, **4**, 6461. DOI: 10.1038/srep06461.
2. Chen, H.-P., Kish, L.B., Granqvist, C.G., Schmera, G. (2014). "On the 'cracking' scheme in the paper 'A directional coupler attack against the Kish key distribution system' by Gunn, Allison and Abbott", *Metrology and Measurement Systems*, **21**, 389–400. http://arxiv.org/abs/1405.2034.
3. Chen, H.-P., Kish, L.B., Granqvist, C.G., Schmera, G. (2014). "Do electromagnetic waves exist in a short cable at low frequencies? What does physics say?", *Fluctuations and Noise Letters*, **13**, 1450016. http://arxiv.org/abs/1404.4664; http://vixra.org/abs/1403.0964.
4. Kish, L.B., Abbott, D., Granqvist, C.G. (2013). "Critical analysis of the Bennett–Riedel attack on secure cryptographic key distributions via the Kirchhoff-law–Johnson-noise scheme", *PLoS ONE*, **8**, e81810. http://www.plosone.org/article/info%3Adoi%2F10.1371%2Fjournal.pone.0081810.
5. Kish, L.B. (2006). "Totally secure classical communication utilizing Johnson(-like) noise and Kirchoff's Law", *Physics Letters A*, **352**, 178–182.
6. Kish, L.B., Granqvist, C.G. (2014). "On the security of the Kirchhoff-law–Johnson-noise (KLJN) communicator", *Quantum Information Processing*, **13**, 2213–2219. DOI: 10.1007/s11128-014-0729-7; http://arxiv.org/abs/1309.4112; http://vixra. org/abs/1309.0106.

7. Smulko, J. (2014). "Performance analysis of the 'intelligent' Kirchhoff's-law–Johnson-noise secure key exchange", *Fluctuations and Noise Letters*, **13**, 1450024.
8. Gingl, Z., Mingesz, R. (2014). "Noise properties in the ideal Kirchhoff-law–Johnson-noise secure communication system", *PLoS ONE*, **9**, e96109.
9. Mingesz, R., Vadai, G., Gingl, Z. (2014). "What kind of noise guarantees security for the Kirchhoff-loop–Johnson-noise key exchange?", *Fluctuations and Noise Letters*, **13**, 1450021. http://arxiv.org/abs/1405.1196.
10. Mingesz, R., Gingl, Z., Kish, L.B. (2008). "Johnson(-like)-noise–Kirchhoff-loop based secure classical communicator characteristics, for ranges of two to two thousand kilometers, via model-line", *Physics Letters A*, **372**, 978–984.
11. Saez, Y., Kish, L.B., Mingesz, R., Gingl, Z., Granqvist, C.G. (2014). "Current and voltage based bit errors and their combined mitigation for the Kirchhoff-law–Johnson-noise secure key exchange", *Journal of Computational Electronics*, **13**, 271–277. http://vixra.org/abs/1308.0113; http://arxiv.org/abs/1309.2179.
12. Horvath, T., Kish, L.B., Scheuer, J. (2011). "Effective privacy amplification for secure classical communications", *EPL (Europhysics Letters)*, **94**, 28002 (1–6). http://arxiv.org/abs/1101.4264.
13. Kish, L.B. (2006). "Protection against the man-in-the-middle-attack for the Kirchhoff-loop–Johnson(-like)-noise cipher and expansion by voltage-based security", *Fluctuations and Noise Letters*, **6**, L57–L63. http://arxiv.org/abs/physics/0512177.
14. Kish, L.B., Granqvist, C.G. (2014). "Elimination of a Second-Law-attack, and all cable-resistance-based attacks, in the Kirchhoff-law–Johnson-noise (KLJN) secure key exchange system", *Entropy*, **16**, 5223–5231. http://www.mdpi.com/1099-4300/16/10/5223.

Chapter 5

Cream on the Cake

There are too many interesting results for all of them to be included in this book. In this chapter, we select and feature a few from the most important outcomes; sometimes only briefly. The fundamental ones are shown in detail.

Figure 5.1. Happy moments before the Wikipedia and arXiv storms began. With Klas Gunnarsson (middle) and Peter Svedlindh (right) in front of the Uppsala Concert Hall, Sweden. I shared an office with Klas during my first stay (1989) in the low temperature condensed matter physics lab at Uppsala University, Sweden; he was Peter's Ph.D. student at that time. Peter and I have had many joint interactions in science (see Section 1.5 of the book for more information). Today, Klas is an Associate Professor, and Peter is the head of the Solid State Physics division in the Department of Engineering Sciences, at Uppsala University.

5.1. General Security Proof of the KLJN Key Exchange

5.1.1. Below is the most fundamental and general security proof of the KLJN key exchange for non-ideal conditions. The proof is based on the already proven perfect security at mathematically ideal (non-practical) conditions; and on the strict continuity of classical physical functions in linear systems and stable non-linear systems. As the whole classical physics since Newton, is built on the same foundations; this security proof of KLJN is, therefore, as solid as the classical physics we know today.

5.1.2. On the contrary, the security proof of quantum key distribution is based on the *no-cloning theorem*, which seems to be far less robust, as evolving quantum measurement theories and methods may serve surprises. See for example, the discussion and links at: "The No Cloning theorem vs weak measurement", on physicsforum.com.

5.1.3. *Correction: see note 5.1.4 under Equation 5.*

Figure 5.1.1. The young Horace Yuen, today's quantum security guru (Courtesy of H. Yuen). He has criticized the quantum key distribution establishment for its lack of satisfactory security proof. He has, together with his co-editor, Osamu Hirota, promoted the paper below to 'invited paper' status in the journal, *Quantum Information Processing*.

Paper: On the security of the Kirchhoff-law–Johnson-noise (KLJN) communicator[1,2]

Abstract. A simple and general proof is given for the information theoretic (unconditional) security of the Kirchhoff-law–Johnson-noise (KLJN) key exchange system under practical conditions. The unconditional security for ideal circumstances, which is based on the Second Law of Thermodynamics, is found to prevail even under slightly non-ideal conditions. This security level is guaranteed by the continuity of functions describing classical physical linear, as well as stable non-linear, systems. Even without privacy amplification, Eve's probability for successful bit-guessing is found to converge towards 0.5 — *i.e.*, the perfect security level — when ideal conditions are approached.

Keywords: Classical physical secure key exchange, Johnson noise.

1. Introduction

In today's normal secure communication, the communicating parties A (Alice) and B (Bob) use software tools to generate and share encryption keys [1]. An eavesdropper (Eve) cannot extract this key because her computational power is limited, but sufficient computing power — for example by using a hypothetical quantum computer or its noise-based logic version — would enable her to fully crack the key. Computing technology progresses at a high pace, which means that today's software-based security is only computationally conditional, and furthermore it does not imply *future-proof security*. Indeed Eve is potentially able to crack a recorded key exchange, and thereby imperil the whole system for data exchange in the near future, even if such a task looks hopeless at the moment.

The inherent deficiencies in the present communication systems have led scientists to explore various physical phenomena for secure key

[1] Kish, L.B., Granqvist, C.G. (2014). "On the security of the Kirchhoff-law-Johnson-noise (KLJN) communicator", *Quantum Information Processing* **13**, 2213–2219; http://arxiv.org/abs/1309.4112; http://vixra.org/abs/1309.0106

[2] Forewords: Hirota, O., Yuen, H.P. (2014). "Introduction to special issue on secure quantum communication", *Quantum Information Processing* **13**, 2153–2154; http://link.springer.com/article/10.1007/s11128-014-0813-z/fulltext.html

exchange, so that the laws of physics would guarantee security [1]. The aim of these endeavors is to implement a key exchange scheme wherein either the exchange cannot be measured/recorded or, if the information is measured/recorded by Eve, the exchange is nil. This situation is referred to as *perfect unconditional security* or *information theoretic security* [2]. However, perfect security is elusive: one can get ever so close but never reach the goal.

Therefore scientists working with physics-based secure key exchange systems have developed special security devises [3–5], such as *statistical distance* measures between the probability distribution characterizing Eve's extracted key and that of a perfectly secure key of the same length. The *trace distance* in *quantum key distribution* (QKD) is a good example even though its particular use is debated in QKD [3–5]. The numerical value provided by these security measures would be zero for a perfectly secure key. *Practically-perfect* unconditional (or information theoretic) security, which is characteristic for QKD, uses measures that exponentially converge towards perfect security for increasing key length.

The impossibility to have a perfectly secure key in a physical system has led to the common practice to speak of *unconditional security* (or information theoretic security) with an understanding that this is only a practically-perfect unconditional security, *i.e.*, the maximum that can be reached in a physical crypto system. We shall proceed in the same way for the rest of this paper.

It was a commonly accepted assumption for years that only QKD would be able to perform unconditionally secure key exchange, and that this scheme can provide unconditional (practically-perfect) security. However this dogma was refuted by the Kirchhoff-law–Johnson-noise (KLJN) secure key exchange scheme [6] which was introduced in 2005 [7] and subsequently experimentally demonstrated [8]. The KLJN scheme is a statistical/physical competitor to quantum communicators; its security is based on the Second Law of Thermodynamics, which implies that the security of the ideal scheme against passive (non-invasive listening/measuring) attacks is as strong as the impossibility to build a perpetual motion machine of the second kind. The security against active (invasive) attacks is — perhaps surprisingly — provided by the robustness of classical physical quantities, which guarantees that these quantities can be monitored *continuously* without destroying their values (which is totally different for the case of quantum physics).

The KLJN communicator has been studied during several years with regard to active (invasive) attacks as well as passive attacks associated with non-idealities of the elements embodied in this system [9]. The scheme was consistently found to be unconditionally secure, with the definition above, against each of these attacks.

The present paper presents a general proof for the unconditional security of the KLJN system against any type of non-ideality-based attack, even against attacks that are yet unknown. The proof is based on the continuity of functions describing classical physical linear, and stable non-linear, systems.

2. The Kirchhoff-law–Johnson-noise secure key exchange system

Figure 1 outlines the KLJN secure key exchange system. For each clock period, *i.e.*, duration of a single bit exchange, Alice and Bob connect their

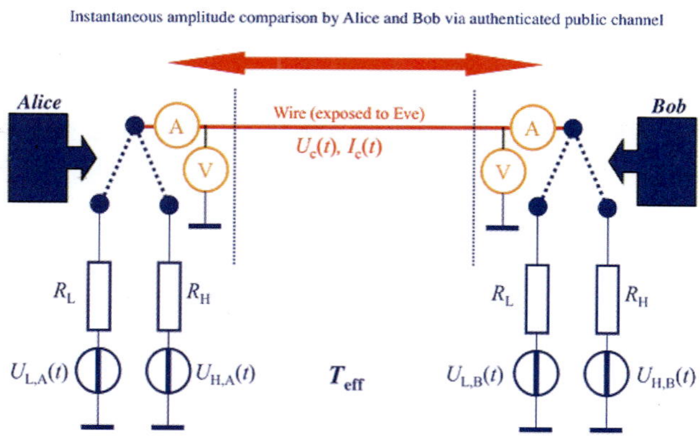

Figure 1. Schematic of the Kirchhoff-law–Johnson-noise secure key exchange system. The various circuit elements are discussed in the main text. Instantaneous voltage $U_C(t)$ and current $I_C(t)$ amplitudes in the wire are measured by Alice and Bob and compared via a public authenticated data exchange (not shown). R, t andT_{eff} denote resistance, time and effective temperature, respectively.

randomly chosen resistor, R_A and R_B, respectively, to the line. These resistors are randomly selected from the set $\{R_0, R_1\}$, $R_0 \neq R_1$, where the elements represent the 0 and 1 bit values. The Gaussian voltage noise generators — imitating the Fluctuation-Dissipation Theorem and delivering band-limited white noise with publicly agreed bandwidth — represent enhanced thermal (Johnson) noise at a publicly agreed effective temperature T_{eff} (typically $T_{eff} \geq 10^9 K$ [9]). Their noises are statistically independent from each other and from the noise during the former clock period.

In the case of secure bit exchange (*i.e.*, the 01 or 10 situation), the power density spectrum $\langle U_c^2 \rangle$ and the mean-square amplitude $\langle U_c^2 \rangle$ of the voltage $U_c(t)$ on the wire, and the same quantities of the current $I_c(t)$ in the wire, are given as

$$\langle U_{c,01/10}^2 \rangle = \Delta f \, S_{uc,01/10}(f) = 4kT_{eff} \frac{R_0 R_1}{R_0 + R_1} \Delta f, \tag{1}$$

$$\langle I_{c,01/10}^2 \rangle = \Delta f \, S_{ic,01/10}(t) = \frac{4kT_{eff}}{R_0 + R_1} \Delta f, \tag{2}$$

respectively, where Δf is the noise bandwidth; further details of the KLJN system are given elsewhere [6, 9]. Eve cannot distinguish between the 01 and 10 situations by measuring mean-square values, because both of them lead to the same results as evident from Eqs. (1) and (2), while the 00 and 11 cases yield distinguishable, and hence insecure, bit arrangements.

The only quantity that potentially could give directional information is the voltage–current cross-correlation, $\langle U_c(t) I_c(t) \rangle$ which gives a directional power flow that is zero in thermal equilibrium. Thus the ultimate security of the ideal system against passive attacks is provided by the Second Law of Thermodynamics implying that, for thermal equilibrium, the power $P_{0\to1}$, by which resistor R_0 is heating resistor R_1, is equal to the power $P_{1\to0}$ by which R_1 is heating R_0 [9]. The net power flow between Alice and Bob is zero, as can be quantitatively derived from the Johnson formula, *viz.*,

$$P_{0\to1} = \frac{S_{0,uc}(f)\Delta f}{R_1} = 4kT_{eff} \frac{R_0 R_1}{(R_0 + R_1)^2} \Delta f, \tag{3}$$

$$P_{1\to0} = \frac{S_{H,uc}(f)\Delta f}{R_0} = 4kT_{eff} \frac{R_0 R_1}{(R_0 + R_1)^2} \Delta f. \tag{4}$$

The equality $P_{H \to L} = P_{L \to H}$, inherent in Eqs. (3) and (4), is in accordance with the Second Law of Thermodynamics, and a violation of this equality would mean going against basic laws of physics and allow Eve to use the voltage–current cross-correlation $\langle U_c(t)\, I_c(t) \rangle$ to extract the bit [9]. This security proof against passive (listening) attacks holds only for Gaussian noise — which has the property that superposition of Gaussian signals remain Gaussian — and its power density spectrum provides the maximum achievable information about the noise. Thus neither higher-order distribution functions nor other tools, such as higher-order statistics, are able to provide additional information.

If Eve is tampering with or changing the system via an active/invasive intervention — such as launching a man-in-the-middle attack [9] — the laws of physics are not enough to guarantee security. Similarly, non-idealities, which represent deviations from the original scheme, pose vulnerabilities. For defending the system against attacks of this kind, the instantaneous voltage and current amplitudes are measured by Alice and Bob, and these quantities are compared via a public authenticated data channel (see Fig. 1). Note, due to the classical physical nature of the scheme, Alice and Bob have a full and deterministic model of the system and continuous monitoring of the quantities is allowed. Then, based on their comparison and publicly set preconditions, Alice and Bob decide to keep or discard the bit having compromised security, see below. This authentication uses only $\log_2(M)$ secure bits of the exchanged ones, where M is the number of bits carrying the current and voltage data in the public channel, which shows that authentication is feasible.

3. General security proof

We now provide a general proof of the unconditional security of the KLJN key exchange. This proof accounts for all non-ideality features, including transients, non-zero propagation time, non-zero cable resistance and capacitance, non-identical temperatures and other inaccuracies. It is important to emphasize that all our analyses are based on a *technically-unlimited Eve* with infinitely accurate and fast measurements limited only by the laws of physics and statistics. For non-ideal situations and at active (invasive) attacks, the *uncertainly principle between measurement duration and statistical errors* makes it impossible for Eve to extract the key regardless of the accuracy or speed of her measurements.

The four steps of the proof are as follows:

(**1**) It has been proven that the ideal KLJN system is perfectly secure [7–9], which means that Eve's probability for making successful guesses of the exchanged key bit is exactly $p = 0.5$, just as for guesses generated by throwing an ideal random coin.

(**2**) Continuous functions describe the variables of passive classical physical systems, such as those in the core KLJN system (and the same holds true even for analog systems in the absence of thresholds and positive feedback). The function $p_\delta(Q)$, describing Eve's probability to make successful guesses for an actual realization of the KLJN system, is also continuous. Here Q is the set of parameters describing the strength of non-ideality, *i.e.*, $Q = \{x_1, x_2, \ldots, x_k\}$, where the variables $x_i \geq 0$ $(i = 1, 2, \ldots, k)$ in Q characterize the non-ideal features, and k is a finite number. The parameter $0 \leq \delta \leq \omega$ of the function $p_\delta(Q)$ characterizes the strength of Eve's eavesdropping measurement output; in the case of a wire resistance attack, for example, δ is the absolute value of the measured difference between the mean-square voltages at Alice's and Bob's side. The upper limit ω is enforced by Alice and Bob who, due to the classical physical nature of the system, also know Eve's measured δ and discard bits with $\delta \geq \omega$. They classify the shared bits with $\delta > \omega$ as high-risk and discard them. Alice and Bob could potentially reach perfect security by choosing $\omega = 0$ at arbitrary Q, but such a choice would imply that virtually all of the bits would be discarded. Thus a practical system must use a balanced choice of Q and ω for the best security at given speed, fidelity and cost.

We now show that the perfect security limit can be approached at arbitrary ω with a proper design of Q. For the sake of simplicity, it is easy to define the quantities in Q so that their values are zero in a mathematically ideal case and $p_\delta(Q, \delta)$ is a non-decreasing function of Q. Such choices, for example, can be $x_1 =$ cable length, $x_2 =$ bandwidth, $x_3 =$ resistivity, x_4

= 1/cable-diameter, x_5 = specific-parasitic-capacitance, x_6 = propagation-time/transient-protocol-duration, *etc.*

Similarly, it is easy to define δ so that $p_\delta(Q)$ is a non-decreasing function of δ.

(3) In the case of $Q = (0, 0, \ldots, 0)$, the probability $p_\delta(Q) = 0.5$ (for arbitrary δ) represents perfect security. By using a Taylor expansion at this point, *i.e.*, in the limit $x_i \to 0$ and for all i, one obtains

$$0.5 \le p_\delta(Q) = 0.5 + \sum_{i=1}^{k} \frac{\partial p_\delta[Q = (0, \ldots 0)]}{\partial x_i} x_i$$
$$\le 0.5 + \sum_{i=1}^{k} \frac{\partial p_\omega[Q = (0, \ldots 0)]}{\partial x_i} x_i. \quad (5)$$

5.1.4. *Correction: In Equation (5), the Taylor polynomial must be further extended to include the second order terms and cross-terms to account for the dependence on cable length. Luckily, the rest of the derivation and its conclusions remain the same.*

Thus

$$0.5 \le p_\delta(Q) = 0.5 + q_\delta(Q) \le 0.5 + q_\omega(Q), \quad (6)$$

where $q \to 0$ when $x_i \to 0$ for all i, and $q \to 0$ when $\delta \to 0$. In other words, for a given $\omega(\omega > 0)$ and by choosing the variables within Q to be non-zero but sufficiently small, it is possible to approach the perfect security limit of $p_a(Q) = 0.5$. A similar type of convergence toward perfect security holds at a given $Q(Q > 0)$ for $\omega \to 0(\omega > 0)$.

It should be noted that the above security proof does not utilize privacy amplification [10], which can be an additional way to go. Because the bit error probability decays exponentially with increasing duration of the bit sharing period, the remarkably low bit error probability — such as 10^{-12} at practical conditions [11] and the resulting extraordinarily high fidelity — allow the use of very high privacy amplification if this is required by economical constrains posed by the values of Q and ω, as further discussed below.

For practical situations, performance aspects (such as speed and wire cost) determine how large magnitudes of δ_{max} and Q are feasible and how much privacy amplification [10] is required. For example, decreasing δ_{max} means larger fraction of discarded bits, *i.e.*, slower key exchange speed, while decreasing Q by increasing cable diameter implies a cable cost that grows by the square of this diameter.

(**4**) The above considerations provide a general proof of information theoretic (unconditional) security. Below, we evaluate the security also by a statistical distance measure, namely the *total variation distance* [12] between the probability distribution characterizing Eve's extracted key and that of a perfectly secure key of the same length. We show that this quantity characterizing the security of the key exchange decays exponentially for increasing length of the key.

In order to assess the security of the shared key, one must compare the probability distribution of successfully guessing each possible key sequence of the N-bit-long key extracted by Eve, encompassing 2^N different sequences, with that of the perfectly secure key having uniform distribution. The total variation distance Δ [12] of the probability distribution characterizing Eve's extracted key and that of a perfectly secure key of the same length is

$$\Delta(E, I) = \max_{j=1,\ldots,2^N} [P(E_j) - P(I_j)], \tag{7}$$

where E and I indicate Eve's extracted key and a perfectly secure key, respectively, and $P(E_j)$ and $P(I_j)$ are the probabilities for correctly guessing the jth version of Eve's key and of the perfectly secure key, respectively. From Eqs. (6) and (7) we obtain

$$\Delta(E, I) = \max_{j=1,\ldots,2^N} [P(E_j) - P(I_j)] = [0.5 + q_\omega(Q)]^N - 0.5^N. \tag{8}$$

For the case of $Nq_\omega(Q) \ll 0.5$, which is a realizable condition according to Eqs. (5) and (6), one then gets

$$\begin{aligned} \Delta &= (0.5 + q_\omega(Q))^N - 0.5^N \\ &= 0.5^N[(1 + 2q_\omega(Q))^N - 1] \cong 2Nq_\omega(Q)0.5^N. \end{aligned} \tag{9}$$

It is thus found that the total variation distance between the probability distribution characterizing Eve's extracted key and that of a perfectly secure key of the same length decays exponentially towards zero for increasing key length provided the condition $q_{\omega}(Q) \ll 0.5/N$ holds. This condition is given by Eqs. (5) and (6).

It should be noted that this security proof is not valid for non-ideality-based hacking attacks against system components containing active electronics with thresholds, such the current/voltage measurement systems at Alice's and Bob's locations which can be overloaded and saturated by a Makarov-type blinding attack via a large voltage spike [1]. However, a defense against such attacks is easily accomplished by use of proper algorithms that discard the bits when the voltage and current are outside their expected ranges, even if they are identical at Alice's and Bob's sides.

4. Conclusion

This paper presents a general proof for the unconditional security of the Kirchhoff-law–Johnson-noise communicator against any type of non-ideality-based attack, even against those that are yet unknown. The proof is based on the continuity of functions characterizing classical physical linear, as well as stable non-linear, systems.

Acknowledgement

LK is grateful to Horace Yuen, Vadim Makarov, Vincent Poor and Renato Renner for helpful discussions about relevant security measures.

References

[1] Gerhardt, I., Liu, Q., Lamas-Linares, A., Skaar, J., Kurtsiefer, C., Makarov V. (2012). "Full-field implementation of a perfect eavesdropper on a quantum cryptography system", *Nature Commun.*, **2**, 349.

[2] Liang, Y, Poor, H.V., Shamai, S. (2008). "Information theoretic security", *Foundations Trends Commun. Inform. Theory*, **5**, 355–580.

[3] Yuen, H.P. (2013). "Essential lack of security proof in quantum key distribution". arXiv:1310.0842.

[4] Hirota, O. (2012). "Incompleteness and Limit of Quantum Key Distribution Theory". arXiv:1208.2106.

[5] Renner, R. (2012). "Reply to recent scepticism about the foundations of quantum cryptography". arXiv:1209.2423v1.

[6] Abbott, D., Schmera, G. (2013). "Secure communications using the KLJN scheme", *Scholarpedia*, **8**, 31157. (open access) http://www.scholarpedia.org/article/Secure_communications_using_the_KLJN_scheme

[7] Kish, L. B. (2006). "Totally secure classical communication utilizing Johnson(-like) noise and Kirchhoff's law", *Phys. Lett. A*, **352**, 178–182.

[8] Mingesz, R., Gingl, Z., Kish, L.B. (2008). "Johnson(-like)-noise-Kirchhoff-loop based secure classical communicator characteristics, for ranges of two to two thousand kilometers, via model-line", *Phys. Lett. A*, **372**, 978–984.

[9] Mingesz, R., Kish, L.B., Gingl, Z., Granqvist, C.G., Wen, H., Peper, F., Eubanks T., Schmera, G. (2013). "Unconditional security by the laws of classical physics", *Metrol. Meas. Syst.*, **XX**, 3–16. (open access) http://www.degruyter.com/view/j/mms.2013.20.issue-1/mms-2013-0001/mms-2013-0001.xml

[10] Horvath, T., Kish, L.B., Scheuer, J. (2011). "Effective privacy amplification for secure classical communications", *EPL*, **94**, 28002.

[11] Saez, Y., Kish, L.B., Mingesz, R., Gingl, Z., Granqvist, C.G. (2013). "Current and voltage based bit errors and their combined mitigation for the Kirchhoff-law-Johnson-noise secure key exchange", *J. Computational Electronics*, **13**, 271–277. arXiv:1309.2179.

[12] Rosenthal, J.S. (1995). "Convergence rates of Markov chains", *SIAM Rev.*, **37**, 387–405.

5.2. Nullifying the (Bergou–)Scheuer–Yariv Attack and the Second-Law Attack

5.2.1. This paper fully eliminates the earliest and most famous attack: the wire-resistance based attack. Even though the practical importance of eliminating that attack is low because the miniscule information leak can easily be fixed by privacy amplification, the fundamental importance of the protocol is high. This is the first-ever solution where the information *leak due to an existing non-ideality* is completely *eliminated by introducing a proper new non-ideality*. We expect more of this class of defense type to emerge. The paper also describes a more fundamental and much more effective wire-resistance attack utilizing the Second Law of thermodynamics. Luckily, this attack is also nullified by the method described below.

Paper: Elimination of a Second-Law-attack, and all cable-resistance-based attacks, in the Kirchhoff-law–Johnson-noise (KLJN) secure key exchange system[1]

PACS 72.70.+m – Noise processes and phenomena

PACS 89.20.Ff – Computer science and technology

PACS 89.90.+n – Other topics in areas of applied and interdisciplinary physics

Abstract. We introduce the, so far. most efficient attack against the Kirchhoff-law–Johnson-noise (KLJN) secure key exchange system. This attack utilizes the lack of exact thermal equilibrium in practical applications and is based on cable resistance losses and the fact that the Second Law of Thermodynamics cannot provide full security when such losses are present. The new attack does not challenge the unconditional security of the KLJN scheme, but it puts more stringent demands on the security/privacy enhancing protocol than for any earlier attack. In this paper we present a simple defense protocol to fully eliminate this new attack by increasing the noise-temperature at the side of the smaller resistance value over the noise-temperature at the at the side with the greater resistance. It is shown

[1]Kish, L.B., Granqvist, C.G. (2014). "Elimination of a Second-Law attack, and all cable-resistance-based attacks, in the Kirchhoff-law–Johnson-noise (KLJN) secure key exchange system", *Entropy*, **16**, 5223–5231. http://arxiv.org/abs/1406.5179; http://vixra.org/abs/1406.0124

that this simple protocol totally removes Eve's information not only for the new attack but also for the old Bergou–Scheuer–Yariv (BSY) attack. The presently most efficient attacks against the KLJN scheme are thereby completely nullified.

Introduction

The Kirchhoff-law–Johnson-noise (KLJN) scheme [1, 2], shown in Fig. 1, is a classical statistical physical competitor to a quantum key distribution for secure communication. For the duration of a single bit exchange, the communicating parties (Alice and Bob) connect their randomly chosen resistor and corresponding noise-voltage generator to a wire line (cable). These resistors are randomly selected from the publicly known set $\{R_L, R_H\}$, $R_L \neq R_H$, where the elements represent low (L) and high (H) bit values. The Gaussian voltage noise generators — mimicking the Fluctuation-Dissipation Theorem and delivering band-limited white noise with publicly agreed bandwidth — produce enhanced thermal (Johnson) noise at a publicly agreed effective temperature T_{eff}, typically being $T_{eff} \geq 10^9$ K [3],

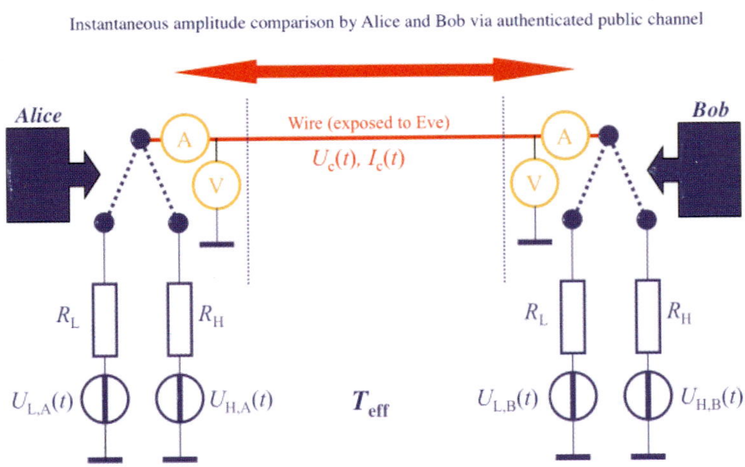

Figure 1. Schematic of the Kirchhoff-law–Johnson-noise secure key exchange system. To defend against active and hacking attacks, the cable parameters and integrity are randomly monitored; the instantaneous voltage $U_c(t)$ and current $I_c(t)$ amplitudes in the cable are measured and compared via public authenticated data exchange; and full spectral and statistical analysis/checking is carried out by Alice and Bob. R, t and T_{eff} denote resistance, time and effective temperature, respectively. Line filters, etc., are not shown.

so the temperature of the wire can be neglected. The noises are statistically independent of each other and from the noise of the former bit period.

In the case of secure bit exchange — *i.e.*, the LH or HL bit situations for Alice and Bob — an eavesdropper (Eve) cannot distinguish between these two situations by measuring the mean-square value of the voltage $U_c(t)$ and/or current $I_c(t)$ in the cable, because both arrangements lead to the same result. In the rest of the paper we assume that one of these secure bit exchange situations (either LH or HL) apply.

To avoid potential information leak by variations in the shape of a probability distribution, the noises are Gaussian [1], and it has been proven that other distributions are not secure [4, 5]. From a physics perspective, the security is provided by the Second Law of Thermodynamics because directional information, due to the direction of power flow, does not exist since the mean power flow is zero even though the LH and HL situations have asymmetric resistance arrangements [1]. In other words, the security of the *ideal* KLJN scheme against passive (non-invasive listening/measuring) attacks is as strong as the impossibility to build a perpetual motion machine of the second kind. The security against active (invasive) attacks is — perhaps surprisingly — provided by the robustness of classical physical quantities, which guarantees that these quantities can be monitored (and their integrity with the cable parameters and model can be checked) *continuously* without destroying their values. We observe, in passing, that the situation is totally different for the case of quantum physics.

The most famous and explored, and so far the most effective, attack against the *non-ideal* KLJN scheme is the Bergou-Scheuer-Yariv (BSY) cable resistance attack [6, 7] which utilizes the fact that, due to the non-zero cable resistance, the mean-square voltage will be slightly less at the cable end with the smaller resistance value than at the other end with the greater resistance. It should be noted that the results (including their physical units) are wrong in Ref. [7], but a correct evaluation of the BSY effect was carried out later by Kish and Scheuer (KS) [8]. Eve's measured absolute difference between the mean-square voltages $\langle U_{\mathrm{cH}}^2(t)\rangle$ and $\langle U_{\mathrm{cL}}^2(t)\rangle$ of the "H" and "L" ends (*cf.* Fig. 2) is given by [8]

$$\Delta_{\mathrm{KS}} = \left|\langle U_{\mathrm{cH}}^2(t)\rangle - \langle U_{\mathrm{cL}}^2(t)\rangle\right| = 4kT_{\mathrm{eff}}\Delta f \left|\frac{R_{\mathrm{c}}^2(R_{\mathrm{H}} - R_{\mathrm{L}})}{(R_{\mathrm{H}} + R_{\mathrm{c}} + R_{\mathrm{L}})^2}\right|, \qquad (1)$$

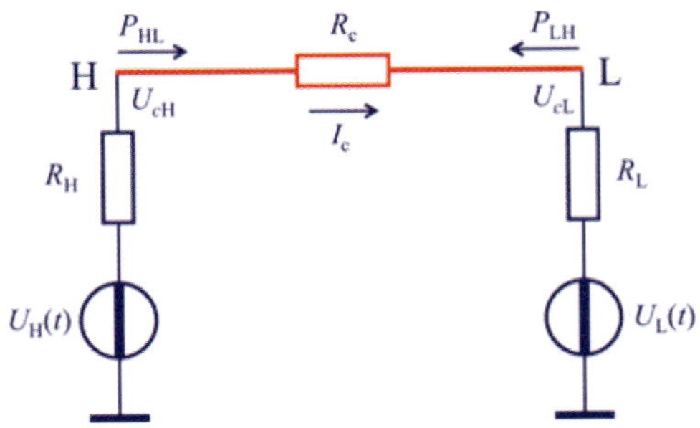

Figure 2. Scheme devised to illustrate the Bergou-Scheuer-Yariv attack and the Second-Law-attack. Alice's and Bob's locations are arbitrary in the figure. During the Second-Law-attack, the powers flowing out from the "H" and "L" ends of the cable are calculated and compared. The temperature of the cable resistor R_c can be neglected because of the high noise temperature of the generators. The notation is consistent with that in Fig. 1.

where k is Boltzmann's constant, Δf is noise bandwidth and R_c is cable resistance. Clearly Δ_{KS} scales with the square of the cable resistance, $i.e.,\Delta_{\mathrm{KS}} \propto R_{\mathrm{c}}^2$.

The Second-Law-attack

In the rest of the paper we use the rules about transformations of noise spectra in linear systems, along with Johnson's formula for thermal noise, and write [1]

$$\langle U_{\mathrm{R}}^2(t)\rangle = 4kT_{\mathrm{eff}}R\Delta f. \tag{2}$$

Here $\langle U_{\mathrm{R}}^2(t)\rangle$ denotes mean-square voltage fluctuations on the resistor, with resistance R, within the bandwidth Δf.

The cable resistance has a non-zero value, and therefore the resistors and their noise generators are not in thermal equilibrium in practical versions of the KLJN system (with T_{eff} much greater than the cable temperature). Consequently the Second Law of Thermodynamics cannot provide full security. The cable-heating powers by the generators at the "H" and "L" ends are different and are given by

$$P_{\mathrm{Hc}} = \langle I_{\mathrm{A}}^2(t)\rangle R_{\mathrm{c}} = \frac{4kT_{\mathrm{eff}}R_{\mathrm{H}}\Delta f}{(R_{\mathrm{H}}+R_{\mathrm{c}}+R_{\mathrm{L}})^2}R_{\mathrm{c}}, \tag{3}$$

and

$$P_{\mathrm{Lc}} = \left\langle I_{\mathrm{B}}^2(t) \right\rangle R_{\mathrm{w}} = \frac{4kT_{\mathrm{eff}} R_{\mathrm{L}} \Delta f}{(R_{\mathrm{H}} + R_{\mathrm{c}} + R_{\mathrm{L}})^2} R_{\mathrm{c}} = P_{\mathrm{Hc}} \frac{R_{\mathrm{L}}}{R_{\mathrm{H}}}. \tag{4}$$

The difference between P_{Hc} and P_{Lc} can be utilized for the Second-Law-attack, because the resistor values R_{H} and R_{L} are publicly known. The implementation of this attack is to measure and compare the net power flows at the two ends of the cable, as illustrated in Fig. 2. The mean power flow P_{HL} from the "H" end toward the "L" end of the cable, and the mean power flow P_{LH} from the "L" end toward the "H" end are, respectively,

$$\begin{aligned} P_{\mathrm{HL}} &= \left\langle U_{\mathrm{H}}^2(t) \right\rangle \left(\frac{R_{\mathrm{c}} + R_{\mathrm{L}}}{R_{\mathrm{H}} + R_{\mathrm{c}} + R_{\mathrm{L}}} \right)^2 \frac{1}{R_{\mathrm{c}} + R_{\mathrm{L}}} \\ &\quad - \left\langle U_{\mathrm{L}}^2(t) \right\rangle \left(\frac{R_{\mathrm{H}}}{R_{\mathrm{H}} + R_{\mathrm{c}} + R_{\mathrm{L}}} \right)^2 \frac{1}{R_{\mathrm{H}}} \\ &= 4kT_{\mathrm{eff}} \Delta f \frac{R_{\mathrm{H}}(R_{\mathrm{c}} + R_{\mathrm{L}}) - R_{\mathrm{L}} R_{\mathrm{H}}}{(R_{\mathrm{H}} + R_{\mathrm{c}} + R_{\mathrm{L}})^2} \\ &= 4kT_{\mathrm{eff}} \Delta f \frac{R_{\mathrm{H}} R_{\mathrm{c}}}{(R_{\mathrm{H}} + R_{\mathrm{c}} + R_{\mathrm{L}})^2} \end{aligned} \tag{5}$$

and

$$\begin{aligned} P_{\mathrm{LH}} &= \left\langle U_{\mathrm{L}}^2(t) \right\rangle \left(\frac{R_{\mathrm{c}} + R_{\mathrm{H}}}{R_{\mathrm{H}} + R_{\mathrm{c}} + R_{\mathrm{L}}} \right)^2 \frac{1}{R_{\mathrm{c}} + R_{\mathrm{H}}} \\ &\quad - \left\langle U_{\mathrm{H}}^2(t) \right\rangle \left(\frac{R_{\mathrm{L}}}{R_{\mathrm{H}} + R_{\mathrm{c}} + R_{\mathrm{L}}} \right)^2 \frac{1}{R_{\mathrm{L}}} \\ &= 4kT_{\mathrm{eff}} \Delta f \frac{R_{\mathrm{L}}\,(R_{\mathrm{c}} + R_{\mathrm{H}}) - R_{\mathrm{H}} R_{\mathrm{L}}}{(R_{\mathrm{H}} + R_{\mathrm{c}} + R_{\mathrm{L}})^2} \\ &= 4kT_{\mathrm{eff}} \Delta f \frac{R_{\mathrm{L}} R_{\mathrm{c}}}{(R_{\mathrm{H}} + R_{\mathrm{c}} + R_{\mathrm{L}})^2} \end{aligned} \tag{6}$$

The power flows P_{HL} and P_{LH} are directly measurable by Eve, and their difference,

$$\Delta P_{\mathrm{HL}} = P_{\mathrm{HL}} - P_{\mathrm{LH}} = 4kT_{\mathrm{eff}} \Delta f \frac{R_{\mathrm{c}}(R_{\mathrm{H}} + R_{\mathrm{L}})}{(R_{\mathrm{H}} + R_{\mathrm{c}} + R_{\mathrm{L}})^2}, \tag{7}$$

gives the difference between the powers supplied by the two cable ends; with the measured cable voltages and current (see Fig. 2) it is

$$\Delta P_{\mathrm{HL}} = P_{\mathrm{HL}} - P_{\mathrm{LH}} = \langle I_{\mathrm{c}}(t)U_{\mathrm{cH}}(t)\rangle - \langle -I_{\mathrm{c}}(t)U_{\mathrm{cL}}(t)\rangle$$
$$= \langle [U_{\mathrm{cH}}(t) + U_{\mathrm{cL}}(t)] I_{\mathrm{c}}(t)\rangle. \tag{8}$$

It should be observed that the opposite current sign at the "L" end expresses the fact that the current flowing *out* from the "H" end is flowing *into* the "L" end (using the same current sign would instead provide the power dissipated in the cable resistance, which is always positive and gives no directional information).

Suppose now that Eve measures the above current–voltage cross-correlations at the two ends and evaluates the pertinent quantities. With the notation introduced in Fig. 3, one finds that

$$\Delta P_{\mathrm{AB}} = P_{\mathrm{AB}} - P_{\mathrm{BA}} = \langle [U_{\mathrm{cA}}(t) + U_{\mathrm{cB}}(t)] I_{\mathrm{c}}(t)\rangle. \tag{9}$$

As an example, suppose that R_{H} has the greater resistance value and R_{L} the smaller one, *i.e.*, $R_{\mathrm{L}} < R_{\mathrm{H}}$. In the *ideal* case, when $R_{\mathrm{c}} = 0$, one obtains $\Delta P_{\mathrm{AB}} = 0$ in accordance with the Second Law of Thermodynamics, which yields $\langle U_{\mathrm{c}}(t) I_{\mathrm{c}}(t)\rangle = 0$. However, in the *practical* case, with $R_{\mathrm{c}} > 0$, one finds

(*i*) if $\Delta P_{\mathrm{AB}} > 0$, then Alice has R_{H} and Bob has R_{L},
(*ii*) if $\Delta P_{\mathrm{AB}} < 0$, then Alice has R_{L} and Bob has R_{H}.

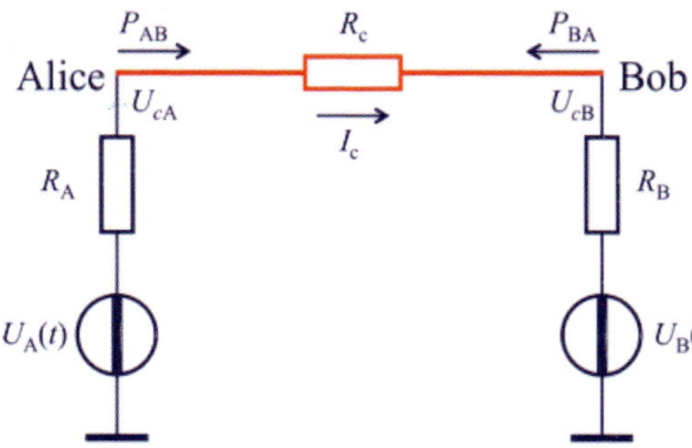

Figure 3. Eve's measurements during the Second-Law-attack. The powers flowing out from the two ends of the cable are measured and compared. The notation is consistent with that in Fig. 1.

The signal inherent in the Second-Law-attack scales linearly with R_c, which provides a much better situation for Eve — especially in the case of vanishing cable resistance — than the square-law scaling of the BSY attack. Moreover, it is also obvious that in a practical case [3, 9, 10], where $R_c \ll R_L \ll R_H$, Eve's signal-to-noise ratio is always greater in the Second-Law-attack than in the BSY attack. This is so because the BSY attack evaluates the dc fraction of $\approx R_c^2/(R_L R_H)$ in the measured (empirical) mean-square channel noise voltage, while the Second-Law-attack evaluates the dc fraction of R_c/R_H in the measured mean power flow. It should be noted that the measured mean-square channel noise voltage, and the measured mean power flow, follow similar statistics because they are the time average of the products of Gaussian processes [11].

The Second-Law-attack is an elegant and efficient one, but it *does not challenge the unconditional security of the KLJN scheme* [2]. Eve's probability p of successful guessing can arbitrarily approach the limit $p = 0.5$ by proper tuning of the parameters inherent in the KLJN scheme, such as resistances and bandwidth, and privacy amplification can be implemented if needed; this was evaluated in detail elsewhere [2], where relationships were reported between security level, cable parameters and communication speed. Nevertheless the new Second-Law-attack is important and may significantly increase the demands on parameter tuning and/or necessitate elaborate privacy amplification [12], which of course come at a cost.

In the rest of this paper we demonstrate two methods capable of fully eliminating the Second-Law attack. The advanced method nullifies the BSY attack as well.

Natural/"Simple" defense

Suppose it is possible to keep the cable and the resistors at the same temperature. This temperature-equilibration method virtually eliminates any Second-Law-attack information for Eve (but not the information in the BSY-attack, albeit its formula for the information leak is changed).

Temperature equilibration constitutes a very simple defense, but the cable temperature and its possible variations cannot be neglected any longer. If the cable temperature is different from that of the resistors, then the KLJN scheme is vulnerable to the Hao-type attack [13]. In principle, with cables of homogeneous temperatures, this attack can be avoided if Alice and Bob are able to monitor the temperature value of the cable by resistance and Johnson noise measurements, since they can then choose T_{eff} to be the same as the cable temperature. While these steps can be taken, the KLJN scheme is no longer simple. Moreover, the mentioned defense method may be unpractical because of the requirement of a homogeneous cable temperature, small noise levels, and since it prohibits the adoption of enhanced KLJN methods wherein Alice and Bob eliminate their own contributions in order to accomplish higher speed and security [9, 14].

Advanced defense, also eliminating all cable resistance attacks

As we have seen, the cable end with the smaller resistance value emits less power toward the other end, and this is the foundation of the Second-Law-attack. This effect, as well as Eve's related signal, can be completely eliminated by properly changing the *ratio* of the noise-temperatures of the generators for the resistors with the smaller and the greater resistance values (see Fig. 4).

Suppose now that we introduce an offset in the noise-temperatures of the generators for the R_{H} and the R_{L} resistors so that the equation

$$\Delta P_{\mathrm{HL}} = P_{\mathrm{HL}}(T_{\mathrm{eff}}) - P_{\mathrm{LH}}(\beta T_{\mathrm{eff}}) = 0 \qquad (10)$$

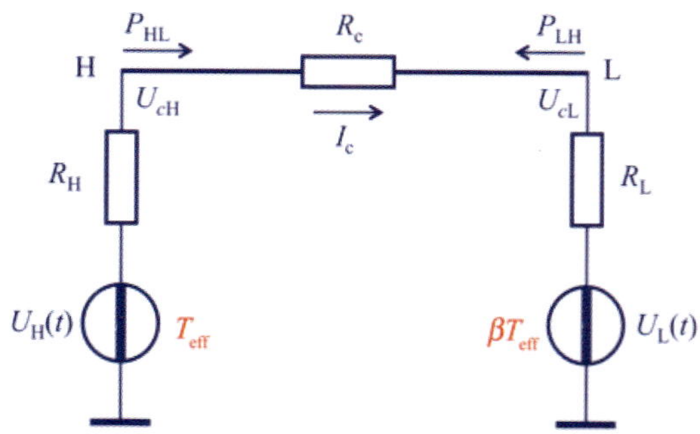

Figure 4. Schematic for illustrating the elimination of the Second-Law-attack and the BSY-attack by introduction of a proper temperature offset. The notation is consistent with that in Fig. 1.

holds, where T_{eff} is the noise temperature at the R_H resistors and βT_{eff} is the noise temperature of the R_L resistors. The solution of the equation is

$$\beta = \frac{1 + \frac{R_c}{R_L}}{1 + \frac{R_c}{R_H}}. \tag{11}$$

This value of β for the temperature-offset consequently eliminates Eve's opportunity to use the Second-Law-attack. One finds $\beta > 1$ for $R_L < R_H$ and $\beta < 1$ for $R_H < R_L$.

The remaining, essential question is whether the defense method delineated above introduces a higher signal for Eve's BSY-attack or not. Reevaluating the analysis of the BSY [8] with the a temperature offset given by Eq. (11), one obtains

$$\begin{aligned}\Delta_{KS}(T_{\text{eff}}, \beta T_{\text{eff}}) &= \left|\langle U_{cH}^2(t)\rangle - \langle U_{cL}^2(t)\rangle\right| \\ &= 4kT_{\text{eff}}\Delta f R_H \left|\frac{R_c^2(1-\alpha\beta) - \alpha R_H R_c(\beta-1)}{(R_H + R_c + R_L)^2}\right|,\end{aligned} \tag{12}$$

where $\alpha = \frac{R_L}{R_H}$. By substituting the above value for β, the nominator becomes zero so that

$$\Delta_{KS}(T_{\text{eff}}, \beta T_{\text{eff}}) = \left|\langle U_{cH}^2(t, T_{\text{eff}})\rangle - \langle U_{cL}^2(t, \beta T_{\text{eff}})\rangle\right| = 0 \tag{13}$$

Hence a modification of the noise temperature of the generators supplying the noise of the R_L resistors by the factor β yields a complete elimination the strongest attacks against the KLJN key exchange scheme, namely the Second-Law-attack and the BSY-attack [6–8].

Conclusions

We introduced the so far most efficient attack against the Kirchhoff-law–Johnson-noise (KLJN) secure key exchanger, *i.e.*, the Second-Law-attack. This attack utilizes the lack of exact thermal equilibrium in practical applications involving cables with non-zero resistance and results in more advantageous scaling and signal-to-noise ratio for Eve.

An advanced defense against this attack, involving a proper increase of the noise-temperature of the noise generator for the smaller resistances compared to that of the generators for the greater resistances surprisingly eliminated not only the Second-Law-attack but also the old Bergou-Scheuer-Yariv attack [6–8]. Removing these attacks can radically reduce Eve's fidelity while increasing Alice's and Bob's ones as a result of the potentially allowed longer bit-exchange periods or higher bandwidths [15].

Finally it should be emphasized that, in order to reduce the risk for hacking attacks or attacks due to possible malfunction, it is important not only to monitor and compare the voltage and current amplitudes at the two cable ends but also to run Gaussianity, spectral and other proper statistical checks on the signals, and to monitor the cable transfer function and signal integrity against hacking.

References

[1] Kish, L. B. (2006). "Totally secure classical communication utilizing Johnson(-like) noise and Kirchoff's law", *Phys. Lett. A* **352** (2006) 178–182.

[2] Kish, L. B., Granqvist, C. G. (2014). "Elimination of a Second-Law-attack, and all cable-resistance based attacks, in the Kirchhoff-law–Johnson-noise (KLJN) secure key exchange system", *Quantum Inf. Process.*, **13**, 2213–2219.
DOI: 10.1007/s11128-014-0729-7.

[3] Mingesz, R., Gingl, Z., Kish, L. B. (2008). "Johnson (-like)–Noise–Kirchhoff-loop based secure classical communicator characteristics, for ranges of two to two thousand kilometers, via model-line", *Phys. Lett. A*, **372** (2008) 978–984.

[4] Gingl, Z., Mingesz, R. (2014). "Noise Properties in the Ideal Kirchhoff-Law–Johnson-Noise Secure Communication System", *PLoS ONE*, **9** e96109.

[5] Mingesz, R., Vadai, G., Gingl, Z. (2014). "What kind of noise guarantees security for the Kirchhoff-Loop-Johnson-Noise key exchange?", *Fluct. Noise Lett.*, **13** (3), E1450021. *arXiv*:1405.1196.

[6] Bergou, J. (2005). Statement in Cho, A.'s interview of him, "Simple noise may stymie spies without quantum weirdness", *Science* **309**, 2148.

[7] Scheuer, J., Yariv A. (2006). "A classical key-distribution system based on Johnson (like) noise — How secure?", *Phys. Lett. A*, **359**, 737–740.

[8] Kish. L. B., Scheuer. J. (2010). "Noise in the wire: the real impact of wire resistance for the Johnson (-like) noise based secure communicator", *Phys. Lett. A*, **374**, 2140–2142.

[9] Kish, L. B. (2013). "Enhanced secure key exchange systems based on the Johnson-noise scheme", *Metrol. Meas. Syst.*, **20**, 191–204. DOI: 10.2478/mms-2013-0017.

[10] Mingesz, R., Kish, L. B., Gingl, Z., Granqvist, C. G., Wen, H., Peper, F., Eubanks, T., Schmera, G. (2013). "Unconditioal Security by the Laws of Classical Physics", *Metrol. Meas. Syst.*, **20**, 3–16. DOI: 10.2478/mms-2013-0001

[11] Kish, L. B., Mingesz, R., Gingl, Z., Granqvist C. G. (2012). "Spectra for the product of Gaussian noises", *Metrol. Meas. Syst.*, **19**, 653–658.

[12] Horvath, T., Kish, L. B., Scheuer, J. (2011). "Effective privacy amplification for secure classical communications", *EPL*, **94**, 28002. DOI: 10.1209/0295-5075/94/28002

[13] Hao, F. (2006). "Kish's key exchange scheme is insecure", *IEE Proc. Inform. Soc.*, **153**, 141–142.

[14] Smulko, J. (2014). "Performance Analysis of the 'Intelligent' Kirchhoff-Law–Johnson-Noise Secure Key Exchange", *Fluct. Noise Lett.*, **13** (3), 1450024. DOI: 10.1142/S0219477514500242

[15] Saez, Y., Kish, L. B., Mingesz, R., Gingl, Z., Granqvist, C. G. (2014). "Current and voltage based bit errors and their combined mitigation for the Kirchhoff-law–Johnson-noise secure key exchange", *J. Comput. Electron.*, **13**, 271–277. DOI: 10.1007/s10825-013-0515-2

5.3. Enhanced KLJN Schemes

Paper: Enhanced secure key exchange systems based on the Johnson-noise scheme[1]

5.3.1. The invention of these seven new types of enhanced KLJN systems — especially my "keyed" protocol — was triggered by my visit to Horace Yuen in October 2012; specifically, by his new (keyed) quantum protocol. With these KLJN schemes and the other ones mentioned in Section 5.8, we have 10 different KLJN systems.

[1] Kish, L.B. (2013). "Enhanced secure key exchange systems based on the Johnson-noise scheme" *Metrology & Measurement Systems*, **20**, 191–204; open access: http://www.degruyter.com/view/j/mms.2013.20.issue-2/mms-2013-0017/mms-2013-0017.xml; http://arxiv.org/abs/1302.3901; http://vixra.org/abs/1302.0055

Abstract. We introduce seven new versions of the Kirchhoff-Law-Johnson-(like)-Noise (KLJN) classical physical secure key exchange scheme and a new transient protocol for practically-perfect security. While these practical improvements offer progressively enhanced security and/or speed for the non-ideal conditions, the fundamental physical laws providing the security remain the same.

In the "intelligent" KLJN (iKLJN) scheme, Alice and Bob utilize the fact that they know exactly, not only their own resistor value, but also the stochastic time function of their own noise, which they generate before feeding it into the loop. By using this extra information, they can reduce the duration of exchanging a single bit, and in this way they achieve not only higher speed but also an enhanced security because Eve's information will significantly be reduced due to smaller statistics.

In the "multiple" KLJN (MKLJN) system, Alice and Bob have publicly known identical sets of different resistors with a proper, publicly known truth table about the bit-interpretation of their combination. In this new situation, for Eve to succeed, it is not enough to find out which end has the higher resistor. Eve must exactly identify the actual resistor values at both sides.

In the "keyed" KLJN (KKLJN) system, by using secure communication with a formerly shared key, Alice and Bob share a proper time-dependent truth table for the bit-interpretation of the resistor situation for each secure bit exchange step during the generation of the next key. In this new situation, for Eve to succeed, it is not enough to find out the resistor values at the two ends. Eve must also know the former key.

The remaining four KLJN schemes are combinations of the above protocols, synergized to enhance the security properties. These are: the "intelligent-multiple" KLJN (iMKLJN), the "intelligent-keyed" KLJN (iKKLJN), the "keyed-multiple" KLJN (KMKLJN) and the "intelligent-keyed-multiple" KLJN (iKMKLJN) key exchange systems.

Finally, we introduce a new transient-protocol offering practically-perfect security without the use of privacy amplification, which is not needed in practical applications but it is shown for the sake of ongoing discussions.

Keywords: Information theoretic security; unconditional security; practically perfect security; secure key distribution via wire; secure smart power grid.

1. Introduction

In this section we briefly define our basic terms of secure key exchange utilizing the laws of physics and introduce the Kirchhoff-Law-Johnson-(like)-Noise (KLJN) secure key exchange protocol.

1.1. *Conditional, unconditional, perfectly and imperfectly secure key exchange*

In private-key based secure communication, the two parties Alice (A) and Bob (B) possess an identical secure key, which is not known to the public, and they utilize this key in a cipher software to encrypt/decrypt the messages they send/receive [1]. Thus, to be able to communicate, Alice and Bob must first generate and share a secure key, which is typically a random bit sequence. The first important problem of secure communication is how to generate and share this key in a *secure* way between Alice and Bob. Note, even if Alice and Bob may already have shared a former secret key to communicate securely and they may be able to share a new key via that secure communication, that is not a *security-growing* method because, if the old key is cracked by an eavesdropper (Eve), the new key will also become compromised, implying that such a simple method cannot be used to share the new key. Of course, Alice and Bob may exchange a secure key by personally meeting or using a mail courier service; however that is neither satisfactory in terms of speed (i.e. not a high speed exchange) nor secure enough (against spying) if they share many keys for future use. It is safest to generate the new key "ad-hoc" when it is needed. Today's internet-based secure communications use software tools to generate and share secure keys; where the reason why Eve (who is monitoring the channel) cannot extract the key is her limited computational power. *However, enough knowledge about the secure key is publicly accessible after the key exchange to crack the key if sufficient computational time and/or power is available* [1]. Thus, these methods offer only conditional security because, with sufficient computing power (for example, by having a hypothetical quantum computer or its noise-based-logic version), the key would instantaneously be cracked by Eve. Due to the unexpected progression of computing technologies, this type of security is not only conditional, but is also *not future-proof* [2]. Eve can potentially crack the recorded key exchange and communication in the near future even if such a task looks hopeless presently.

Due to these facts, scientists have been exploring various physical phenomena for secure key exchanges, where the laws of physics could offer the security. The goal is to have a key exchange where either the exchange cannot be measured/recorded, or when the information measured/recorded by Eve is zero (a situation called perfect information theoretic security); or where this information is practically miniscule (a situation that is *practically-perfect* information theoretic security). If the extracted information by Eve is zero or small, and *its amount does not depend on Eve's accessible resources when she is approaching the limits imposed by the laws of physics*, the security is called *unconditional* [2, 3]. In practical cases "unconditional" and "information theoretic" security are interchangeable terms [1]. Thus the security classification can be classified as *perfect unconditional* (perfect information theoretic), *imperfect unconditional* (imperfect information theoretic) or conditional, where "perfect" with practical elements can only be "practically perfect", and only simplified mathematical models of the system may offer really perfect security at the theoretical level. In other words, perfect unconditional security can only be defined at the (simplified) conceptual level in any physical system, while imperfect unconditional security is the level that *any* real physical system can aim to reach due to the limitations posed by non-ideal elements and situations [2,3]. Perfect security is like infinity, it can be approached but never reached.

1.2. *The Kirchhoff-Law-Johnson-(like)-Noise (KLJN) secure key exchange scheme*

The KLJN scheme is a statistical/physical competitor to quantum communicators and its security is based on Kirchhoff's Loop Law and the Fluctuation-Dissipation Theorem. More generally, it is founded on the Second Law of Thermodynamics, which indicates that the security of the ideal scheme is as strong as the impossibility to build a perpetual motion machine of the second kind.

Until 2005, it was a commonly accepted that only quantum key distribution (QKD) is able to perform information theoretic (unconditional) secure key exchange. While QKD can, theoretically, provide perfect security, practically, it is always imperfect. In 2005, the Kirchhoff-Law-Johnson-(like)-Noise (KLJN) secure key exchange [2–16] scheme was introduced [4]; it was later built and its security demonstrated [7]. These ideas have

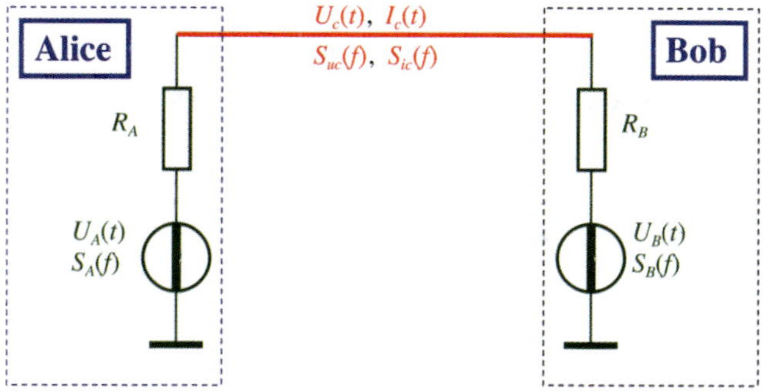

Figure 1. Outline of the core KLJN system. Parasitic elements leading to non-ideal features and defense circuit block (current/voltage monitoring/comparison) against invasive attack are not the topics of this paper thus they are not shown/discussed here. The resistors R_A and R_B are randomly selected from the $\{R_L, R_H\}$ set.

also inspired new concepts in computing, particularly noise-based logic and computing [17–24], where the complexity of data processing has been the issue, instead of the security of data.

The core KLJN system, without the defense circuitry (current–voltage measurement/comparison, filters, etc.) against invasive and non-ideality attacks, is shown in Fig. 1.

We first briefly survey the foundations of the ideal KLJN system [2, 4, 9]. Fig. 1 shows a model of the idealized KLJN scheme designed for secure key exchange [4]. At each KLJN-clock period, which is the duration of a single bit exchange, Alice and Bob connect their randomly chosen resistor, R_A and R_B, respectively, to the line. These resistors are randomly selected from the set $\{R_L, R_H\}$, $R_L \neq R_H$, where the elements represent the low, *L* (0), and high, *H* (1), bits, respectively. Alice and Bob randomly choose one of the resistors and connect it to the wire line. The situations *LH* and *HL* represent secure bit exchange [4], because Eve cannot distinguish between them through measurements, while *LL* and *HH* are insecure. The Gaussian voltage noise generators—delivering white noise with publicly agreed bandwidth—represent an enhanced thermal (Johnson) noise at a publicly agreed effective temperature T_{eff} (typically $T_{eff} \geq 10^9 K$ [7]) where their noises are statistically independent from each other or from the noise during the former KLJN-clock period. According to the Fluctuation-Dissipation Theorem, the power density spectra $S_{u,L}(f)$ and $S_{u,H}(f)$ of the voltages $U_{L,A}(t)$ and $U_{L,B}(t)$ supplied by the voltage generators in R_L and R_H are given by

$$S_{u,L}(f) = 4kT_{eff}R_L \quad \text{and} \quad S_{u,H}(f) = 4kT_{eff}R_H, \tag{1}$$

respectively.

In the case of secure bit exchange (*i.e.*, the *LH* or *HL* situation), the power density spectrum $S(f)$ and the mean-square amplitude $\langle U_{ch}^2 \rangle$ of the channel voltage $U_{ch}(t)$ and the same measures of the channel current $I_{ch}(t)$ are given as

$$\langle U_{ch,HL/LH}^2 \rangle = \Delta f \; S_{u,ch,HL/LH}(f) = 4kT_{eff}\frac{R_L R_H}{R_L + R_H}\Delta f, \tag{2}$$

$$\langle I_{ch,HL/LH}^2 \rangle = \Delta f \; S_{i,ch,HL/LH}(t) = \frac{4kT_{eff}}{R_L + R_H}\Delta f, \tag{3}$$

respectively, where Δf is the noise-bandwidth; and further details are given elsewhere [4, 9]. It should be observed that during the *LH* and *HL* cases, due to linear superposition, the spectrum given by Eq. (2) represents the sum of the spectra at two particular situations, *i.e.*, when only the noise generator of R_L is running one gets

$$S_{L,u,ch}(f) = 4kT_{eff}R_L\left(\frac{R_H}{R_L + R_H}\right)^2, \tag{4}$$

and when the noise generator of R_H is running one has

$$S_{H,u,ch}(f) = 4kT_{eff}R_H\left(\frac{R_L}{R_L + R_H}\right)^2. \tag{5}$$

The ultimate security of the system against passive attacks is provided by the fact that the power $P_{H \to L}$, by which the noise generator of resistor R_H is heating resistor R_L, is equal to the power $P_{L \to H}$ by which the noise generator of resistor R_L is heating resistor R_H [4, 9]. Thus the net power flow between Alice and Bob is zero, which is required by the Second Law of Thermodynamics. A proof of this can easily be derived from Eqs. (4–5) for the noise-bandwidth ofΔf:

$$P_{L \to H} = \frac{S_{L,u,ch}(f)\Delta f}{R_H} = 4kT_{eff}\frac{R_L R_H}{(R_L + R_H)^2}\Delta f, \tag{6a}$$

$$P_{H \to L} = \frac{S_{H,u,ch}(f)\Delta f}{R_L} = 4kT_{eff}\frac{R_L R_H}{(R_L + R_H)^2}\Delta f. \tag{6b}$$

The equality $P_{H \to L} = P_{L \to H}$ (*cf.* Eq. (6)) is in accordance with the Second Law of Thermodynamics; violating this equality would mean not only going against basic laws of physics (the inability to build a perpetual motion machine) but also allowing Eve to use the voltage-current cross-correlation $\langle U_{ch}(t)\ I_{ch,}(t)\rangle$ to extract the bit [4]. However the only quantity that could provide directional information is zero, $\langle U_{ch,HL/LH}(t)\ I_{ch,HL/LH}(t)\rangle = 0$, and hence Eve has no information to determine the bit location during the *LH* and *HL* situations. This security proof against passive (listening) attacks holds only for Gaussian noise, which has the well-known property that its power density spectrum or autocorrelation function already provides the maximum achievable information about the noise, and no higher order distribution functions or other tools (such as higher-order statistics) are able to serve with additional information.

The error probability of the bit exchange between Alice and Bob is determined by the following issue. In the case of the *LL* bit status of Alice and Bob, which is not secure situation, the channel voltage and current satisfy:

$$\langle U_{ch,LL}^2\rangle = \Delta f\ S_{u,ch,LL}(f) = 4kT_{eff}\frac{R_L}{2}\Delta f$$

and

$$\langle I_{ch,LL}^2\rangle = \Delta f\ S_{i,ch,LL}(t) = \frac{2kT_{eff}}{R_L}\Delta f, \tag{7}$$

while in the case of the other non-secure situation, the *HH* bit status, the channel voltage and current satisfy:

$$\langle U_{ch,HH}^2\rangle = \Delta f\ S_{u,ch,HH}(f) = 4kT_{eff}\frac{R_H}{2}\Delta f$$

and

$$\langle I_{ch,HH}^2 \rangle = \Delta f\, S_{i,ch,HH}(t) = \frac{2kT_{eff}}{R_H} \Delta f \tag{8}$$

During a key exchange in this classical way, Alice and Bob must compare the predictions of Eqs. (2, 3, 7, 8) with the actually measured mean-square channel voltage and current to decide if the situation is secure (LH or HL) while utilizing the fact that these mean-square values are different in each of these three situations (LL, LH/HL and HH). If the situation is secure, Alice and Bob will know that the other party has the inverse of his/her bit, which means, a secure key exchange takes place. To make an error-free key exchange, Alice and Bob must use sufficiently large statistics, which means a long-enough KLJN-clock period. However, the length of the KLJN-clock period determines the speed of the exchange of the whole key and, most importantly, Eve's statistics when utilizing non-ideal features to extract information. Thus protocols that can reduce the necessary duration of the KLJN-clock period for satisfactory statistics (fidelity) for Alice and Bob would enhance not only the speed but also the security. The new ("intelligent") KLJN protocol described in Sec. 2 offers this kind of improvement.

1.3. *On invasive attacks and non-idealities*

It should be observed [2, 3, 4, 6, 9, 10, 11, 12] that deviations from the shown circuitry — including invasive attacks by Eve, parasitic elements, delay effects, inaccuracies, non-Gaussianity of the noise, *etc.* — will cause a potential information leak toward Eve. The circuit symbol "line" in the circuitry represents an ideal wire with uniform instantaneous voltage and current along it. Fortunately the KLJN system is very simple, implying that the number of such attacks is strongly limited. The defense method against attacks utilizing these aspects is straightforward and is generally based on the comparison of instantaneous voltage and current data at the two ends via an authenticated communication between Alice and Bob.

These attacks are not the subject of the present paper. Instead, we refer the reader to our relevant papers where they have been analyzed [2, 3, 4, 6, 9, 10, 11, 12] and where misconceptions (or errors) of other papers written by others are also corrected. In surveys [2, 3], existing invasive attacks by other authors and us have been surveyed.

It is important to emphasize that, if the security of a certain bit is compromised, that is *also known by Alice and Bob*; therefore they can decide to discard the bit to have a clean secure key. The price for that is a reduced key exchange speed, while approaching the perfect security. Alternatively, privacy amplification can be executed [13], which also results in a slower key exchange. In conclusion, in the KLJN system, Alice and Bob can always protect themselves against the eavesdropping of the key, though they remain vulnerable to Eve jamming the key exchange (the same situation exists in QKD too, because the single photons are the easiest objects to jam).

2. The "intelligent" KLJN (iKLJN) key exchange protocol

The important characteristics of all passive attack types against practical KLJN systems is that Eve's bit-guessing success rate is strongly limited by poor statistics [4, 6, 9, 10, 11, 12]. In most cases, Eve has a very weak signal-to-noise ratio due to the limited KLJN-clock period, which is the time window available for Eve to make those statistics [11, 12]. Thus, if we could further limit Eve's time window, her success rate would further decrease. As we have already pointed out above, the minimum duration of the clock period in the original KLJN scheme is set by Alice and Bob, by their need to successfully classify the measured mean-square channel voltage and/or current levels by comparing them with the predictions of Eqs. (2, 3, 7, 8) in order to identify which one of the LL, LH/HL, HH situations does the actual status correspond to [4].

The Intelligent KLJN (iKLJN) system allows using shorter KLJN-clock periods, thus further weakening Eve's statistics. It has the same hardware as the original KLJN system but the protocol is more calculation-intensive. Alice and Bob utilize the fact that they exactly know, not only their own resistor value, but also the stochastic time function of their own noise, which they generate before feeding it into the loop. In the iKLJN method, Alice and Bob, by utilizing the superposition theorem on the channel noise, subtract their own contribution to generate a *reduced-channel-noise* that does not contain their own noise component. Because they don't know the resistance value at the other end, they must run two alternative *computational-schemes* simultaneously to calculate the reduced-channel-noises to account for the possible resistance situations (in total, four time functions, two voltage and two current noises, corresponding to the two possible resistance situations at the other end), as shown below. They then analyze at which one of these situations, the reduced-channel-noise does not contain their noise contribution. The reduced-channel-noise that does not contain their noise component is calculated with the correct assumption about the actual resistance value used by the other party. Thus the nature of the decision Alice and Bob makes is changed in the iKLJN: instead of evaluating mean-square noise amplitudes, they must assess the independence of two noise processes. Note that they also continue to assess the channel noise situation in the classical way by evaluating the mean-square of the channel noise amplitudes, which has partially independent information; thus combining the new and old information sources in the guessing process significantly shortens the KLJN-clock period needed for a given error probability.

At the same time, Eve can only use her old way, the parasitic elements, to extract any information. Because Eve's available observation time window (the KLJN-clock period set by Alice and Bob) becomes shorter, the information that she can extract is also significantly reduced. She may not even be sure during the shortened KLJN-clock period if a secure bit exchange took place or not; therefore her related error rate will increase. Thus, in the non-ideal situation, when information-leak exists, the

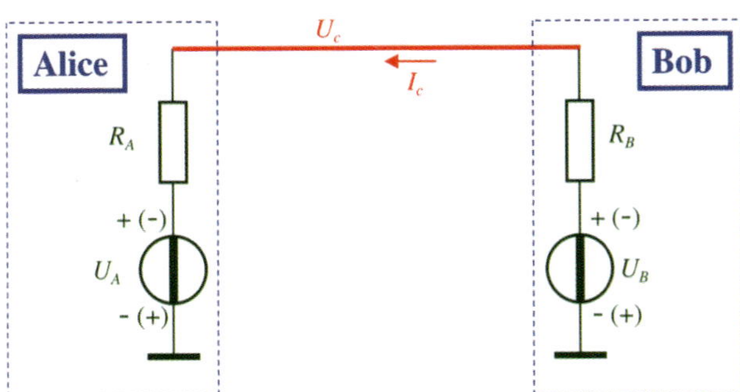

Figure 2. Snapshot of the current and the voltages in the KLJN system: at a given time moment, the polarities of Alice's and Bob's voltages and the resulting current are shown.

reduced observation time window progressively worsens Eve's probability of successfully guessing not only the key bits but also guessing which KLJN-clock periods had secure bit exchanges.

2.1. *Analysis of the "intelligent" KLJN (iKLJN) key exchange protocol*

To analyze the system with this new approach and to illustrate its way of functioning, first, let us assume that Bob's resistance is R_B and Alice's one is:

$$R_A = \alpha R_B \tag{9}$$

where $\alpha \neq 1$. We analyze Bob's protocol to demonstrate the process. Alice is acting in a similar way, which results in the same type of features.

According to Kirchhoff's Loop Law, the channel noise current $I_c(t)$ and noise voltage $U_c(t)$ at a given instant of time are given as:

$$I_c = \frac{U_B - U_A}{R_B(1+\alpha)} \tag{10}$$

$$U_c = \frac{U_A + \alpha U_B}{1+\alpha}, \tag{11}$$

where, for convenience, we skipped the time variable from the equations. Bob's calculation of the reduced-channel-noise currents and voltages is carried out in the following way.

Hypothesis-1: Reduced-channel-noise with the incorrect assumption

In one of the computational-schemes, Bob supposes that the resistance value of Alice is the same as his, that is, $R_A = R_B$, which is the incorrect assumption. Then the "incorrect" reduced-channel-current and reduced-channel-voltage amplitudes, $I^*_{c,1}$ and $U^*_{c,1}$, are:

$$I^*_{c,1} = I_c - \frac{U_B}{2R_B} = \frac{U_B - U_A}{R_B(1+\alpha)} - \frac{U_B}{2R_B} = -\frac{1}{R_B}\frac{2U_A - U_B(1-\alpha)}{2(1+\alpha)} \tag{12}$$

$$U^*_{c,1} = U_c - \frac{U_B}{2} = \frac{U_A + \alpha U_B}{1+\alpha} - \frac{U_B}{2} = \frac{2U_A - U_B(1-\alpha)}{2(1+\alpha)} \tag{13}$$

Hypothesis-2: Reduced-channel-noise with the correct assumption

In the other computational-scheme, Bob supposes that the resistance value of Alice is different from his, that is, $R_A = \alpha R_B$, which is the correct assumption. Then the "correct" reduced-channel-current and reduced-channel-voltage amplitudes, $I^*_{c,2}$ and $U^*_{c,2}$, are:

$$I^*_{c,2} = I_c - \frac{U_B}{R_B(1+\alpha)} = \frac{U_B - U_A}{R_B(1+\alpha)} - \frac{U_B}{R_B(1+\alpha)} = \frac{-U_A}{R_B(1+\alpha)} \quad (14)$$

$$U^*_{c,2} = U_c - \frac{\alpha U_B}{1+\alpha} = \frac{U_A + \alpha U_B}{1+\alpha} - \frac{\alpha U_B}{\alpha+1} = \frac{U_A}{1+\alpha} \quad (15)$$

It is obvious from our approach and the results in Eqs. (12–15) that, in the case of the incorrect assumption, the reduced-channel-noises contain both the noise contribution of Alice ($U_A(t)$) and that of Bob ($U_B(t)$) while, in the case of the correct assumption, they only contain Alice's noise ($U_A(t)$). Thus, Bob, by using a proper statistical tool to compare the reduced-channel-noises with his noise ($U_B(t)$) and checking for the independence, he can identify the "correct" assumption and, in this way, learn the actual resistor value of Alice. One of the simplest ways of that is the evaluation of the cross-correlations between his noises and the reduced-channel-noises:

With the incorrect assumption:

$$\langle U_B I^*_{c,1} \rangle = \frac{U_B^2(1-\alpha)}{2(1+\alpha)R_B} \neq 0 \quad (16)$$

$$\langle U_B U^*_{c,1} \rangle = \frac{\langle U_B^2 \rangle(\alpha - 1)}{2(1+\alpha)} \neq 0 \quad (17)$$

With the correct assumption:

$$\langle U_B I^*_{c,2} \rangle = \frac{-\langle U_A U_B \rangle}{R_B(1+\alpha)} = 0 \quad (18)$$

$$\langle U_B U^*_{c,2} \rangle = \frac{\langle U_A U_B \rangle}{1+\alpha} = 0 \quad (19)$$

While evaluating and comparing these cross-correlations is a new and independent source of information for Bob about Alice's resistance; in the field of statistics, there are more efficient ways of estimations, which will be explored in the future.

Naturally, Alice is proceeding in the same way as Bob. Then, combining this "intelligent" method with the original assessments based on the voltage and current noises will significantly enhance the information of Alice and Bob, and thus they can use shorter KLJN-clock periods to reach the same error probability as earlier. And because Eve does not have access to the "intelligent" way of information extraction, her successful guessing probability of the resistor situation in the non-ideal KLJN system will drop significantly compared to the original KLJN situation, while in the ideal system her information remains at zero. Moreover, due to the reduction of KLJN-clock duration, Eve's error probability of guessing if a secure bit exchange took place or not will also decrease, which further enhances her uncertainty.

Finally, an important question: *What is the price of this "intelligent" enhancement of the original KLJN protocol*? A higher computational capacity is needed for Alice and Bob to carry out this task, which implies higher electrical power requirements too. Thus, when computational performance is limited or low power requirements are essential, the classical KLJN protocol (and its "multiple" and/or "keyed" versions, see below) is the way to go.

3. An educational problem about the KLJN system

This section is not essential for the rest of the paper; thus Readers who are not interested in the deeper understanding of the foundations of the KLJN system can jump to the next section. A natural question arises about the iKLJN theory: *Do we need statistical evaluation? Or perhaps we could quickly determine the correct assumption by just using the reduced voltage and current values*? Similar questions arose in 2005 before Johnson(-like) noise was introduced into the KLJN system and a random DC voltage generator pair version of it was explored. The answer was that Johnson noise and statistical analysis were needed. That study was not published. We will now show that the similar situation in the iKLJN system leads to the same conclusion; statistics cannot be avoided.

If we use Ohm's law between the correctly deduced reduced-channel-voltage and reduced-channel-current components, we get

$$\frac{U^*_{c,1}}{I^*_{c,1}} = -\frac{1}{R_B}, \tag{20}$$

which is the expected result where the negative sign is due to the direction of current component into Bobs resistor from Alice's voltage generator, see Fig. 2.

Our naive expectation may be that, if we do the same derivation with the *incorrectly deduced* current and voltage components then the result will be different and then Bob can instantaneously find out that *Hypothesis-2* is the valid assumption for Alice's resistor.

Unfortunately, even the incorrect assumption yields the same result:

$$\frac{U^*_{c,2}}{I^*_{c,2}} = -\frac{1}{R_B} \tag{21}$$

This surprising result is due to the degeneracy of the system of equations describing the channel voltage and current, which also prohibits Bob from deducing Alice's resistor even though he knows his own resistor and voltage (a situation that led the author to test thermal noise in this system in 2005). No "instantaneous" way to find out Alice's resistance (that is, the value of α) exists due to this degeneracy. In conclusion, only statistical methods can provide the necessary information for Bob and Alice.

4. The "multiple" KLJN (MKLJN) key exchange protocol

In the "multiple" KLJN (MKLJN) system, Alice and Bob have publicly known identical sets of different resistors $\{R_1 < R_2 < \cdots < R_n\}$. For each KLJN-clock period, they randomly choose a resistor from this set and connect it (with a corresponding independent noise generator) to the line. There is a publicly known truth table about the bit-interpretation of the different combinations of the chosen $[R_i, R_j]$ resistor pair, whenever $R_i \neq R_j$, with the condition that the bit interpretation of $[R_i, R_j]$ is the inverse of the bit interpretation of $[R_j, R_i]$. It is designed so that, when the estimation of one of the resistors is missed at one of the sides and the neighboring resistor value is estimated instead, the bit-interpretation reverses in order to make Eve's guessing statistics worse.

In this new situation, for Eve to succeed, it is *not enough* to find out which end has the higher resistance. Eve must *exactly identify* the actual resistor values of both sides (while Alice and Bob only have to identify the value on the other side) to know which $[R_i, R_j]/[R_j, R_i]$ situation is the relevant one in the truth table and, in accordance with the original KLJN principle, even then Eve is unable to decide if $[R_i, R_j]$ or $[R_j, R_i]$ is the case. The result of modification is, again, an enhanced security in the non-ideal case.

5. The "keyed" KLJN (KKLJN) key exchange protocol

This enhancement is inspired by Horace Yuen's "keyed" quantum key exchange (called KCQ) [25] to enhance the security of his new quantum key exchange protocol. It works after a secure key is already generated/shared by Alice and Bob in the KLJN protocol. Then, by using secure communication with the shared key, they share a time-dependent truth table for the bit-interpretation of the $[R_L, R_H]_i$ versus $[R_H, R_L]_i$ resistor situation at the i-th secure bit exchange step during the generation of the next key. Note, the $[R_L, R_H]_i$ and $[R_H, R_L]_i$ situations must always mean opposite bit values.

It is obvious that KKLJN is a security growing technique because, even if Eve would succeed with correctly guessing the former key and learn the truth table, the security of the new key is still information-theoretically (unconditional) secure and it only "falls back" to the security level of the original KLJN key exchange. In the non-ideal case, if Eve has no information about the former key, the information of Eve about the key is progressively less than at the standard KLJN, and in the ideal case, her information remains zero.

6. The "keyed-multiple" KLJN (KMKLJN) key exchange protocol

Naturally, the KKLJN protocol can be enriched by using multiple resistor sets, $n > 2$, in the same fashion as the MKLJN system is doing but, instead of a publicly known truth table the bit-interpretation of the $[R_L, R_H]_i$ versus $[R_H, R_L]_i$ resistor situations is randomly changed for the subsequent key exchanges and the relevant truth table is shared by secure communication utilizing the former key. The KMKLJN protocol synergically combines the security enhancement of the MKLJN and KKLJN protocols.

7. Three more protocols: iMKLJN, iKKLJN and iKMKLJN

The "multiple", "keyed" and "keyed-multiple" KLJN protocols can be combined with the "intelligent" method of accessing the resistors at the other end by Alice and Bob to reduce the KLJN-clock duration and Eve's information, and to increase the speed. This enrichment should always be made whenever calculation power is enough. These are: the "intelligent-multiple" KLJN (iMKLJN), the "intelligent-keyed" KLJN (iKKLJN), the "keyed-multiple" KLJN (KMKLJN) and the "intelligent-keyed-multiple" KLJN (iKMKLJN) key exchange systems.

8. Transient-protocol for practically-perfect security

Finally, we introduce a new transient-protocol offering *practically-perfect* security without privacy amplification (which is not needed in practical applications, but shown for the sake of ongoing discussions). At the beginning of each bit-exchange period (KLJN-clock period), until the thermalization, which is the mixing and equilibration of noise-propagation (not waves) in the cable takes place, there is potential for an information leak. Even though, so far, no concrete method has been shown that would be able to extract the key bit, precautions have been made to reduce this effect with low-pass line-filters (which must always be present to keep away wave-generating frequency components of the wave-limit, and to prevent hacking attacks), and by ramping-up the generator voltages or, instead of ramping, starting the generators from zero voltage. However, even in these cases, one can expect some small information leak at the beginning because the system starts with noises that have not been mixed. This information leak is supposed to be very small with these stochastic signals (still needs a concrete attack to see how much it is) because the relaxation time constant of the low-pass filters is much longer than the propagation time in the cable (this is required by the no-wave limit). Thus, we show the "ultimate transient protocol" below for the sake of discussion, and not for practical applications.

The protocol is as follows. First Alice and Bob randomly decide if they want to use R_L or R_H during the next KLJN-clock period for their R_A and R_B, respectively. Then, to execute the key exchange, they use continuously variable resistors (such as potentiometers, etc.). If noise generators are also used to enhance the noise-temperature, then the band-limited white noise spectra of the noise-generators are also variable in a synchronized fashion, so that the noise-temperature stays constant at the publicly agreed value T_{eff}. At the beginning of the KLJN-clock period, both Alice and Bob start with:

$$R_A(0) = R_B(0) = \frac{R_L + R_H}{2} \tag{22}$$

and they stay at this value until the noises equilibrate in the wire. Thus no informative transients can be observed after connecting the resistors to the line. Then Alice and Bob execute independent slow continuous-time random-walks with their resistor values (and in a synchronized fashion with the spectral parameter of their noise generators). The random walks are executed so slowly that, from a thermodynamic point of view, the system is changing in the adiabatic limit: there is virtually a thermal equilibrium in the line during the whole random-walk process.

There is a publicly pre-agreed time period t_r to execute these independent random walks. If within this time period Alice and Bob reach their randomly preselected R_A and R_B value, they stop the random walk and stay at this value. Then, after the t_r time period, they start the KLJN protocol, in the regular way or a proper advanced fashion described in this paper. In this way, the transient effects and the information leak they may cause, are virtually kept away.

If by the end of the t_r time period, either the random-walk of Alice or Bob (or both) does not reach the randomly preselected resistance value, he/she (or both) submits a cancellation signal via an authenticated channel and the bit exchange process is terminated and a new KLJN-clock period starts in the way described above.

Note, it is also possible to introduce alternative protocols where there is no preliminary random decision by Alice and Bob and they use the random value they get by the random walk at the end of the t_r time period. If the difference of obtained resistance values is large enough they can use it for secure key exchange. The fact of whether the difference is large enough will only turn out at the end of the KLJN bit exchange period; they can then decide (based on publicly known rules) if they keep the exchanged bit.

9. Summary

While the KLJN secure key exchange has perfect security at idealized (mathematical) conditions, in the case of non-ideal devices (including non-zero range with finite speed), there is some information leak, similar to that which occurs in a quantum key exchange. To reach *practically-perfect* security, Alice and Bob can either do privacy amplification [13] or discard the bits [9] that provide information to Eve beyond a certain threshold. In this paper, we show additional ways to enhance the security in the non-ideal cases by using one of the seven new extended protocols introduced above.

In Table 1, we summarize the hardware and computational requirements of the various protocols.

It should be noted that the foundations of the security remain the same: the Second Law of Thermodynamics, the particular properties of Gaussian stochastic processes, and Kirchhoff's Law.

Finally, we should emphasize that, due to the exchange and comparison of the current/voltage data at the two ends of the line, Alice and Bob know Eve's exact information [9]. This is a new and unique situation in physical secure key exchange systems because QKD does not have this advantage.

Table 1. Comparison of the standard KLJN hardware requirements with that of the improved versions.

	KLJN	iKLJN	MKLJN	KKLJN	KMKLJN	iMKLJN	iKKLJN	iKMKLJN
Number n of resistors	2	2	$n > 2$	2	$n > 2$	$n > 2$	2	$n > 2$
Loops to compute	—	2	—	—	—	$n(n-1)$	2	$n(n-1)$

Though the exact implications of this fact have yet to be explored, it is obvious that it can offer ways for enhancing security not thought of before by properly discarding risky bits.

Acknowledgements

Discussions with Horace Yuen about his KCQ system and other general issues of security, and with Robert Mingesz and Zoltan Gingl about the KLJN system are appreciated.

References

[1] Liang, Y., Poor, H.V., Shamai. S. (2008). "Information theoretic security", *Foundations Trends Commun. Inform. Theory* **5**, 355–580. DOI: 10.1561/0100000036.

[2] Mingesz, R., Kish, L.B., Gingl, Z., Granqvist, C.G., Wen, H., Peper, F., Eubanks, T., Schmera, G. (2013). "Unconditional security by the laws of classical physics", *Metrology & Measurement Systems* **20**, 3–16. open access: http://www.metrology.pg.gda.pl/full/2013/M&MS_2013_003.pdf.

[3] Mingesz, R., Kish, L.B., Gingl, Z., Granqvist, C.G., Wen, H., Peper, F., Eubanks, T., Schmera, G. (2013). "Information theoretic security by the laws of classical physics" in: Balas, V.E. *et al.* (Eds.) *Soft Computing Applications, AISC 195*, pp. 11–25. Springer, Berlin.

[4] Kish, L.B. (2006). "Totally secure classical communication utilizing Johnson(-like) noise and Kirchhoff's law", *Phys. Lett. A* **352**, 178–182.

[5] Cho, A. (2005). "Simple noise may stymie spies without quantum weirdness", *Science* **309**, 2148. Available online at: http://www.ece.tamu.edu/~noise/news_files/science_secure.pdf.

[6] Kish, L.B. (2006). "Protection against the man-in-the-middle-attack for the Kirchhoff-loop-Johnson(-like)-noise cipher and expansion by voltage-based security", *Fluct. Noise Lett.* **6**, L57–L63. http://arxiv.org/abs/physics/0512177.

[7] Mingesz, R., Gingl, Z., Kish. L.B. (2008). "Johnson(-like)-noise-Kirchhoff-loop based secure classical communicator characteristics, for ranges of two to two thousand kilometers, via model-line", *Phys. Lett. A* **372**, 978–984.

[8] Palmer, D.J. (2007). "Noise encryption keeps spooks out of the loop", *New Scientist* **2605**, 32. Available online at: http://www.newscientist.com/article/mg19426055.300-noise-keeps-spooks-out-of-the-loop.html.

[9] Kish, L.B., Horvath, T. (2009). "Notes on recent approaches concerning the Kirchhoff-law-Johnson-noise-based secure key exchange", *Phys. Lett. A* **373**, 901–904.

[10] Kish, L.B., Scheuer, J. (2010). "Noise in the wire: The real impact of wire resistance for the Johnson(-like) noise based secure communicator", *Phys. Lett. A* **374**, 2140–2142.

[11] Kish, L.B. (2006). "Response to Scheuer–Yariv 'A classical key-distribution system based on Johnson (like) noise — How secure?'", *Phys. Lett. A* **359**, 741–744.

[12] Kish, L.B. (2006). "Response to Feng Hao's paper 'Kish's key exchange scheme is insecure'", *Fluct. Noise Lett.* **6**, C37–C41.

[13] Horvath, T., Kish, L.B., Scheuer, J. (2011). "Effective privacy amplification for secure classical communications", *Europhys. Lett.* **94**, 28002. http://arxiv.org/abs/1101.4264.

[14] Kish, L.B., Saidi, O. (2008). "Unconditionally secure computers, algorithms and hardware", *Fluct. Noise Lett.* **8**, L95–L98.

[15] Kish, L.B., Mingesz, R. (2006). "Totally secure classical networks with multipoint telecloning (teleportation) of classical bits through loops with Johnson-like noise", *Fluct. Noise Lett.* **6**, C9–C21.

[16] Kish, L.B., Peper, F. (2012). "Information networks secured by the laws of physics", *IEICE Trans. Commun.* **E95-B**, 1501–1507.

[17] Kish, L.B. (2009). "Noise-based logic: Binary, multi-valued, or fuzzy, with optional superposition of logic states", *Phys. Lett. A* **373**, 911–918.

[18] Kish, L.B., Khatri, S., Sethuraman, S. (2009). "Noise-based logic hyperspace with the superposition of 2^N states in a single wire", *Phys. Lett. A* **373**, 1928–1934.

[19] Bezrukov, S.M., Kish, L.B. (2009). "Deterministic multivalued logic scheme for information processing and routing in the brain", *Phys. Lett. A*, **373**, 2338–2342.

[20] Gingl, Z., Khatri, S., Kish, L.B. (2010). "Towards brain-inspired computing", *Fluct. Noise Lett.* **9**, 403–412.

[21] Kish, L.B., Khatri, S., Horvath, T. (2011). "Computation using noise-based logic: Efficient string verification over a slow communication channel", *Eur. J. Phys. B*, **79**, 85–90. http://arxiv.org/abs/1005.1560.

[22] Peper, F., Kish, L.B. (2011). "Instantaneous, non-squeezed, noise-based logic", *Fluct. Noise Lett.* **10**, 231–237. Available on: http://www.worldscinet.com/fnl/10/1002/open-access/S0219477511000521.pdf.

[23] Wen, H., Kish, L.B., Klappenecker, A., Peper, F. (2012). "New noise-based logic representations to avoid some problems with time complexity", *Fluct. Noise Lett.* **11**, 1250003. http://arxiv.org/abs/1111.3859.

[24] Mullins, J. (2010). "Breaking the noise barrier", *New Scientist* **2780**. Available online at: http://www.newscientist.com/article/mg20827801.500-breaking-the-noise-barrier. html?full=true.
[25] Yuen, H.P. (2009). "Key generation: Foundation and a new quantum approach", *IEEE J. Selected Topics in Quantum Electronics* **15**, 1630.

5.4. Physical Unclonable Function (PUF) Hardware Keys

5.4.1. In May 2013, Chiman Kwan, the Chief Technology Officer (CTO) at Signal Processing, Inc., contacted me with a call for research proposals about Physical Unclonable Functions (PUF) and asked if KLJN could be used. The results of the quick collaboration are the schemes described in the paper below. Though our grant proposal was turned down, we had, at least, this paper for our efforts. The best PUF system shown below is exponentially stronger than any other existing PUF alternatives. Only quantum key distribution could be a competitor if it can be integrated on a chip — but that is a long shot.

Figure 5.4.1. Chiman Kwan (Ph.D.). He is the Chief Technology Officer of Signal Processing, Inc. (Maryland). Many years back, Chiman sponsored my research on fluctuation-enhanced sensing, when he had related grants. He was also helping by seeking funds for KLJN research, though there was never success. He has a pleasant personality, is open-minded and very intelligent — the best sponsor I have ever had.

Paper: Physical Unclonable Function Hardware Keys Utilizing Kirchhoff-Law–Johnson-Noise Secure Key Exchange and Noise-based Logic[1]

Abstract. A weak unclonable function (PUF) encryption key means that the key can can be cloned by the hardware's manufacturer, and only the manufacturer. Nobody else is able to so that. Strong unclonable function (PUF) encryption key means that even the manufacturer of the hardware is unable to clone the key. In this paper, we first introduce an "ultra"-strong PUF with intrinsic dynamical randomness that is not only unclonable, but is also renewed into an independent key (with fresh randomness) during each use via the unconditionally secure key exchange. The solution utilizes the Kirchhoff-law–Johnson-noise (KLJN) method for dynamic key renewal, and a one-time-pad (OTP) secure key for the challenge/response process. The secure key is stored in a flash memory on the chip to provide tamper-resistance and non-volatile storage with zero power requirements in standby mode. Simplified PUF keys are shown, such as a strong PUF utilizing KLJN protocol during the first run, and then the noise-based logic (NBL) hyperspace vector string verification method for the challenge/response during the rest of its life. or until it is re-initialized. Finally, the simplest PUF utilizing NBL without KLJN (and thus can be cloned by the manufacturer but not by anybody else) is discussed.

1. Introduction

Intellectual Property (IP) protection of hardware designs such as field-programmable gate arrays (FPGA) and other integrated circuits (IC) has become important in recent years. It has been estimated that US$100 billion of the global IT industry revenue has been lost due to counterfeiting [1]. Attackers can easily tap the bit stream containing hardware configuration information and clone the hardware system using the information obtained.

To protect the IP of hardware systems, researchers developed a physical unclonable function (PUF) that is embodied in a physical structure, and that is easy to evaluate but hard to predict even if given the exact manufacturing process that produced it [2]. The earliest research into PUFs started in 1983 [3] and 1984 [4, 5]. The term PUF was coined in 2001 [6] and 2002 [7].

[1]Kish. L.B., Kwan, C., (2013). "Physical Unclonable Function Hardware Keys Utilizing Kirchhoff-Law–Johnson-Noise Secure Key Exchange and Noise-based Logic", *Fluctuation and Noise Lett.* **12**, 1350018. http://vixra.org/abs/1305.0068; http://arxiv.org/abs/1305.3248

In PUF, a challenge C_i and a response R_i, $(i = 1, 2, \ldots, Q)$ called a challenge–response pair (CRP) is needed. That is, a PUF is considered as a function that maps challenges to responses [1,2]. For a PUF to work, three conditions must be in place [1]:

(*i*) It is required that a response, R_i (to a challenge, C_i), gives only a negligible amount of information in another response, R_j (to a different challenge, C_j), with $i \neq j$;
(*ii*) Without having the corresponding PUF at hand, it is impossible to come up with the response, R_i, corresponding to the challenge, C_i, except with an exponentially low probability (versus the bit-number characterizing the PUF); and
(*iii*) *It is assumed that PUFs are tamper-evident.*

We note in passing that there is a frequent misunderstanding of the meaning of a Physical(ly) Unclonable Function (or PUF). Particularly, inspired by PUF systems utilizing random optical media, *it is often believed that the PUF is defined by the technical difficulties of accessing and characterizing the physical structure of a PUF device*. That means tamper-proof design.

Thus, the technical difficulties of cloning a PUF by physical means belongs to condition (*iii*), which is guaranteed per-definition, and the no-cloning condition is strictly defined by conditions (*ii*) and (*iii*). This does not mean that tamper-evidence is not a serious problem in PUF development.

Concerning the randomness they utilize, there are two types of PUFs. The first category uses explicitly-introduced randomness. Examples of these include Optical PUF and Coating PUF [2]. Two key advantages of this type of PUFs are the much greater ability to distinguish devices from one another, and their minimal environmental variations, compared to the second type of PUFs: PUFs using intrinsic randomness. Typical examples of the latter include Delay PUF, SRAM PUF, Butterfly PUF, Bistable Ring PUF, and Magnetic PUF.

The advantages of PUFs include the following. First, there is no need for a dedicated power-source (e.g., a battery) to keep the key memory alive. Second, data remanence issues are eliminated. Third, additional physical security measures to protect the memory device are minimized.

However, it has been pointed out in [8] that all existing PUFs in the literature still lack statistical uniqueness and robustness with respect to environmental conditions. In addition, the ability of the PUF to reliably and repeatedly generate a unique PUF of a minimum bit-length of 256 bits still needs significant research. Finally, [9] points out that many PUFs, including standard Arbiter PUFs and Ring Oscillator PUFs of arbitrary sizes, and XOR Arbiter PUFs, Lightweight Secure PUFs, and Feed-Forward Arbiter PUFs of up to a given size and complexity, are all vulnerable to numerical modeling attacks.

PUFs can be classified also according to their level of strength. A w*eak* PUF key can be cloned by its maker but not anybody else. A *strong* PUF cannot be cloned even by its maker. However, an even stronger PUF, ultra-strong ones, can be realized — which we will show below.

In this paper, we will introduce an *ultra-strong*, a *strong* and a *weak* PUF hardware key. These systems use two tools of noise-based informatics: the Kirchhoff-law–Johnson-noise (KLJN) unconditionally secure key exchange method [9–14], and noise-based logic (NBL) [16–26].

In Section 2, we briefly describe the KLJN system [9–14] and the version of NBL tool [26] required to build this "ultra-strong PUF".

In Section 3, we introduce a novel PUF key with intrinsic randomness. It may be called an "*ultra-strong*" PUF not only because it cannot be copied by its maker, but during each use, it is randomly changing by generating and sharing a new independent unconditional secret with the lock; thus, if a counterfeiter accidentally succeeds to make a clone (this has an exponentially low probability), that clone will be useless the next time the legal user applies the PUF key. The generation and regeneration of the intrinsic randomness in the key utilizes the unconditionally secure key distribution based on the Kirchhoff-law–Johnson-Noise (KLJN) method [9–14], which offers comparable or higher security than quantum key exchange (QKD) schemes do [13]. To generate the challenge/response, a one-time-pad (OTP) is used, which gets renewed during each use of the PUF key.

In Section 4, we present a simplified version of the PUF key described above, which is still "strong", that even the manufacturer is unable to clone it after its first use. The key refreshment is used only on the first time the

PUF key is applied to the lock. Then during the subsequent applications the secure key does not act as a one-time-pad anymore. The challenge-response system utilizes noise-based logic (NBL) [16–25] hyperspace vectors [19] and the string verification method [26] based on that.

In Section 5, we present the simplest PUF utilizing noise-based informatics: a key that can be cloned by the manufacturer (this means it is not a strong version) however it cannot be cloned by anybody else. This system contains a secret key installed by the manufacturer. The challenge-response system utilizes the hyperspace vector [19] of noise-based logic (NBL) [16–25] and the string verification method [26] based on that.

2. Brief description of the tools: the KLJN secure key exchange and noise-based logic

2.1. *The KLJN key exchange method with information theoretic (unconditional) security*

The Kirchhoff-law–Johnson-noise (KLJN) secure key exchange scheme [9–14] was first proposed in 2005 [9, 10]. It is a statistical/physical competitor to quantum key exchange, and its security is based on Kirchhoff's Loop Law and the Fluctuation-Dissipation Theorem. More generally, it is founded on the Second Law of Thermodynamics, which indicates that the security of the ideal scheme is as similarly strong as the impossibility to build a perpetual motion machine (of the second kind).

The basic KLJN system [9, 13] is shown in Fig. 1. Two identical pairs of resistors (R_L and R_H) and two binary switches with corresponding states (representing bit values) are used; one at Point A (Alice), and one at Point B (Bob). Depending on the state of a switch, which is randomly selected at the beginning of each clock cycle, one of resistors (with its own Johnson noise generator) is connected to the cable. Thus there are four possible states in which the two binary switches can be in: *LL*, *LH*, *HL*, and *HH*. The two mixed states, *LH* and *HL*, have identical mean-square channel noise voltage U_{ch} and current I_{ch} amplitudes, which cannot be distinguished by the eavesdropper (Eve). However, Alice and Bob know, not only the noise levels in the channel, but also the resistance value of

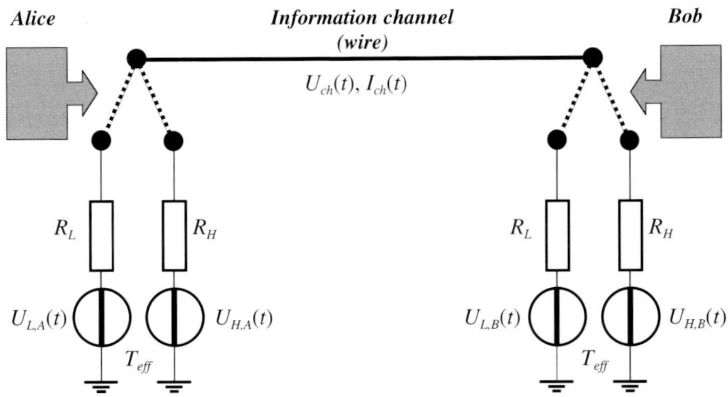

Figure 1. Basic circuitry of the KLJN secure key exchange system [9, 10]. This simple version is not protected against active (invasive) attacks or attacks that utilize non-ideal features of the building elements.

their own resistor connected to the line. So, they can logically deduce the state at the other side from the resultant of the resistance values (deduced from channel noise intensity) and from the state of their own switch.

This system, when ideal, offers unconditional (information theoretic) security against passive (listening) attacks only. To achieve unconditional security against active (invasive) attacks and/or attacks that utilize non-ideal features of the building elements, the following extra elements are essential [9, 13, 14]:

(*a*) External noise generator to have stable noise temperature with high accuracy.
(*b*) Filters to limit the frequency range to the measurement bandwidth and to the no-wave limit.
(*c*) Measurement of the instantaneous voltage and current values at the two ends (by Alice and Bob) and to compare them via an authenticated public channel. When they differ, those situations may indicate invasive eavesdropping and those bits must be discarded. For the number of secure bits used up by the authentication, see Section 3.

While there were several attacks published against the KLJN system by utilizing non-ideal building elements; the above described defense method has been able to protect the system and maintain the unconditional security at the practically-perfect level [13]. Recently, seven new KLJN schemes which achieve the practically-perfect unconditional security with higher bit exchange rates have been proposed.

The KLJN system has been proposed to provide unconditional security for hardware in computers, games and instruments [15].

2.2. *Noise-based logic, its hyperspace, and the string verification method*

Noise-based logic [16–26] is a new class of deterministic multivalued logic schemes where the information carrier is a system of random orthogonal time functions, as discussed below. For example, the stochastic time functions can have continuum amplitude distribution; can be random spikes; or be random telegraph signals. The universality of the binary versions of these logics schemes is proven.

Here we focus on the random telegraph wave scheme [22, 23, 26] and briefly introduce it based on [26]. There is a set of a simple type of random telegraph waves (RTW), $R_i(t_j)$, with discrete amplitudes ± 1 and discrete time, where the index i stands for the i-th RTW, and the index j for the j-th clock cycle. At the beginning of each clock period, a given random telegraph wave takes a 1 or -1 amplitude with 50% probability. There is no memory in the system, except that the chosen amplitude is held until the end of the clock period where a new random selection takes place. Thus, for the sake of simplicity, we drop the continuum time parameter in the notation for an RTW, and refer only to the amplitude of the ith RTW during the jth clock cycle by $R_i(j)$. In other words, an RTW can be generated by a discrete random number generator that generates $+1$ or -1 with a 50% probability each.

The product of an arbitrary number of independently generated RTWs is:

$$W_x(j) = \prod_{i=1}^{N} R_i(j) \tag{1}$$

and is called a hyperspace vector — which is also an RTW with the same statistical properties, and it is orthogonal to the original RTWs or any other

RTW generated independently, i.e.:

$$\langle W_x(j) R_k(j) \rangle = \left\langle R_k(j) \prod_{i=1}^{N} R_i(j) \right\rangle = 0, \tag{2}$$

where either $k \in \{1, \ldots, N\}$ or $k > N$, which means that R_k is an RTW generated independently of $R_1, R_2, \ldots, R_N$.

The string verification method utilizes the fact that, if two RTWs are generated independently, the probability $P(m)$ that their amplitudes are identical over m clock steps is

$$P(m) = 2^{-m} \tag{3}$$

To represent an N-long bit string by such a product (hyperspace vector) we need $2N$ independently generated RTW signals to assign a different RTW to the H and L values of the N bits.

Suppose, both Alice and Bob have an N-bit long string and they would like to detect if one or more of their bit values are different [26]. Then they use the same publicly agreed set of $2N$ independent random number generators and assign each of them to the publicly agreed relevant bit and its value. Then using the N RTWs (random number generators) that represent the bit values in the string, they make the RTW product described by Eq. (1). Let us call Alice's product $W_A(t)$, and Bob's $W_B(t)$.

Suppose that the bit strings of Alice and Bob differ by at least a single bit value. Then at least one RTW in these products (see Eq. (1)) is different.

Obviously, if the $W_A(t)$ and $W_B(t)$ products contain the same RTW elements, which means that the two strings are identical, the following relations hold:

$$W_A(t_j) - W_B(t_j) = 0 \tag{4}$$

$$W_A(t_j) W_B(t_j) = 1 \tag{5}$$

Thus, a single comparator device checking for non-zero values in the first case; or a multiplier and a comparator checking for negative values in the second case can verify if the assumption about identical strings have been violated [26].

In conclusion, if the strings are different, the generated hyperspace vectors will also be different or deviate in a short time. Thus the comparison of a short sample of $W_A(t)$ and $W_B(t)$ is enough to detect any difference with high probability. The probability that the difference of the strings is not detected is given by Eq. (3). This error probability gets infinitesimally small very quickly. For example, to reach the theoretical error probability of approximately 10^{-25} of the logic gates in computer chips at idealistic conditions, the required clock steps of comparison m is only 83 (because $0.5^{83} \approx 10^{-25}$). Thus, Alice and Bob are able to detect that there is an arbitrary difference between their bit strings *of arbitrary length N* by communicating only 83 bits [26].

Note, an analogous representation of RTWs and their string verification can be achieved by using 0 and 1 amplitudes in the RTW and applying the XOR logic operation instead of multiplication.

After the above preparations, the description of the three different PUF systems is straightforward.

3. Ultra-strong PUF hardware key with KLJN and one-time-pad

The system is illustrated in Fig. 2. Both the Key and the Lock are integrated on chips. The manufacturer prepares the PUF with a secure key, which is clonable. When the user uses the key the first time (initialization) the Lock recognize the key by this clonable secure key and then the Lock and the Key performs a KLJN key exchange and generates a new key that is unknown even to the manufacturer. The new secure key is saved in a

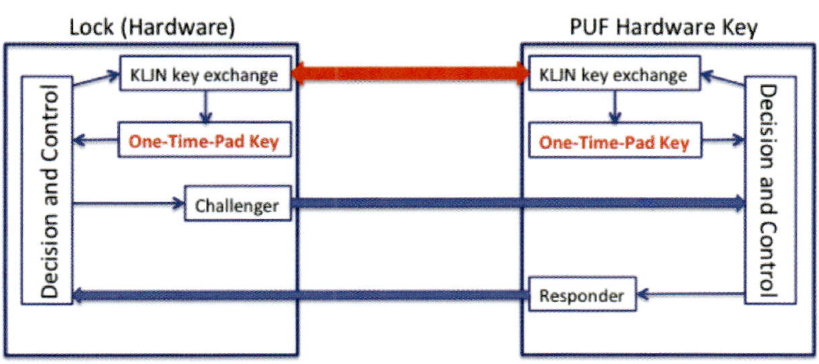

Figure 2. The Ultra-strong PUF hardware key with unconditionally secure KLJN key exchange and one-time pad.

flash memory in the chip; thus making it virtually impossible to access this information.

After this initialization procedure, the unclonable key is ready to be used and is *ultra-strong* because at each new application of the key, the Lock and the Key exchange a new, independent, unconditionally-secure key. This PUF is ultra-strong because, if a counterfeiter accidentally succeeds in making a clone (this has an exponentially low probability), that clone will be useless the next time the legal user applies the PUF key.

The challenge/response process is extremely simple: the Lock requests for the PUF key to send the secure key they exchanged during the last use of the key. It is important to note that the exchanged secure-key is used as a *one-time pad*, which is the most secure way of secure communications, because during the next application of the PUF, a new, independent, secure key will be used.

If a counterfeiter is trying to open the lock, his chances of succeeding with a randomly generated N-bit long key is 2^{-N}, where N is the length of the secure key, that is exponentially small. It is also important to note that there is no information about the next key, and no modeling/simulation of system behavior can provide any information.

Note: minimum $N + \log_2 F$ key bits must be exchanged, where F is the number of authenticated public bits exchanged for current–voltage comparison against active/invasive attacks, because the authentication of F bits in a public channel requires using at least $\log_2 F$ secure bits [27]. The logarithmic dependence guarantees that the load by authentication stays negligibly small.

4. Strong PUF hardware key with KLJN and noise-based logic

Here we present a simplified version of the PUF key described above, which is still "strong", that is, even the manufacturer is unable to clone it after its first use (see Fig. 3). The key refreshment is only used the first time, during the initialization (which can be repeated if needed). Then, during subsequent applications, the secure key cannot be used as a one-time-pad because the same secure key must be used to provide an exponentially difficult job for

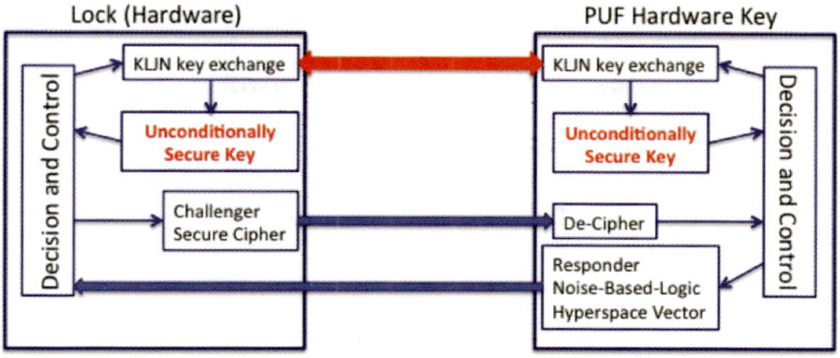

Figure 3. The Strong PUF hardware key with unconditionally secure KLJN key exchange and noise-based logic.

the counterfeiter. The challenge must proceed via a secure communication, where the secure key is used in the cipher. The Lock generates a random string, which then gets encrypted by the cypher and sent to the PUF key. The response system in the PUF key utilizes noise-based logic (NBL) and generates and sends the product RTW described by Eq. (1), that corresponds to the given bit string. A counterfeiter has again 2^{-N} chance to succeed if he tries with his only available tool: brute force.

Then the Lock will use the string verification method described above: it also generates the required RTW product and uses Eqs. (4) and (5) to check if the sequence coming from the Key is satisfactory. If the Key sends N clock steps of the product, the chance that a brute force counterfeiter escapes detection is, again, 2^{-N}.

The outline of the Responder is shown in Fig. 4. The secure key bits shared by the unconditionally secure KLJN protocol determine for each bit if its pseudo-random-number generator A or B is assigned to its high or low value. All the $2N$ generators are independent and publicly known — except for their secret assignment.

5. Simple PUF hardware key with noise-based logic

The simplest PUF utilizing noise-based informatics is shown in Fig. 5. It has a secure key that is implemented by the manufacturer, and can thus be cloned by the manufacturer (this means it is not a strong PUF); however, this PUF cannot be cloned by anybody else. Like to the former system, the

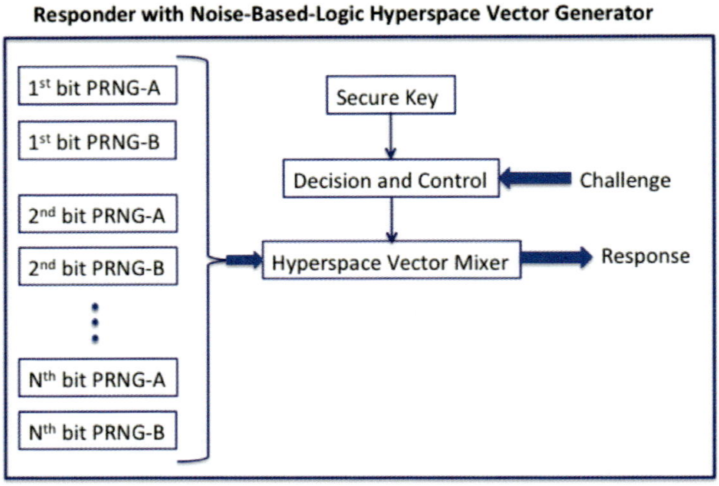

Figure 4. The Responder with RTW-based noise-based logic.

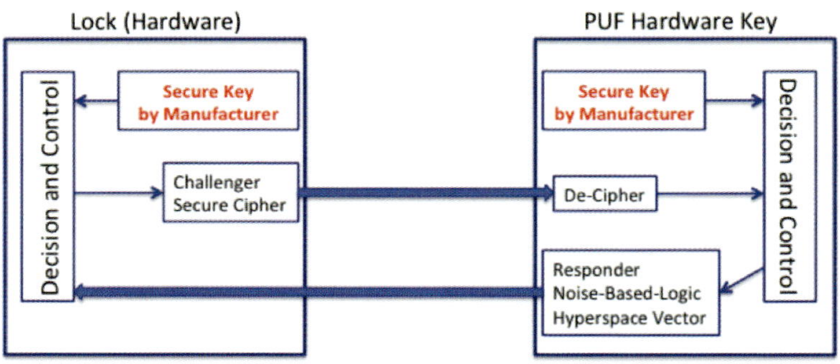

Figure 5. Simple PUF hardware key with noise-based logic.

challenge–response system utilizes secure communication, and the NBL scheme shown in Fig. 4. is described in Section 2.2.

6. Conclusion

We showed three ways to realize physical unclonable function (PUF) encryption keys where noise-based informatics is utilized. The key lengths can easily be much longer than 256 bits. All these systems can be integrated on a chip, which provides robustness against tampering and environmental effects.

References

[1] Guajardo, J., Kumar, S., Schrijen, G.-J., Tuyls, P. (2007). "Physical Unclonable Functions and Public-Key Crypto for FPGA IP Protection", *International Conference on Field Programmable Logic and Applications,*. Aug. 2007, pp. 189–195.

[2] Bauder, D.W. (1983). "An anti-counterfeiting concept for currency systems", Research report PTK-11990. Sandia National Labs. Albuquerque, NM.

[3] Simmons, G. (1984). "A system for verifying user identity and authorization at the point-of sale or access," *Cryptologia* **8**, 1–21.

[4] Simmons, G. (1991). "Identification of data, devices, documents and individuals", *IEEE International Carnahan Conference on Security Technology*, pp. 197–218.

[5] Pappu, R., Recht, B., Taylor, J., Gershenfeld, N. (2002). "Physical one-way functions", *Science* **297**, 2026–2030. http://dx.doi.org/10.1126/science.1074376.

[6] Gassend, B., Clarke, D., van Dijk and, M., Devadas, S. (2002). "Silicon Physical Random Functions", *Proceedings of the Computer and Communications Security Conference*, November 2002.

[7] http://www.dodsbir.net/sitis/display_topic.asp?Bookmark=44320. Not available anymore.

[8] Rührmair, U., Dror, G., Sehnke, F., Devadas, S., Sölter, J., Schmidhuber, J. (2010). "Modeling Attacks on Physical Unclonable Functions," CCS'10, October 4–8, 2010, Chicago, Illinois, USA.

[9] Kish, L.B. (2006). "Totally secure classical communication utilizing Johnson(-like) noise and Kirchhoff's law", *Phys. Lett. A* **352**, 178–182.

[10] Cho, A. (2005). "Simple noise may stymie spies without quantum weirdness", *Science* **309**, 2148. Available online at: http://www.ece.tamu.edu/~noise/news_files/science_secure.pdf.

[11] Palmer, D.J. (2007). "Noise encryption keeps spooks out of the loop", *New Scientist* **2605**, 32. Available online at: http://www.newscientist.com/article/mg19426055.300-noise-keeps-spooks-out-of-the-loop.html.

[12] Kish, L.B. (2013). "Enhanced secure key exchange systems based on the Johnson-noise scheme", *Metrol. Meas. Syst.* **20**, 191–204; open access: http://www.degruyter.com/view/j/mms.2013.20.issue-2/mms-2013-0017/mms-2013-0017.xml

[13] http://arxiv.org/abs/1302.3901, http://vixra.org/abs/1302.0055

[14] Mingesz, R., Kish, L.B., Gingl, Z., Granqvist, C.G., Wen, H., Peper, F., Eubanks, T., Schmera, G. (2013). "Unconditional Security by the

Laws of Classical Physics," *Metrol. Meas. Syst.* **XX** (1), 3–16. Open access: http://www.degruyter.com/view/j/mms.2013.20.issue-1/mms-2013-0001/mms-2013-0001.xml

[15] Kish, L.B. (2006). "Protection against the man-in-the-middle-attack for the Kirchhoff-loop–Johnson(-like)-noise cipher and expansion by voltage-based security", *Fluct. Noise Lett.* **6**, L57–L63.

[16] Kish, L.B., Saidi, O. (2008). "Unconditionally secure computers, algorithms and hardware", *Fluct. Noise Lett.* **8**, L95–L98.

[17] Kish, L.B. (2009). "Noise-based logic: Binary, multi-valued, or fuzzy, with optional superposition of logic states", *Phys. Lett. A* **373**, 911–918. http://arxiv.org/abs/0808.3162.

[18] Mullins, J. (2010). "Breaking the noise barrier", *New Scientist* **2780**, 29 September 2010. Available online at: http://www.newscientist.com/article/mg20827801.500-breaking-the-noise-barrier. html?full=true.

[19] Wen, H., Kish, L.B. (2012). "Noise based logic: why noise? A comparative study of the necessity of randomness out of orthogonality", *Fluct. Noise Lett.* **11**, 1250021. http://arxiv.org/abs/1204.2545.

[20] Kish, L.B., Khatri, S., Sethuraman, S. (2009), "Noise-based logic hyperspace with the superposition of 2^N states in a single wire", *Phys. Lett. A* **373**, 1928–1934. http://arxiv.org/abs/0901.3947.

[21] Bezrukov, S.M., Kish, L.B. (2009). "Deterministic multivalued logic scheme for information processing and routing in the brain", *Phys. Lett. A* **373**, 2338–2342. http://arxiv.org/abs/0902.2033.

[22] Gingl, Z., Khatri, S., Kish, L.B. (2010). "Towards brain-inspired computing", *Fluct. Noise Lett.* **9**, 403–412. http://arxiv.org/abs/1003.3932.

[23] Kish, L.B., Khatri, S., Peper, F. (2010). "Instantaneous noise-based logic", *Fluct. Noise Lett.* **9**, 323–330. http://arxiv.org/abs/1004.2652.

[24] Peper, F., Kish, L.B. (2011). "Instantaneous, non-squeezed, noise-based logic", *Fluct. Noise Lett.* **10**, 231–237. DOI: 10.1142/S0219477511000521.

[25] Wen, H., Kish, L.B., Klappenecker, A., Peper, F. (2012), "New noise-based logic representations to avoid some problems with time complexity", *Fluct. Noise Lett.* **11**, 1250003. http://arxiv.org/abs/1111.3859.

[26] Wen, H., Kish, L.B., Klappenecker, A. (2013). "Complex noise-bits and large-scale instantaneous parallel operations with low complexity", *Fluct. Noise Lett.* **12**, 1350002. http://vixra.org/abs/1208.0226.

[27] Kish, L.B., Khatri, S., Horvath, T. (2011). "Computation using Noise-based Logic: Efficient String Verification over a Slow Communication Channel", *Eur. J. Phys. B.* **79**, 85–90. http://arxiv.org/abs/1005.1560.

[28] Hjelme, D.R., Lydersen, L., Makarov, V. (). "Quantum cryptography". http://arxiv.org/abs/1108.1718.

5.5. Unconditionally Secure Smart Grids[1,2]

5.5.1. This result is the only KLJN scheme which has been patented, thanks to the efforts of Elias Gonzalez, a PhD student, and Nicholas Chremos, senior licensing manager at Texas A&M, USA.

Figure 5.5.1. Elias Gonzalez was a PhD student at Texas A&M University. He graduated in 2016. He has a B.Sc. in Physics, and is a clever model creator and analyzer. He is most interested in security applications and, after graduation, he found a job in that sector.

5.5.2. There was an interesting but sad episode with Elias. He had received an invitation to go to China for a summer research program focusing on KLJN, and had won a National Science Foundation (NSF) fellowship for that (he was one

(*Continued*)

[1]Gonzalez, E., Kish, L.B., Balog, R., Enjeti, P. (2013). "Information theoretically secure, enhanced Johnson noise based key distribution over the smart grid with switched filters", *PLoS ONE* **8**, e70206. Open access: http://journals.plos.org/plosone/article?id=10.1371/journal.pone.0070206

[2]Gonzalez, E., Kish, L.B., Balog, R., Granqvist, C.G., Enjeti, P. (2014). "Encryption Key Distribution System and Method", U.S. Patent Application Serial No. 14/489,025, September 17, 2014.

(*Continued*)

of 30 winners, in the whole Nation). He was then invited to Washington DC to complete the necessary preparation; following which, he acquired the visa and fight tickets. A few days before his scheduled departure to China, he received a phone call from the NSF program manager. She said that Elias's fellowship had been canceled because the research topic was of high national interest, and therefore could not be shared with China. (Note: it is true that there is an export ban on certain encryption schemes but international research and publications on encryption schemes are not banned). It was a funny but sad way of recognizing the importance of KLJN by NSF...

This concept is based on the idea that power lines and power networks (also known as Smart Grids), can be used for KLJN key exchange, with filters that separate the KLJN frequency band from the 50/60 Hz power frequency. In this way, single KLJN loops can be isolated from the non-KLJN loads in the networks and from the other KLJN loops. Thus no new wires are necessary for users to adopt KLJN. However, switched filter boxes must be installed at each unit on the smart grid, even at units that do not have a KLJN device. These filters must be controlled by a central server, for example, via wireless means. Due to the lack of space for practical applications of KLJN in this first edition, we refer the Reader to the figures below and to the open access paper.[1]

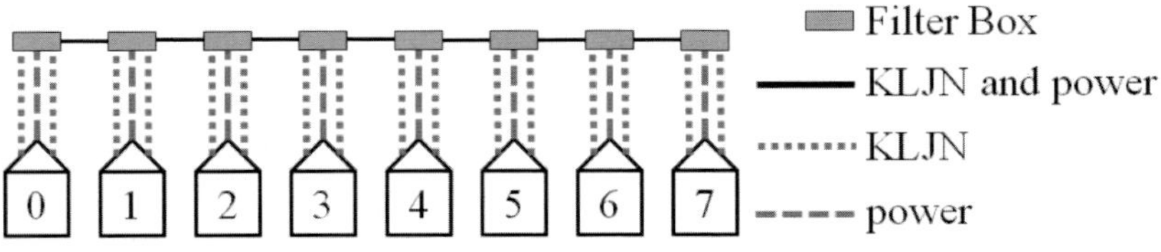

Figure 5.5.2. Line network example for network size $N = 7$. Each host is connected to a filter box and the filters boxes are connected to the power grid. Note how each host has three wire connections to its filter box: KLJN to the left; power; KLJN to the right.

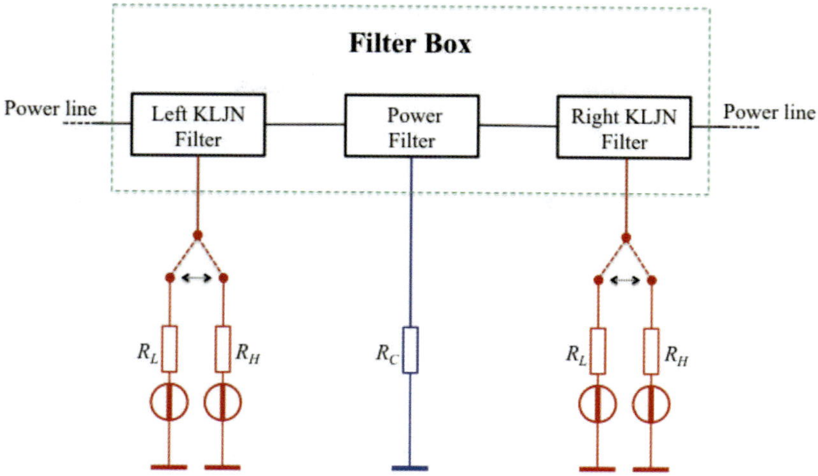

Figure 5.5.3. A filter box example. *Rc* represents the consumer electronics load (50/60 Hz).

5.6. Unconditionally Secure Vehicle Communication Networks[1,2]

5.6.1. This short project is a simple example of the potential applications of KLJN in areas of high strategical importance; where (information-theoretic) unconditional security is essential.

5.6.2. In the last few years, vehicular communication networks have become an emerging research topic. The main motivation for the deployment of a more intelligent vehicular system is the need to enhance transportation safety and efficiency. In this type of network, vehicles will be equipped with advanced

(*Continued*)

[1] Saez, Y., Cao, X. Kish, L. B., Pesti, G. (2014). "Securing vehicle communication systems by the KLJN key exchange protocol", *Fluctuation and Noise Lett.* **13**, 1450020; http://arxiv.org/abs/1404.1900

[2] Cao, X. Saez, Y., Pesti, G., Kish, L.B. (2015). "On KLJN-based Secure Key Distribution in Vehicular Communication Networks", *Fluctuation and Noise Lett.* **14**, 1550008; http://arxiv.org/abs/1409.5911

(Continued)

sensing and computing capabilities where communication protocols will enable them to share information with each other and the roadside infrastructure. The incorporation of this new range of technology will create a smart network where every vehicle is aware of its surrounding environment. In fact, a great number of applications are under development to improve traffic safety and mobility, and to perform financial transactions (e.g. toll collection) securely and efficiently.

5.6.3. A malicious hacker breaking into such a system could cause large-scale accidents, for example. Thus making the key exchange unconditionally secure is an important factor for enhancing not only data security, but also vehicular and human safety. The project resulted in two papers (not shown here). To understand the schematics of such a system, see Fig. 5.6.2.

Figure 5.6.1. Xiaolin Cao. She is an Associate Professor at Jilin University, China. She came to Texas A&M University, USA, for a one-year visiting scholar position (sponsored by a Chinese foundation) to study the potential of KLJN for applications in Vehicular Ad Hoc Networks.

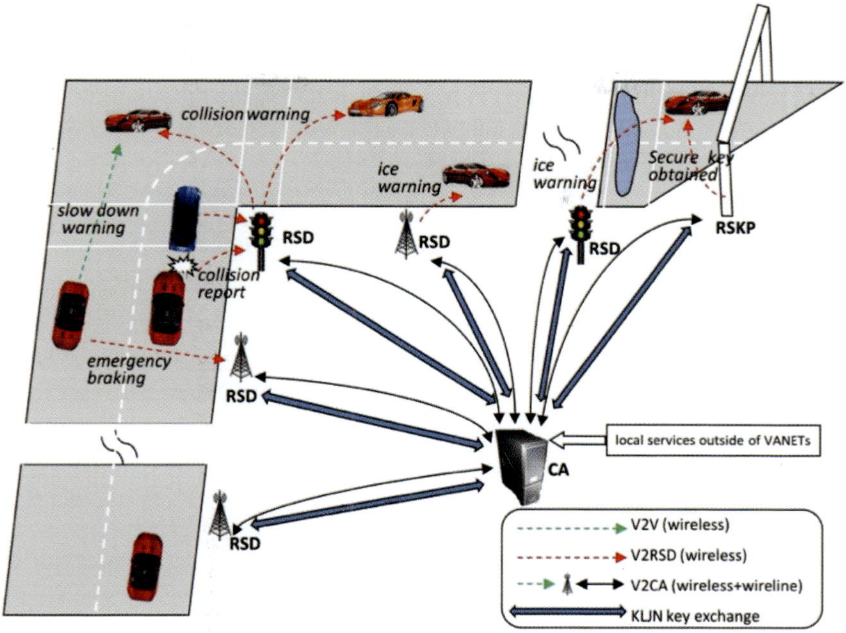

Figure 5.6.2. A possible way to implement KLJN, to secure the key exchange in an Ad Hoc Vehicular Network unconditionally (see references on page 411). In this system, the Central Authority (CA) has wire connections to the Road Side Key Providers (RSKP) and the Road Side Devices (RSD, traffic lamps, wireless towers, etc.). Thus unconditionally secure keys can be continuously generated and shared between them via the KLJN method. The RSKP units communicate part of the key to the cars via near-field communication; for example, by magnetic dipole communicators under the lanes. Then the shared, unconditionally-secure keys are used for high-speed wireless communications between the cars and the RSDs, and between the RSDs and the CA. Simple analysis shows that, with reasonable economic effort, and on roads full of cars, the key refreshing time would span less than 20 minutes. This duration can easily be shortened further by using multiple cables (e.g. existing phone cables with 500 wire pairs) and chip technology for KLJN.

5.7. Design Rules-of-thumb for Security

So far, known non-idealities have resulted only in miniscule information leaks (Section 5.1) that can be *i*) sometimes left untouched, *ii*) nullified by introducing compensating effects (Section 5.2), or *iii*) managed by privacy amplification[1] and/or by discarding high-risk bits.[2] The few reported larger information leaks were due to flawed concepts and/or poor experimental design (see Sections 4.2, 4.4, 4.6, 4.6.1–4.6.3). The real danger is the Unknown — possible situations or non-idealities that have not yet been recognized and, without relevant defense measures, may result in a major information leak. Good examples are the quantum hacking attacks by Vadim Makarov[3] and his associates (also see references in Section 4.4), which fully cracked even commercial quantum communicators (a pair of which had been priced at around US$100,000)! In this section, I list some of the essential security measures I would use if my neck was on the line.

The following are security measures for you, the Designer — supposing you want to approach the ultimate security, even for unexpected attack types — with your practical KLJN key exchanger:

5.7.1. *General directives*

This is a classical physics system. *Utilize this advantage in all aspects*! You should construct a high-accuracy model of the whole system (a computer model with professional-quality cable simulators, such as LTSPICE, is fine), including propagation, transient and possible reflection effects (at high-frequency disturbance), and use this model in all signal monitoring and procedures of continuous comparisons of the measured quantity with the simulated values.[4] At random times, the KLJN system must be checked with probe signals and noises to see that the model is still an accurate

[1]Horvath, T., Kish, L.B., Scheuer, J. (2011). "Effective Privacy Amplification for Secure Classical Communications", *EPL* (formerly *Europhysics Letters*) **94**, 28002; http://arxiv.org/abs/1101.4264

[2]See our 2009 paper in Section 4.2 and also R. Mingesz (2014) "Experimental study of the Kirchhoff-Law-Johnson-Noise secure key exchange", *International Journal of Modern Physics: Conference Series*, **33**, 1460365.

[3]http://www.vad1.com/lab/

[4]H.P. Chen, M. Mohammad, L.B. Kish (2016), "Current Injection Attack against the KLJN Secure Key Exchange", submitted for publication. http://arxiv.org/abs/1512.03685, http://vixra.org/abs/1512.0274

Figure 5.7.1. Vadim Makarov the mighty quantum hacker, probably after he had cracked all the commercial QKD systems on the market. (Courtesy of IQC/Peter Kovacs). While he offered patches to fix those vulnerabilities in order to keep those companies on the playfield, he was still the feared security-killer "Godfather" of quantum-hardware (see Section 1.6 also). To strongly reduce the opportunities of an "*electronics-based-Vadim*" executing high-profile cracks, some advice is given in this section. Note, these may not always be enough! Basic rules of thumb: "*Vadims*" are always watching, they never sleep and they enter through the narrowest gap/negligence. Security is not the place for pride but for humility. See also Gunn-Allison-Abbott's self-implicated self-cracks due to poor KLJN design in Sections 4.6.1–4.6.3.

description of the system. If not, hacking or another type of damage is in effect. In details:

(*a*) Naturally, the basic voltage and current comparison checks by Alice and Bob must continuously be performed.
(*b*) Classical physical information is extremely robust: you can continuously measure it without destroying it. Utilize that fact! Continuously monitor it for any discrepancies, spurious signals, unexpected statistics or spectral component, and design an alarm for bit removal protocol. The smallest deviation from Gaussianity and/or any unexpected spurious signal component is a call for alarm.

(*c*) For frequency ranges you cannot monitor, implement proper filters.

(*d*) For amplitude ranges out of the regular working regime, install an alarm. This can be handy even though Makarov's blinding attack does not work with KLJN due to the opposite current directions of injected currents at Alice's and Bob's ends.

(*e*) Test the wire and its linearity at random times using random, *high-amplitude* signals, including DC and high-frequency AC excitation; to monitor for any sign of non-linearity and also the noise of standalone cable without signal (circulator, etc. attacks). Note any unexpected features and shut down the communication until the problem is clarified.

(*f*) The above defense operations are against hacking/active attacks. However, operations against passive (listening) attacks should also be carried out by evaluating all the known passive attack protocols. Bits with higher risk than a threshold value should be discarded (see Section 5.1 of this book).

(*g*) Use a healthy amount of privacy amplification against possible unknown attack types, see subsection 3 below.

5.7.2. *Random numbers and noises*

You must use virtually-perfect physical random noise sources and privacy amplification tools on the obtained raw random bits to approach the situation of truly random numbers. It is quite easy to pass "randomness tests" designed for pseudorandom number generators (PRNG); however PRNGs do not offer acceptable security levels.

The problem is information theoretic security. Pseudorandom number generators do not create new secrets. A PRNG has a finite number of parameters/seeds (in the case of chaotic one, an initial and boundary conditions) as secret. These parameters are a finite number of bits; themselves a kind of secure key.

In accordance with Kerchhoff's principle (see Section 1.6 of the book) all these parameters/seeds are known by Eve, provided they are fixed. To avoid this problem, if you want to use a PRNG for a new key exchange, you will need a new set of truly random parameters. Actually, you will probably need to refresh these random parameters/seeds even during a single-bit exchange.

How to do that? You can use a properly designed physical random number generator driven by a thermal noise source. Those, if properly designed, far outperform any quantum random generators.

Chip technology offers an option of putting a large number of thermal noise based random number generators on a single chip. Such systems would satisfy even the highest expectations.

5.7.3. *Privacy amplification*

For the most pessimistic expectations — if my neck were put on the line — I would add the simple Exclusive OR (XOR) operation based privacy amplification scheme (see the reference on the first page of this section); XOR-ing the key bit pairs to have a half as long key with higher security. I would repeat this procedure 5 times. The resulting key exchange will be 32 times slower. However, it will be practically clean even if there is a catastrophically high security breach with $p = 0.85$. The resulting short key will have a probability (of information leak) of only $p = 0.50001$; which means a less than 10^{-9} relative information leak toward Eve. The resulting 20-fold increase of bit-error rate between Alice and Bob will still be astronomically small, see Section 4.3.

5.7.4. *Enhanced usage of the shared key (new method)*

See the example[5] describing how to make conditionally secure software-based key exchanges unconditional by running them via a one-time-pad based cipher with an unconditionally secure KLJN key. The *known-plain-text* part of the key exchange should go without encryption. Then the same KLJN key can be used for an extraordinarily large number of times without compromising security. This method can also be used for QKD.

[5]L.B. Kish (2015), "Enhanced usage of keys obtained by physical, unconditionally secure distributions", *Fluct. Noise Lett.* **14**, 1550007. also at: http://arxiv.org/abs/1408.5800 (May 5, 2015).

5.8. The Most Recent Developments

The passing of one year since the completion of the first draft of the book has made it essential to add a list of the most important developments during that time (2015).

5.8.1. *The cable capacitance attack*

This is a passive attack utilizing the information leak due to one of the non-idealities, the parasitic (cable) capacitance. It was proposed at the very beginning of KLJN research but it was never calculated or tested. In his paper,* Barry Chen used the industrial cable and circuit simulator LTSPICE to validate the attack. Simulation results showed that the attack works but privacy amplification and/or capacitor killer (capacitance compensation) arrangements can effectively eliminate the leak. There is also a new, simple, very efficient defense method not shown in the published paper: adding a proper shunt capacitor.

5.8.2. *The current injection attack*

This active attack† and a simple defense were already mentioned in the first KLJN paper, and later analyzed in the response to Bennett–Riedel (see Sections 3.4 and 4.4). Barry used the LTSPICE to emulate the current injection attack against the ideal KLJN system, and a practical KLJN system, respectively. We showed that two security enhancement techniques, namely, the instantaneous voltage/current comparison method, and a simple privacy amplification scheme, independently and effectively eliminated the information leak and successfully preserved the system's unconditional security. It was also demonstrated the first time that the most efficient defense method is to run an *in-situ*, full simulation of the system, including the cable, where the measured voltages at the two ends are inputted (see

*H.P. Chen, E. Gonzalez, Y. Saez, L.B. Kish, "Cable Capacitance Attack against the KLJN Secure Key Exchange", *Information* **6** (2015) 719–732. http://www.mdpi.com/2078-2489/6/4/719/htm

†H.P. Chen, M. Mohammad, L.B. Kish, "Current Injection Attack against the KLJN Secure Key Exchange", accepted for publication (2015) in *Metrology and Measurement Systems* (2016). http://arxiv.org/abs/1512.03685, http://vixra.org/abs/1512.0274

Section 5.7). Then the simulated quantities are compared with the other measured quantities. Any discrepancy is a potential attack.

5.8.3. *Vadai–Mingesz–Gingl: perfect security at non-zero power flow*

Gergely Vadai, Robert Mingesz and Zoltan Gingl (VMG) published a revolutionary discovery in a Nature journal[‡]: they used two different resistor and noise voltage generator pairs with four different temperatures to provide unconditional security at non-zero power flow!

In this case, the HL and LH bit situations at Alice and Bob results in not only in the same mean-square voltage and current noises, respectively, but also in the same mean power flow which is greater than zero. These new types of KLJN systems are not protected by the Second Law but by the Fluctuation-Dissipation-Theorem in the classical physical regime (just

Figure 5.8.1. Gergely Vadai. He was a PhD student of Zoltan Gingl at the University of Szeged, Hungary, when this research was carried out. After defending his PhD he is now an Assistant Professor at the same place: the Department of Technical Informatics chaired by Zoltan.

[‡]G. Vadai, R. Mingesz and Z. Gingl, "Generalized Kirchhoff-law–Johnson-noise (KLJN) secure key exchange system using arbitrary resistors", *Sci. Rep.* 2015 (2015) 13653. http://www.nature.com/articles/srep13653

like it was said in the first KLJN paper [see Section 3.4], when the non-zero power flow was not on the horizon yet). The highest impact (so far) of the VMG discovery is that it inspired the Random-Temperature–Random-Resistor KLJN scheme, which is naturally robust against all existing attacks (see below).

5.8.4. *The Random-Resistor–Random-Temperature KLJN scheme*

Inspired by the VMG discovery, a new KLJN scheme is created: the Random-Resistor–Random-Temperature (RRRT) KLJN scheme,[§] which together with the known but formerly unpublished Random-Resistor (RR) scheme, is the 10th version of KLJN.

The RRRT-KLJN scheme is shown in Fig. 5.8.2. *As of January 2016, all existing former attacks, including GAA's new transient attack, are invalid against this scheme.* No doubt, attack-attempts will come sooner or later.

5.8.5. *More light on the flaws in the Gunn–Allison–Abbott (GAA) "directional coupler" attack*

As we have seen in Section 4.6, the GAA attack in *Science Reports* (2014) looks like an interesting idea; however it has proven to be a total *failure* — from its *concept*, through the *physics* and *experiments*. See the considerations in Subsection 5.8.8 below, which explains the origin of these types of flaws in today's science publications; when researchers with good intentions and hard work often fail to achieve their goals and results.

Yet, history shows that *the GAA paper was not fully in vain*. It actually triggered some interesting discussions and debates between physicists and engineers about the non-wave nature of signals in a short wire at low frequencies (quasi-static limit). It is proven in Section 4.6 that such signals are not electromagnetic waves; otherwise basic laws of physics are violated. However, contrary to this fact, electrical engineers often use calculations based on wave equations even in the no-wave (quasi-static) situation, because that simplifies the math. Engineering, in general, uses a

---

[§]L.B. Kish, C.G. Granqvist, " L.B. Kish, C.G. Granqvist, "Random-resistor–random-temperature Kirchhoff-law-Johnson-noise (RRRT-KLJN) key exchange", *Metrology and Measurement Systems* 23 (2016), pp. 3–11. http://www.degruyter.com/view/j/mms.2016.23.issue-1/mms-2016-0007/mms-2016-0007.xml; http://arxiv.org/abs/1509.08150.

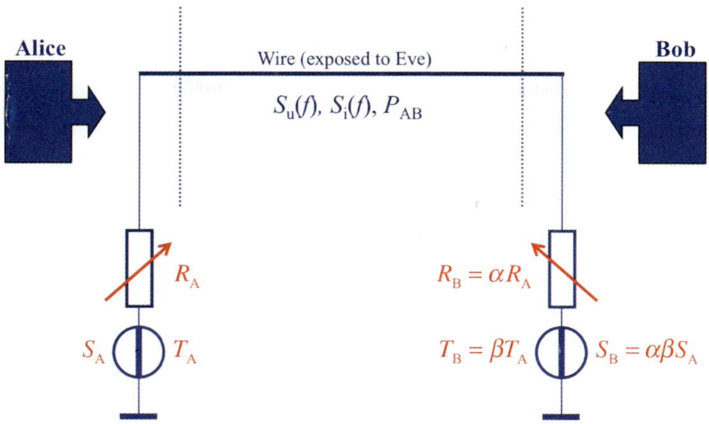

Figure 5.8.2. The Random-Resistor–Random-Temperature (RRRT) KLJN scheme, which is the 10th KLJN scheme. The temperatures and the resistors at Alice (A) and Bob (B), and their corresponding voltage noise spectra, are continuum random variables with a new random choice made at the beginning of each KLJN period. The Low (L) and High (H) bit-values at Alice and Bob are determined by the relative resistance values; for example, the party with the higher resistance has the high bit. In the hypothetical but non-practical situation when the resistance distribution is a continuum and the inaccuracies of Alice's and Bob's estimation results are zero, 100% of secure bit exchange is accomplished. In the practical case with finite accuracy (finite bit exchange duration) and quasi-continuum discrete distribution, the secure bit exchange efficiency is less than 100%, because the bits with singular noise and power levels must be discarded, but 100% can be approached with proper design.

large number of simplifications for easier handling of calculations and this is one of those. What is interesting is that certain results, such as the cable impedance, comes out correctly even in this limit, with the unphysical calculation. We have investigated** what is behind this paradox. It is a simple math trick, which is unphysical, but works fine — if only a limited part of the result is used (see below).

In linear time-invariant network analysis, the weighting function $h(t)$ is the response to a Dirac pulse (zero-width and infinite height) at the input, as shown in Fig. 5.8.3.

**L.B. Kish, S.P. Chen, C.G. Granqvist, J. Smulko, "Waves in a short cable at low frequencies, or just hand-waving? What does physics say?", invited talk, Proc 23rd International Conference on Noise and Fluctuations (ICNF 2015), Xian, China, June 2–5, 2015. DOI: 10.1109/ICNF.2015.7288604

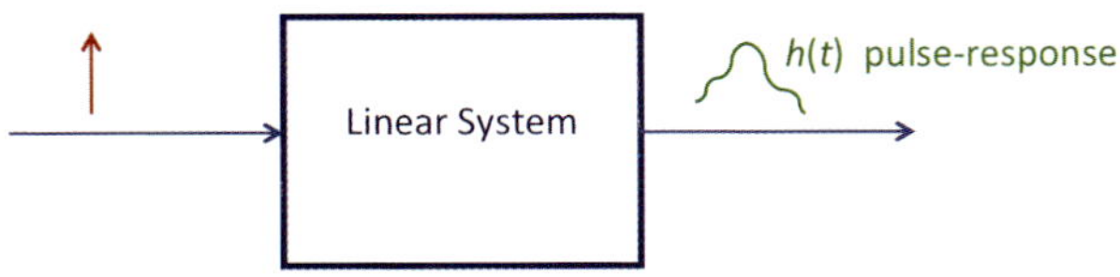

Figure 5.8.3. The weighting function $h(t)$ is the response to a Dirac pulse at the input.

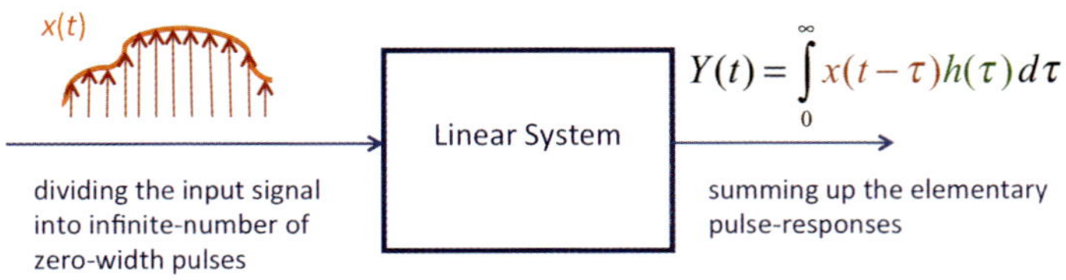

Figure 5.8.4. The output signal $Y(t)$ is the convolution of elementary responses due to the abstract input pulses with zero width and infinite bandwidth.

Any signal $x(t)$ with limited bandwidth at the input is then divided into such abstract, zero-width (Dirac) pulses, and the time response to these input pulses at the output are calculated and summed up by a convolution to get the output signal $Y(t)$; see Fig. 5.8.4.

The zero-width input pulses have infinite bandwidth, even though the total input signal may be very slow. Thus, even if the linear system is a short cable with low signal bandwidth and thus the system is physically quasi-static (no waves); these zero-width pulses will mathematically still propagate and reflect as waves in the cable. The propagation of these abstract elements can mathematically be described as waves and, at the end, during the output-summation, all wave components disappear as a consequence of their destructive interference so that we regain the slow, non-wave output signal $Y(t)$.

It is obvious that these abstract pulses and their propagation are *non-physical* because, for example, we cannot detect X-rays, electron emissions, or impacts of the quantum uncertainty principle due to the high bandwidth and short time scales involved. In engineering and physics, many similar mathematical contrivances are used to calculate results in an easy way. Clearly, mathematics is a much broader field than physics, and physics will ultimately make use of those mathematical solutions that are physical. It is important not to assign physical characteristics to elements that are only non-physical mathematical constructs.

If it is so, and the math can work without valid physics for the signal, the question is where did GAA make a related mistake? Their approach is based on the d'Alambert equation, which is actually not even a wave equation because many non-wave type propagation (such as laminar flow) satisfies it. This equation cannot be used for a steady-state signal, as GAA did, because the propagation velocity (phase velocity) will depend on the terminating resistance, due to the reflections of the abstract Dirac spikes described above. This was clearly shown by Barry Chen's LTSPICE simulations, see Section 4.6.1. However, GAA used a single, direction-independent propagation velocity of the slow signal, which was a deadly mistake.

Ultimately, this effect kills the whole concept because, to know the velocity in the different directions, it requires the knowledge of Alice's and Bob's secret: the resistor values what they use!

To see more examples about violating physics when claims about laws of physics are made by simplified electrical engineering pictures, see subsection 5.8.7 below.

5.8.6. *A new, successful transient attack (Gunn–Allison–Abbott)*

Recently, Lachlan Gunn, Andrew Allison and Derek Abbott (GAA) published an interesting paper[††] with the historically first transient attack, which is utilizing transients at the beginning of the bit exchange. The idea is extremely simple: monitoring the mean-square voltage before the switching transient reaches the other end of the cable. This is a very short sampling time, less than 10% of the correlation time of the noise, thus the change of the voltage is typically small. Thus the easier understanding of this is: monitoring the voltage on the cable capacitance while the cable capacitance is charged up by a DC voltage via the resistor. While the higher resistance has a higher noise voltage, the charging will be slower by a factor that is the square root of the ratio of the resistance values. Naeem Farokhnia validated the attack under real conditions,[‡‡] without ramping,

[††] L. Gunn, A. Allison, D. Abbott, A New Transient Attack on the Kish Key Distribution System, *IEEE Access* **3** (2015) 1640–1648.
[‡‡] N. Farokhnia unpublished (2016).

using LTSPICE simulations and indeed found some information leak, which could easily be fixed by privacy amplification, thus the unconditional security was not challenged. Moreover, if Alice and Bob properly choose the starting voltage amplitude and velocity values, respectively (they need recorded noise data), the information leak can be eliminated even without privacy amplification.

Figure 5.8.5. Bob (J.R. Biard) is one of the rare scientists who not only understands noise, but who can also execute virtuoso calculations or build pioneering devices about it. He has retired from the industry and Texas A&M University and is the member of the National Academy of Engineering. Bob has over 75 US and a few international patents, and his patent story supports our earlier statements that breakthrough research cannot be planned even if research foundations require such plans. From Bob's 75+ patents, three by far made the highest industrial profit, and those profits were huge: the first (infrared) LED diode, the MOS Read-Only-Memory (MOS ROM), and the Shottky-clamped TTL logic circuit. Most importantly the three breakthrough inventions were not in the official work plan of the industry while the rest of the patents were the results of officially planned R&D activities. The breakthroughs were conceived during relaxed lunch and coffee chats instead of during intensive work-rush following written plans. The Nobel Prize for the LED was unfortunately given for the blue LEDs instead of Bob's; though Bob's vision to realize that gallium-arsenide junctions can emit photons was the fundamental breakthrough. Most interestingly, at the time of creating the LED patent, even Bob did not believe that such a LED can be used for anything useful, or that the ROM will be used in the BIOS of every computer. Indeed, it is impossible to immediately see the long-term impact of a scientific discovery. Yet, in grant applications scientists must justify the potential impact of their planned research! Very often such a "justification" is nothing more than technically advanced science fiction.

5.8.7. *Deficiencies in the Gunn–Allison–Abbott transient attack paper*

While the simple transient attack in the IEEE Access paper mentioned in Subsection 5.8.6 above is valid; the paper has many problems. Here we show a few:

(i) First of all, like their 2014 submission to *Science Report* (see Section 4.6) details of the simulation were excluded from the paper. It is known from personal communications with Derek Abbott that it was not a professional cable simulation, such as the ones Barry used in the papers outlined in Sections 5.8.1 and 5.8.2 above, thus it is particularly important to see how the cable was simulated, how was the noise generated, etc.; because these are all potential traps.

(ii) Secondly, GAA is using the quantity "*secrecy rate*" to characterize security. In this way, the bit error probability q of Alice and Bob also enters into the result: the higher the q, the less is GAA's secrecy rate. However, comparing security (secrecy) and secrecy rate, are like comparing apples and oranges: both are fruits, but are also completely different. The basic definition of (perfect) secrecy of a key exchange is as follows[§§]:

The probability p of Eve's successful guessing of the key-bit is not improved by Eve's monitoring of the key exchange.

When Eve's knowledge about a uniformly generated key is zero, $p = 0.5$. In the case of perfect secrecy (security), p stays at 0.5, even when Eve is eavesdropping. The bit error probability q of key exchange between Alice and Bob is absolutely irrelevant in this definition. While, for KLJN, the error probability q can be as low as 10^{-20}, to mix Alice/Bob error probability with the security of key exchange is a fundamental mistake. To illustrate how inadequate secrecy rate is:

It is possible to have two different KLJN systems with identical secrecy rates, where one of the systems has a very poor security (p close to 1) while the other has a very strong security (p close to 0.5).

[§§]C.E. Shannon, "Communication theory of secrecy systems", *Bell Systems Tech. J.* **28** (1949) 656; G. Gilbert, M. Hamrick, "Secrecy, Computational Loads and Rates in Practical Quantum Cryptography", *Algorithmica* **34** (2002) 314–339.

5.8.1. The publication of this paper[‡‡] in *IEEE Access* has its own funny story. Robert Mingesz and I were the reviewers. While we saw the technical weaknesses of the paper and initially forced the authors to make major revisions; in the second refereeing run, we saw that a complete fix would take ages. After some discussion, Robert and I agreed to accept the paper for publication inspite of its deficiencies because we appreciated Lachlan's simple attack idea, which was obviously working. However, when the paper came out, we saw with horror that a *whole extra page that we had not seen during our review had been added* to the text in the publication! This page was the Appendix, which claimed that, so far, the most general security proof of the KLJN scheme (see Section 5.1) is incorrect. Their arguments had errors about physics, electrical engineering and security (see below). We were outraged and asked Derek (who is not only a co-author of the paper, but who was also an editor of *IEEE Access*) how that had happened. He responded that they had decided to make a non-refereed addition, and that the Acting Editor had accepted it. To me, this was not acceptable. I thus contacted *IEEE Access*'s Editor-in-Chief, Michael Pecht, and described the case. He shared our worries and offered us the option to publish a rebuttal in *IEEE Access* with the $1,750 open access publication fee waived. We submitted our rebuttal in February, 2016, but received mostly negative referee reports that were totally illiterate both in security and physics, which led to us withdrawing the paper without debating the reports. We concluded that there was no point attempting to teach the multiplication table for those who were unwilling to learn. With Claes Granqvist, we published our criticism (open-access) in *Metrology and Measurement Systems* **23** (2016); http://arxiv.org/abs/1606.02697. We have added an Appendix to it, where we address some of the worst physics/security mistakes in the *IEEE Access* reports about our paper to prepare the non-expert reader for similar invalid arguments, which may emerge in future papers or on the internet.

(iii) GAA (correctly) argues that, when the parameters are tuned toward the ideal values to match security requirements, there are limits to doing that. They use the example of cable length, which can rarely be close to zero (except in intra-instrument chip-to-chip communications). This is true but the goal of proving unconditional security is to show that there is a practical limit, where the required security level is reached. That we proved via the continuity of function in stable classical physical systems. Moreover, the impact of long cable length can be eliminated by a small bandwidth and the required security level reached.

The positive side of this comment by GAA is that, by re-investigating our treatment, we have realized that some of the parameters (such as the cable length) imply zero first derivative at zero value. This fact requires the following correction in Equation (5) of the paper in Section 5.1. In the Taylor expansion shown by Equation (5), the Taylor series must be expanded to at least its second order terms! Then the higher-order cross terms take care of reducing the information leak to zero when the time allowed for the bit exchange goes to infinity. This situation does not change the main results and the conclusion about unconditional security but it has direct relevance for practical operation and (among others) privacy amplification.

Then the expanded Equation (5) explains why perfect security can be reached at zero cable length even if the other parameters (bandwidth, reciprocal cable diameter, etc.) are not at their ideal zero value.

(iv) Finally, the most important point: security versus physics. Using a circuit example, *GAA claims that functions in stable classical physical systems are not always continuous*. This is a very serious claim and must be incorrect — otherwise the whole theoretical framework and education of classical mechanics, electrodynamics and statistical physics are flawed. Therefore let us investigate GAA's claim. We will see that there are three different types of errors in their argumentation related to electrical circuit theory, security and physics.

Their system is shown in Fig. 5.8.6. It is public knowledge that Alice and Bob have 1 Ohm resistors but their DC voltages are secret. Eve measures the voltages U_{AE} and U_{BE}. If $R_{\mathrm{E}} > 0$, she can exactly determine the voltages U_{A} and U_{B} from the measured quantities; thus Eve has perfect eavesdropping.

In the case of $R_{\mathrm{E}} = 0$ (see Fig. 5.8.7), Eve cannot determine the secret voltages because she does not have enough information. However, ignoring the possibility of current measurement is a circuit-theoretical mistake yielding an *absurdly-limited-Eve*, which *contradicts the basic rules of security analysis* that Eve must utilize all the information she has access to! If she also measures the current, she can exactly determine the voltages: $U_{\mathrm{A}} = U_{\mathrm{E}} - I * 1\Omega$ and $U_{\mathrm{B}} = U_{\mathrm{E}} + I * 1\Omega$. Obviously the non-zero R_{E} is to use the current as extra information via the voltage drop on R_{E}, in which information is lost at zero R_{E}.

In conclusion, there is no discontinuity as GAA claims. Eve's eavesdropping ability is actually constant. Finally, as an educational example let

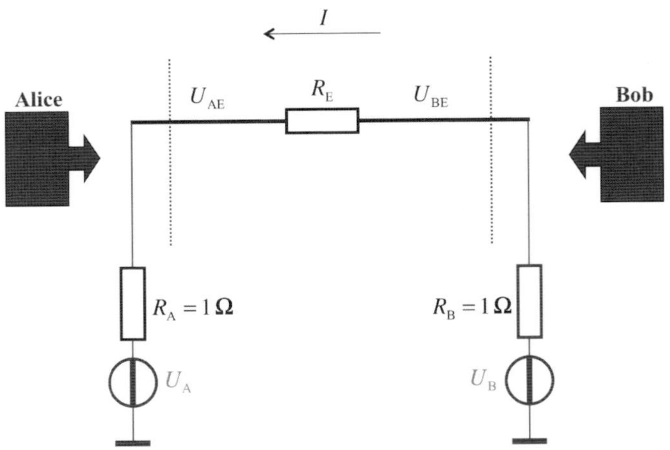

Figure 5.8.6. Eve knows the resistor values and measures the voltages U_{AE} and U_{BE}. If $R_{\mathrm{E}} > 0$, she can determine the voltages U_{A} and U_{B}.

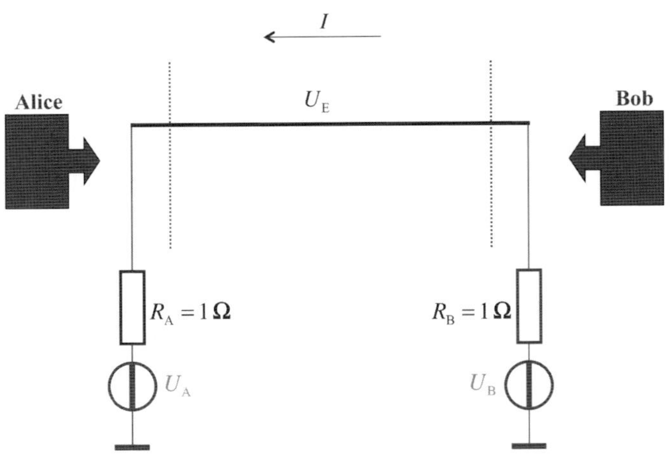

Figure 5.8.7. Eve measures the voltage and knows the resistor values. Yet, she cannot find out the voltages. However, this is a strongly-limited-Eve because she does not measure the current, which contradicts the rules of security that Eve is utilizing all the information she has access to! If she also measures the current, Eve can determine the voltages.

us make the above unphysical, simplified, electrical engineering models physical by realizing that all the resistors have Johnson noises, $U_{\mathrm{An}}(t)$, $U_{\mathrm{Bn}}(t)$ and $U_{\mathrm{En}}(t)$ (see Fig. 5.8.8). Neglecting Johnson noise creates the unphysical assumption of absolute zero temperature, which is forbidden by the laws of physics. The impact of the non-zero noise is significant:

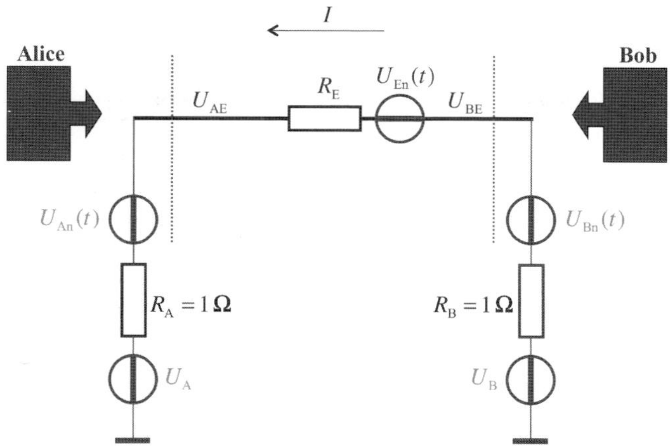

Figure 5.8.8. The correct physics represented by the non-zero Johnson noise voltages of the resistors. The impact of $U_{An}(t)$ and $U_{Bn}(t)$ is significant: guaranteeing a non-perfect eavesdropping and continuous transition toward $R_E = 0$. Perfect eavesdropping can be achieved only after infinite long time averaging, which is an unphysical situation. Neglecting Johnson noise creates the unphysical assumption of absolute zero temperature.

it guarantees a non-perfect eavesdropping and continuous transition of eavesdropping information toward $R_E = 0$. This is true for both a correct, unlimited Eve and the absurdly-limited-Eve. Perfect eavesdropping can be achieved only by infinitely long time averaging, which is again an unphysical situation.

In conclusion, GAA's "proof" of the existence of a discontinuous function in a stable, linear, classical physical system is invalid. As soon as the system is made physical by realizing the unavoidable thermal fluctuations, the law of continuum functions stands. *Simplified engineering pictures are insufficient to prove or disprove laws of physics.*

5.8.8. *The ultimate cause of all my debates about the physics of noise and information*

Currently, I have several active scientific debates going on, which are actually quite tiresome, but hopefully the truth will be revealed.

Gijs Bosman (Professor at the University of Florida, USA) said, when I met him last summer in Barcelona: "I see you are still continuing your windmill fights". He was referring to the very first public debate of my carrier (Montreal, 1987), where I showed that the — at that time celebrated — "quantum 1/f noise" effect does not exist. Already then, he told me that such debates were 'windmill fights' because the other party could not be convinced and the energy was only wasted. *Max Planck, the originator of quantum theory, had a similar claim saying that deep scientific truth rarely wins directly; rather, its opponents must die first.*

When one realizes that one is fighting against most of the world on multiple issues, one must then ask the question: "Am I perhaps crazy?" Luckily (*hopefully*) the answer is *no* and, in my understanding, the main cause of these debates is the *extraordinary specialization in today's science and engineering while the research problems in question are strongly interdisciplinary.*

Eager specialists, *with good intention*, carry out studies utilizing the understanding, methods and tools of their own field and, while doing so, they reach beyond the validity region of their methods — and yet they make claims about areas where their tools break down. A few examples:

(i) Using electrical engineering tools for conditions where the tools produce artifacts or unphysical results. The Reader can find plenty of examples about such in this book; for a quick look, see Subsections 5.8.5 and 5.8.7 above. The cause of the mistake is a flaw in the physics/physical aspects — yielding erroneous conclusions in security.

(ii) Mathematical-physical abstract models are used to determine the energy dissipation in a memory during erasure versus the amount of erased information. However, information theory says that a memory with known data has zero information, thus the erased information cannot be a measure to describe energy dissipation — a flaw in the informatics aspect that yields an erroneous conclusion (e.g. Landauer's principle) in its physics.

(iii) Maxwell's demon or the Szilard engine. Simplified mathematical physical models leave out the energy dissipation of the control units (switches, valves, etc.): a flaw at the engineering side that also yields an erroneous conclusion in physics.

Luckily, the author of this book has evolved through various backgrounds (experimental physics, theoretical physics, electronics), which have provided tremendous help in identifying and avoiding these kinds of traps. The painful implication is, however, the existence of these debates, which may indeed be called "windmill fights" for the truth.

5.8.2. Concerning the recent debates around KLJN, Max Planck's comment above may seem to be relevant. Instead of continuing public debates, I plan to focus on new science. Thus it will be up to other colleagues and the Reader to filter out pseudoscience. Concerning the Scholarpedia entry on KLJN: at the moment (as of January 7, 2016), the article is fine (with my last correction made in February 14, 2014). I advise the Reader to be extremely cautious if new changes appear, and to check all claims and results against the laws of physics and security. Similarly, the Reader is advised to be cautious if another book on the subject, by different authors, emerge in the future! Observe the potential flaws and, while you correct the errors, please remember Einstein's relevant comment: "Anyone who has never made a mistake has never tried anything new".

Chapter 6
Epilogue

In this book, the Reader was presented with some of the scientific and personal background stories behind the unconditionally secure KLJN key exchange system; its fallouts, such as Noise-based Logic and Computing; some debates; and some scientific/personal history. It should be noted that all these developments have been done without financial support, except some minor items such as paying for parts of a trip, which is a strong contrast to the multi-billion dollar support that quantum informatics has received during the same years.

However, the true purpose and focus of scientific research is not to seek grants, citations or popularity, but to seek the truth.

***My message to young scientists*:** If you want to achieve a real breakthrough, you must leave the noisy mainstream and buzz, and go far away, until you reach the frontiers — the real frontiers that differ from the *claimed perimeters* of mainstream science. You may be completely alone, perhaps nobody will hear or cite you — people may even think you are crazy or lost. But, if you are persistent, you can have breathtaking views there, at the *frontiers of knowledge*.[1] This is why we are scientists, isn't it?

Laszlo
College Station, Texas.

[1] Where is the frontiers of knowledge? Here's an example: if you terminate a well-accepted, but incorrect dogma of science, you are certainly standing there even if nobody believes you at first. . .

Figure 6.1. With Zoltan Gingl (left) and Mark Dykman (right) in Barcelona, on July 16, 2015, after the Authors' paper[2] presentation on the non-validity of the celebrated Fluctuation-Dissipation-Theorem (Johnson noise) in the quantum regime and the follow-up discussion, which was still going on when this picture was taken. The reception of the talk was mixed, which was expected because the celebrated equation with "Nobel-Prize-level" importance was challenged. Hopefully, this book will bring to the Reader some flavor of these dynamic discussions and debates and will inspire the Reader to seek the truth and filter out the myths from claims about science.

---

[2]L.B. Kish, G.A. Niklasson, C.G Granqvist, "Zero-point term and quantum effects in the Johnson noise of resistors: A critical appraisal", *J. Stat. Mech.* (2016) 054006. doi:10.1088/1742-5468/2016/05/054006. http://arxiv.org/abs/1504.08229

Subject Index

A

Abbott, Derek: 41, 50, 51, 80, 192, 193, 252, 307, 308, 327, 426
active attack: 30, 34, 41, 46, 98, 100, 110, 117, 132, 207, 209, 210, 245, 256, 263, 267, 271, 276, 291, 293, 356, 366, 382, 404, 418
Antilli, Davide: 113, 273, 274
artifacts (experimental and modeling): 12, 21, 206, 207, 216, 218, 219, 221, 308, 328, Section 4.6.3
arXiv: *xiv*, 100, 165, 299, 353
authentication: 39, 184, 187–189, 209, 230, 263, 270, 290, 291, 357, 359, 366, 382, 391, 400, 404

B

bandwidth: 15, 18, 19, 27, 55, 59, 64, 90, 92, 111, 112, 121, 122, 145, 161, 158, 169, 172–174, 181, 201, 207, 211, 215, 220, Sections 4.3, 4.4, 5.1, 5.2, 5.8
Barcelona: 44, 429, 434
Bennett, Charles: 38, 179, 191, 304
Bennett-Riedel (BR) attacks: 2, 19, 142, 193, Section 4.4, 418
Bennett-Riedel thermodynamics-free system: 113, Section 4.4
Bergou, Janos: 77, 78, 85, 127, 151
(Bergou)-Scheuer-Yariv attack: see Scheuer-Yariv attack
Bezrukov, Sergey: 37, 204, 205
Bhattacharyya, Shankar: 197, 200
Biard, J.R. (Bob): 424
Bosman, Gijs: 429
brain: 33, 54, 196, 198
brainwashing: 102
Brillouin's principle: 199, 284, 302

C

cable capacitance attack: 112, 136, 418
Cao, Xiaolin: 412
Chappell, James: 78, 79
Chen, Hien-Pu (Barry): 310, 418, 423
circulator-based attack: 207, 220, 227
classical physical quantities: 356, 427
communism: 116
conditional security: 35
control (dissipation of, in physics): 253, 300, 303, 304, 305, 430, Section 4.5
cracking the Bennett-Riedel (BR) system: 256, 284, Section 4.4
current injection attack: 147, 177, 179, 418

D

debates (physical informatics, general origin): 429, Section 5.8.8

defense directives (general): 414–417
Detailed Balance Principle: 310, 313, Section 4.6.1
Diamanti, Eleni: 36
Doering, Charlie: 128
dogmas (popular but wrong): *xii*, *xix*, 51, 54, 101, 356, 433
Dykman, Mark: 434

E

eavesdropping: see passive attack
Energy Conservation Law: 3
energy dissipation: 8, 51, 54, 58–60, 198, 199, 300, 304–306, 430
Energy Equipartition Theorem: 24, 310, 313, Section 4.6.1
enhanced KLJN schemes: Sections 5.3, 5.8.3 and 5.8.4
enhanced usage of unconditionally secure keys: 417

F

Ferry, Dave: 192, 302
Fluctuation-Dissipation-Theorem (FDT): 88, 105, 258, 358, 366, 419, 434
Fluctuations and Noise Symposium (SPIE): *xx*, 68

G

Gaussian noise: 4, 5, 14, Section 1.3, 103, 121, 172, 207, 209, 236, 241, 244, 246, 262–264, 268, 276, Section 5.1, 371, 374, 381, 382, 392, 415
Gea-Banacloche, Julio: 8, 50, 68, 69
Gingl, Zoltán: 22, 42, 44, 45, 166, 168, 342, 419, 434
Global Warming: 300
Gonzalez, Elias: 247, 409, 410
Granqvist, Claes-Göran: 4, 49, 253, 426
Gunn-Allison-Abbott (GAA) attack: 2, 253, 415, Subsections in 4.6; Section 5.8.5
Gunn-Allison-Abbott transient attack: Section 5.8.6
Gunn, Lachlan: 79, 307, 308, 327, 342, 423, 426
Gunnarson-Klas: 353

H

Hacking attack 47, 248, 249, 253, 258, 270, 295, 296, 363, 366, 414, 415
Hajdu, Janos: 31, 82
Hao attack 109, 136, 138, 142, 177, 179, 182, 372
Hao, Feng: 138, 139
Heszler, Peter: 17
Hirota, Osamu: 37, 40, 46, 47, 143, 235, 252, 257, 267, 354
Horvath, Tamas: 44, 47, 129, 206
Hrasko, Peter: 17, 18

I

information leak: 1, 21, 31, 41, 127–129, 131, 136, 138, 142–145, 173, 177–179, 181, 182, 209–211, 263, 268, 288, 289, 310, 343, 344, 346, 347, 350, 351, 365, 367, 371, 382, 384, 390–392, 413, 414, 417, 418, 424, 427
information theoretic security: *xiii*, 35, 36, 38, 40, 42, 102, 135, 138, 142, 143, 185, 210, 234, 254, 256, 267, 355, 356, 362, 378, 389, 399, 411, 416, Section 4.4, Section 5.1
invasive attack: see active attack

J

Johnson, John Bertrand: 26
Johnson-Nyquist noise: 1–3, 12, 25, 26, Section 1.4

K

Kak, Subhash, 78
Kirchhoff's loop law: 2
Kerckhoffs's principle: 35, 46, 47
Khatri, Sunil: 10

Kish attack (resistance inaccuracy): 177, 179
- defense against: see Vadai-Mingesz-Gingl KLJN version
Kish cypher books: *xii*
Kish cypher Wikipedia page (deleted): *xi*, *xii*, *xiv*, 41, 166, 192–195, 252, 353
Klappenecker, Andreas: 37, 127, 204
KLJN key exchange (definition): 40
KLJN versions and names: 41
known plain text attack: 40, 46, 417
Kwan, Chiman: 5, 395

L

Landauer's principle/theorem (erasure, non-validity): 299, 300, 302
- experimental "proof" of: 303
LED diode: 424
listening attack: see passive attack
Liu attacks: 206
Liu, Pao-Lo: 44, 45, 124, 142, 206, 207, 223, 226, 255, 281–283, 307, 325

M

Makarov, Vadim: 38, 40, 48, 253, 258, 293, 363, 414, 415, 416
Makra, Peter: 168
Manhattan Project: 165
man-in-the-middle-attack: 46, 100, 117, 149, 152, 184, 186–189, 207, 210, 220, 225–228, 230
Maxwell demon: 253, Section 4.5
Mingesz, Robert: 22, 42, 130, 147, 158, 166, 168, 342, 419, 426
Mingesz attack: 154, 158
MITM attack, see man-in-the-middle-attack
MIT Technology Review: 2, 299

N

National Science Foundation (NSF), USA: *xiv*, 5, 50–52, 75, 125, 126, 165, 409, 410
National Security Agency (NSA), USA: 42, 147
National Security Agency, Germany: 147
New Scientist magazine: 2, 166
no-cloning theorem: *xiii*, 38, 39, 101, 354
Nobel Prizes: 25, 424, 434
noise-based logic: 195, 355, Section 4.1, Section 5.4
non-ideality attack: see passive attack and/or hacking attack
Norton, John, D.: 299, 300, 302
no-wave limit: see quasi static limit
Nyquist, Harry: 26

O

oscillator: 21–23, 88, 199, 309, 398
Orlando system: 113, 273

P

passive attack: 77, 100, 207, 209, 256, 259, 261–263, 268, 269, 271, 272, 275, 282, 284, 290, 356, 418
Pentagon phonecall: 147
Peper, Ferdinand: 43
perfect security: 21, 28, 43, 46, 113, 136, 145, 178, 257, 264, 265, 290, 354–356, 360, 361, 376, 378, 419, 427
perpetual motion machine: 105, 258, 261, 293, 308, 309
physical informatics: 300
Plaga, Reiner: 291
Poor, Vincent: 363
Porod, Wolfgang: 87, 192, 302
privacy amplification: 41, 43–47, 129, 131, 136, 142, 143, 145, 177, 178, 181, 209, 227, 264, 269, 270, 290, 293, 332, 346, 355, 361, 362, 365, 376, 390, 416, 417, Section 5.7.3

Q

quantum 1/f noise (non-existent): 17, 18, 430
quantum communicator: *xi*, 38, 43, 45, 48, 91, 92, 112, 136, 140, 143–145, 149,

158, 163, 167, 170, 182, 209, 210, 253, 258, 356, 378, 414
quantum demons: 303
quantum encryption: see quantum key distribution
quantum key distribution (QKD): 21, 26, 34, 37, 38, 41, 42, 46, 47, 104, 113, 131, 142, 182, 235, 256–258, 264, 356, 378, 383, 392, 398, 415, 417
- cracking its security: 38, 41, 48, 235, 258, 414, 415
quantum key exchange: see quantum key distribution
quasi static limit of electrodynamics: 24, 151, 172, 210, 212–214, 216, 235, 252, 271, 277–281, 283, 285, 288, 310, 313, 315–318, 320, 328, 330, 420, 422

R

random numbers (true randomness): 416, 417
Renner, Renato: 37, 46, 136, 235, 257, 294, 363
Ribbing, Carl-Gustaf: 300

S

Saez, Yessica: 176, 233, 237
Saidi, Olivier: 186
sampling theorem (Nyquist): 18, 266
Scheuer, Jacob (Koby): 100, 127–130
Scheuer-Yariv attack: 2, 28, 113, 128–130, 136, 142, 145, 167, 169, 175–178, 182, 365, 372, Section 3.6, Section 5.2; (called also wire resistance attack)
- total elimination of it: Section 5.2
Schmera, Gabor: 193
Schneier, Bruce: 41, 115, 116, 117
Scholarpedia: 193, 431
science: *xvi*, 11, 37, 50, 52, 54, 99, 101, 126, 192–194, 299, 300, 350, 353, 420, 429–431, 433
- frontiers of: 433
Science magazine: *xi*, 2, 100, 115, 125, 127
Second Law attack 130, 368, Section 5.2
Second Law of Thermodynamics: 3, 21, 24, 28, 85, 105, 199, 259, 261, 265, 269, 274, 280–282, 293, 304, 309, 313, Section 4.6.1
security proof of KLJN (general): Section 5.1, 359, 427
Sethuraman, Swaminathan (Swami): 75, 76, 80
Siemens patent of KLJN (in 2011): 113–114
single photon: 38
Smart Grid (unconditionally secure): section 5.5
Smulko, Janusz: 5, 117, 342
Stacho, Laszlo: 204
statistical distance: 40, 43, 45, 46, 136, 266, 267, 275- 277, 290, 356, 362
stealth communication: Section 3.2, and truly thermal noise based KLJN
Svedlindh, Peter: 32, 353
switches (energy dissipation of): 300, 304
Szeged: 126
Szegedin Whisper: 165, 166
Szilard, Leo: 126
Szilard engine: 253, Section 4.5

T

Taylor, Henry: 94, 96–98, 116
telecloning (of information by KLJN): Section 3.8
thermal noise: see Johnson-Nyquist noise
transient attack: 275, 288, 390, Section 5.8.6

U

unconditional security: see information theoretic security
Unsolved Problems of Noise (UPoN): xx, 78, 434
Uppsala: 116

V

Vadai, Gergely: 419
Vadai-Mingesz-Gingl (VMG) KLJN version: 41, Section 5.8.3, Section 5.8.4

vehicle communication systems (unconditionally secure): Section 5.6
Voice of the Dolphins: 126

W

wave in the wire debate: Section 4.6.1, Section 5.8.5, Section 5.8.8
Wen, He: 197, 200
white noise: 6, 7, 19, 27, 30, 31, 135, 207, 236, 259, 289, 348, 391
Wikipedia: *xi*, *xii*, *xiv*, 41, 166, 192–195, 252, 353
"windmill fights" (for clarifying the truth): 429, 430, 431
wire-resistance-thermal-noise attack: 287

Y

Yuen, Horace: 37–39, 40, 46, 47, 104, 136, 142, 143, 235, 252, 257, 266, 291, 354, 375, 389

Z

zero-point energy/fluctuations: see zero-point noise
zero-point noise: 27, 69, 86–89, 94, 95, 434
zero-signal-power communication: Section 3.2, and truly thermal noise based KLJN
Zubairy, Suhail: 85, 87, 257